Darwin, Geodynamics and Extreme Waves

Sh. U. Galiev

Darwin, Geodynamics and Extreme Waves

Springer

Sh. U. Galiev
Department of Mechanical Engineering
University of Auckland
Auckland, New Zealand

Special gratitude goes to Garry Tee for editing the language in this monograph.

ISBN 978-3-319-16993-4 ISBN 978-3-319-16994-1 (eBook)
DOI 10.1007/978-3-319-16994-1

Library of Congress Control Number: 2015942288

Springer Cham Heidelberg New York Dordrecht London

Cover illustration: Stock Photo © Reniw-Imagery, photo 11716641

Printed on acid-free paper

Springer International Publishing AG Switzerland is part of Springer Science+Business Media (www.springer.com)

Preface

Strongly nonlinear waves can be excited by natural causes in sediment and water basins, ridges, realms of the ocean, and a scalar field in which resonant effects can cause very large amplitude growth. This book particularly focuses on the reasons for the resonant amplification of seismic and ocean waves that have the capacity to destroy cities and ocean-going vessels. In this regard, Charles Darwin's observations of the catastrophic results of the 1835 Chilean earthquake are discussed. Darwin explained the results of earthquake-induced vertical shocks and described the earthquake aftermath in his book *Journal of Researches into the Geology and Natural History of the Various Countries Visited by H.M.S. Beagle, under the Command of Captain FitzRoy, R.N. from 1832 to 1836*. Darwin was the first researcher to observe and analyze local amplifications of seismic effects connected with the topography of the Earth's surface and the Earth's geology. He described the interaction of earthquakes with volcanoes, seaquakes, and tsunami formation.

The year 2009 marked the 170-year anniversary of the publication of this book, which also displays Darwin's outstanding literary talent. It may be compared with another remarkable work, *A Voyage Round the World in His Britannic Majesty's Sloop Resolution, Commanded by Capt. James Cook, during the Years, 1772, 3, 4, and 5* (1777) (Johann Georg Adam Forster), with respect to geographic and literary achievements, but of course, Darwin's book is immeasurably superior in providing this outstanding researcher's view on various natural phenomena.

In the following, I analyze the parts of Darwin's book devoted to the abovementioned 1835 earthquake and related catastrophic natural phenomena. The ongoing mysteries of vertically induced earthquakes are discussed using modern data. I ask how Darwin perceived these phenomena and whether his ideas are endorsed by the discoveries of modern science and whether the results of catastrophic earthquakes can be modeled by means of mathematical methods. These results are studied using strongly nonlinear wave equations. I also propose that similar equations can simulate the dynamics of many objects on the surface of the Earth. The origin of the Universe, dark matter, and dark energy are modeled as strongly nonlinear wave phenomena. The results are demonstrated by a large number of experiments and observations. Thus, the issues considered in this book embrace the areas of mechanics, physics, and mathematics.

The book was written for Master's and Ph.D. students as well as for researchers and engineers in the fields of geophysics, seismology, nonlinear wave studies, cosmology, physical oceanography, and ocean and coastal engineering. It will be of use to those who are interested in the phenomena of natural catastrophes as well as those who want to learn more about the life and work of Charles Darwin.

Auckland, New Zealand Sh. U. Galiev

Acknowledgments

I thank Professors M. Ilgamov, R. Ganiev, R. Nigmatulin, E. Pelinovsky, and T. Talipova (Russian Academy of Science) and Professors V. Troshchenko and V. Astanin (Ukraine Academy of Science) for many useful discussions on the results of this book. Some of these results were obtained in cooperation with Professors N. Akhmediev, Y. Kivshar, and Dr. A. Ankiewicz (Australian National University) and Professor I. Gabitov (Los Alamos National Laboratory) during my visits to Canberra and Los Alamos. I would also like to thank Professors D. Bhattacharyya, R. Flay, B. Mace, and G. Mallinson (University of Auckland), with whom I studied strongly localized waves, extreme waves, and tsunamis. I am most grateful to my friends Professor Stoyko Fakirov (University of Auckland) and Professor Victor Lazarev (Centre Modern Education, Russia) for their assistance at various stages in the production of this work.

I would further like to acknowledge Professor Garry Tee (University of Auckland), the Scientific Editor of this book. His long-time collaboration and contribution in terms of comments, thoughts, and ideas has been most valuable.

Finally, I am grateful for the contribution of all those outstanding scientists and authors of scientific articles at the various websites which have been used in the book.

The occasion of Darwin's bicentenary provided me with the impetus to start writing this book. Over the years, I have been particularly inspired by *The Voyage of the Beagle*, and it was this work that determined the direction of my research. In the course of my rereadings and reflections on this book, it came upon me to explain Darwin's observations on extreme waves and geodynamics on the basis of my own knowledge of mechanics, mathematics, and physics. I leave it to the reader, of course, to judge my success in this undertaking.

Galiev

Contents

Chapter 1
Prologue: A Few Notes About Charles Darwin, His Researches and the Contents of the Book

> *..., I look at the natural geological record, as a history of the world imperfectly kept, and written in a changing dialect; of this history we possess the last volume alone,... (**Charles Darwin.** On the Origin of Species by Means of Natural Selection, London, John Murray, 1859, p. 311.)*

1.1 Darwin's Youth and the Voyage of the *Beagle*

Charles Robert Darwin was born in Shrewsbury, Shropshire, England on 12th February 1809 at his family home, the Mount. He was the fifth of six children of the wealthy society doctor and financier Robert Darwin F.R.S., and Susannah Darwin.

The family was not aristocratic, but belonged to an elite class of highly intellectual English families. According to all accounts the young Darwin was a polite, talkative and affectionate boy with an obvious interest in natural science. However, he had not been keen on the basic subjects on the curriculum at the classical school where he was taught. He recalled that ... *When I left the school I was for my age neither high nor low in it: and I believe that I was considered by all my masters and my Father as a very ordinary boy, rather below the common standard in intellect,... I took much pleasure in watching the habits of birds,.... In my simplicity I remember wondering why every gentleman did not become an ornithologist.*

Towards the close of my school-life, my brother worked hard at chemistry ... and I was allowed to aid him as a servant in most of his experiments. ... This was the best part of my education at school, for it showed me practically the meaning of experimental science...

The 16-year old Darwin's obvious interest in science was stimulated when he began to study at the Medical School of Edinburgh University. There he began his researches and made his first two discoveries in science, connected with soft small sea organisms. During those years Darwin became acquainted with a very pleasant and intelligent black man who had been freed from slavery, who taught him the art of taxidermy. He also had friends who were interested in natural sciences and anatomy. However, Darwin did not wish to become a doctor. The painful experience of watching a surgical operation on a child, performed without anaesthetics,

Sh.U. Galiev, *Darwin, Geodynamics and Extreme Waves*,
DOI 10.1007/978-3-319-16994-1_1

haunted him for many years, and he chose to look for his future profession in another field. His father was probably not happy about this. He had complained that Charles seemed to "care for nothing but shooting, dogs, and rat-catching," and declared "you will be a disgrace to yourself and all your family," to the deep affliction of his son, who held him in the highest esteem [1]. Charles' father decided that he should enter the priesthood. Darwin was not set against this. While he had experienced some doubts about his religion, the idea of being a rural priest was appealing to him. He began to study theology, among other subjects, at Christ College at Cambridge University. This did not take up all of his time, and he was also able to enjoy a social life and pursue his own interests.

By all reports, he was an outstanding student, and a few bright young professors started to involve him in their research on a casual basis. These contacts had very important and unexpected consequences, leading him to be offered a place on a naval ship, the *Beagle*, upon the completion of his studies. The aim of the ship's global voyage was the production of navigational charts. Its captain was Robert FitzRoy, a brilliant aristocrat of royal descent. (The particle 'Fitz' in a person's name signifies that he is descended from an English king.) FitzRoy decided to take a scientist on the voyage who would undertake the study of the flora and fauna of different countries, and Darwin was recommended for this research work. At first sight, the choice of a young graduate for the difficult and responsible mission of studying the natural science of many little-known countries may seem surprising. However, by that time Darwin possessed an extensive stock of knowledge in various areas and was familiar with the techniques which were used in zoological, botanical and geological research. Most importantly, he was able to work independently, observing various phenomena and assessing whether or not they were new discoveries.

On the other hand, it should be stressed that the work was not official – he was not paid – and Darwin could only accept FitzRoy's offer because he came from a wealthy family. Of course, he also needed his father's consent. Initially his father regarded the voyage as unacceptable ... *A wild scheme ..... Disreputable to my character as a Clergyman ... That you should consider it as again changing my profession* (letter to R.W. Darwin, 31st August, 1831). Later, however, he had a change of mind, and Charles was allowed to embark on this global voyage that would change the course of his life.

The voyage lasted for 5 years (1831–1836). At every opportunity Darwin went ashore to do observations and to collect specimens of geological materials, plants and animals. He made long trips deep into territories of South America. In fact, Darwin spent more time on land than he did on the *Beagle*. He brought home a huge volume of specimens, impressions and observations, which he subsequently used for writing his outstanding books and articles.

In 2009, Darwin's bicentenary was celebrated worldwide. That year also marked the 150th anniversary of the publication of **On the Origin of Species by Means of Natural Selection, or the Preservation of Favoured Races in the Struggle for Life**. In the 6th edition of 1872, the title of this book was changed to **The Origin of Species** [2]. Many papers and books were published to commemorate these events (see, for example, [3–10]). 2009 year was also the 170th anniversary of the publication of **Journal of Researches**, published in 1839 and commonly referred to as **The Voyage of the *Beagle*** [11]. In that book Darwin first formulated many of his great ideas, based on his extensive knowledge of different peoples, birds, animals, and plants (Fig. 1.1).

Besides being a brilliant naturalist, Darwin could write with great expressiveness and clarity. This made his books accessible to a wide readership, scientists and non-scientists alike. Among his numerous publications, **The Voyage of the *Beagle*** had particularly universal appeal [11, 13, 14]. English literary critics have always recognized this book as representing, in its language and manner of expression, an excellent example of English prose of the second quarter of the nineteenth century, showing Darwin to be an outstanding stylist (Fig. 1.2).

On this point, let us mention that Osip Mandel'shtam, one of the best Russian poets of the twentieth century, had devoted two articles to Darwin: 'About a problem of Darwin's scientific style' and 'About the literary style of Darwin'. He wrote: 'The books of Darwin and Dickens were read by the same public. The scientific success of Darwin was in some part also literary'. This is high praise indeed, if we bear in a mind that Charles Dickens is one of the most widely-read authors in the world; in terms of great literary figures, he follows closely behind Homer, Miguel de Cervantes and William Shakespeare. Certainly, Mandel'shtam's words apply to **The Voyage of the *Beagle***.

Darwin did not limit himself to discussing scientific problems as a naturalist; his interests were much wider. He provided a broad picture of life as it was lived in the various countries visited by the *Beagle*. Perhaps, after learning this, some readers of the present work will decide to read – or re-read – **The Voyage of the *Beagle***. It is not only the diary of a researcher, but to a greater degree, the diary of an exceptionally courageous, enterprising, noble and humane person. The author reproduces here a passage from that book in illustration of his qualities. *...Picture to yourself the chance, ever hanging over you, of your wife and your little children – those objects which nature urges even the slave to call his own – being torn from you and sold like beasts to the first bidder! And these deeds are done and palliated by men who profess to love their neighbours as themselves, who believe in God, and pray that his will be done on earth! It makes one's blood boil, yet heart tremble, ...* [13, p. 597].

Probably, these words most fully characterize Darwin in the years before his great mission. Throughout his life, he remained a good, modest and kind person, always acting in accordance with the spiritual values which had been instilled in his childhood. Darwin was not only a meticulous scientist; he was a passionate humanitarian [15–17].

During the voyage Darwin developed a rare combination of scientific strengths: a dedication to careful fact-gathering and a propensity to theorize about the facts. The success of his geological ideas encouraged him to search for other universal laws. Indeed, the isolation of the voyage, combined with the exposure to new phenomena, taught him to think independently and in an original way, although within the confines of the scientific culture of his time.

Fig. 1.1 The chart shows the *Beagle's* route and the territories which were visited by Darwin during his travels around South America [12]

Fig. 1.2 Hunting with the bolas as described in **The Voyage of the *Beagle*** [13]

1.2 The Baconian Principles and Darwin's Method

Although Darwin spent long periods of time ashore, he was on the ship for nearly half of the voyage. Prone to seasickness, he would rest lying in a hammock, as the waves surged and snarled against the rounded hull of the ship. He spent his time reading books and trying to understand the data he had collected. He recalled in his **Autobiographies** [18] ... *The investigation of the geology of all the places visited was far more important, as reasoning here comes into play ... Everything about which I thought or read was made to bear directly on what I had seen and was likely to see; and this habit of mind was continued during the five years of the voyage. I feel sure that it was this training which has enabled me to do whatever I have done in science...*

I worked on true Baconian principles, and without any theory collected facts on a wholesale scale, more especially with respect to domesticated productions, by printed enquiries, by conversation with skilful breeders and gardeners, and by extensive reading...., I am surprised at my industry .... Darwin's creative activity began from early on in the voyage, and involved the data he had started to collect from those first days. From this data he could quickly construct working hypotheses, which he would then check. He recalled .... *Moreover, as a number of isolated facts soon become uninteresting, the habit of comparison leads to generalization ...* [13, p. 603]. He hoped to find simple but general ideas which would allow him to describe these data. For example, a brilliant idea about the subsidence of some landmasses and corresponding elevation allowed him to explain much data concerning ocean reefs and atolls.

He wrote to his father on 8th February 1832: ... *I find to my great surprise that a ship is singularly comfortable for all sorts of work.- Everything is so close at hand, & being cramped, make one so methodical, that in the end I have been a gainer.-* .... Darwin described the methodical research techniques which he had developed during the first days of the voyage; in his articles [19, pp. 166–204] and [20, pp. 227–250] he wrote ... *A few cautions may be here inserted on the method of collecting. Every single specimen ought to be numbered with a printed number ... and a book kept exclusively for their entry. As the value of many specimens entirely depends on the stratum or locality whence they were procured being known, it is highly necessary that every specimen should be ticketed on the same day when collected. If this be not done, in after years the collector will never feel an absolute certainty that his tickets and references are correct. It is very troublesome ticketing every separate fossil from the same stratum, yet it is particularly desirable that this should be done; for when the species are subsequently compared by naturalists, mistakes are extremely liable to occur; and it should always be borne in mind, that misplaced fossils are far worse than none at all. Pill-boxes are very useful for packing fossils. ... To save subsequent trouble, it will be found convenient to pack up and mark outside, sets of specimens from different localities. These details may appear trifling; but few are aware of the labour of opening and arranging a large collection, and such have seldom been brought home without some errors and confusion having crept in ....*

In conclusion, it may be re-urged that the young geologist must bear in mind, that to collect specimens is the least part of his labour. ... But let his aim be higher: by making sectional diagrams as accurately as possible of every district which he visits (nor let him suppose that accuracy is a quality to be acquired at will), by collecting for his own use, and carefully examining numerous rock-specimens, and by acquiring the habit of patiently seeking the cause of everything which meets his eye, and by comparing it with all that he has himself seen or read of, ...

Darwin followed these precepts throughout the voyage and indeed throughout his life, which he spent trying to understand 'the mystery of mysteries' .... *Hence, both in space and time, we seem to be brought somewhat near to that great fact – that mystery of mysteries – the first appearance of new beings on this earth. ...* [11, p. 454]. In one of the comments in his notebook, dated 1837, Darwin wrote that ... *the intimate relationship between the vital phenomena with chemistry and its laws makes the idea of spontaneous generation conceivable* ... In a letter of 1871 to his friend Joseph Hooker, he pondered ... *it is often said that all the conditions for the first production of a living being are now present, which could ever have been present. But if (and oh what a big if) we could conceive in some warm little pond with all sort of ammonia and phosphoric salts, — light, heat, electricity present, that a protein compound was chemically formed, ready to undergo still more complex changes, at the present such matter would be instantly devoured, or absorbed, which would not have been the case before living creatures were formed...* In other letters, Darwin would talk of his theory that the origin of life was a result of the self-organizing and evolution of matter, but the time for this idea was yet to come. Indeed, it was only in 1922 that the biochemist Aleksandr Oparin put forward a theory of the origin of life [21, 22]. Oparin believed that in the presence of methane and other gases in the atmosphere of the young Earth, the appearance of life was a fairly unremarkable biochemical process. He theorized that in the extreme conditions presented by the young planet's volcanism and electrical storms, these elements could have been readily 'cooked up,' so to speak, into the precursor chemicals for life. Oparin's ideas were not far from the ideas that Darwin had formulated much earlier, which he could now qualitatively support with a great many observations and experiments [23, 24].

Darwin could only speculate on the origin of life, although throughout his career he studied certain aspects of its evolution. Such theorizing demanded huge efforts. Indeed, over the millennia some of the greatest minds had tried to understand how life, and different species, came into being. In the past these occurrences had usually been ascribed to some instant act of powerful will; and this will of the gods (or a God) was considered as the source of Nature! Many religious dogmas were constructed on the basis of this idea, and it was very difficult to reconcile these dogmas with modern science. Often, scientists would try to develop science within the framework of traditional official ideologies, so as to avoid conflict with society.

The great scientific discoveries of Galileo and Newton had not essentially changed this situation; the Christian picture of the Creation had remained more or less untouched. Isaac Newton, in particular, being a deeply religious person, devoted much time to his attempts to reconcile scientific knowledge and Bible dogmas. At the end of his book **Opticks** (1704) he posed the following questions: '...What is there in places almost empty of matter, and whence is it that the sun and planets gravitate towards one another without dense matter between them ?... what hinders the fixed stars from falling on one another? How came the bodies of animals to be contrived with so much art,... Was the eye contrived without skill in Opticks, and the ear without knowledge of sounds?... and whence is the instinct in Animals?...'.

Newton proceeded to give an answer which was compatible with fundamental world religious doctrines '...does it not appear from phaenomena that there is a Being incorporeal, living, intelligent, omnipresent, who in infinite space,...?'. In one of his letters to Richard Bentley (12th October 1692), he repeated this thought: '...To make this system, therefore, with all its motions, required a cause ... that cause to be, not blind and fortuitous, but very well skilled in mechanics and geometry...' [25].

In contrast with Newton, Darwin was rather more rational. A realist and an atheist, he focused less on hypotheses and more on Nature itself. Only Darwin fully introduced time into the history of Nature. Of course, he knew about the *dialectic* of the ancient Greeks, and he used the global geological ideas of James Hutton and Charles Lyell [26]. He hoped that after **The Origin of Species**, the book of Genesis would be considered only as a literary history. He threw Man off the Throne of the creator God [27].

Darwin's theory of natural selection basically attempts to explain Nature's beautiful design in terms of physical, engineering and natural laws – without invoking the interventions of a 'Grand Designer'. At the same time he did not try, as some scientists and philosophers had done before him (for example, René Descartes) to embrace all natural phenomena. He selected *variability of species* and studied it intensively for more than 20 years ... *patiently accumulating and reflecting on all sorts of facts...* [2, Introduction]. This research method opened the door to great discoveries. Generally speaking, Darwin's method reminds us of Galileo's doctrine that physical principles must rest only on experience and observations.

In spite of its great success, the idea of natural selection was attacked from several sides. Sometimes the criticism was just. All great theories have gaps in physical or philosophical senses, and Darwin understood this very well. For example, it is possible to ask 'Selection by whom or by what?' [28]. He tried to avoid such questions by saying that he had as much right to use metaphorical language as physicists did. In his own words: ... *who objects to an author speaking of the attraction of gravity as ruling the movements of the planets? Everyone knows what is meant and is implied by such metaphorical expressions* ... [2, Chapter 4].

Indeed, up until the time of Darwin, natural phenomena had simply been *described* in precise language which *explained* nothing. In his '**Discourses and Mathematical Demonstrations Concerning Two New Sciences,**' Galileo said of the phrase *the acceleration of natural motion* '...The present does not seem to be the proper time to investigate the cause of the acceleration of natural motion concerning which various opinions have been expressed by various philosophers,.... At present it is the purpose of our Author merely to investigate and to demonstrate some of the properties of accelerated motion (whatever the cause of this acceleration may be)-meaning thereby a motion,....'. Initial reactions to this idea were negative. Even Descartes protested at Galileo's decision to seek descriptive formulas: '...Everything that Galileo says about bodies falling in empty space is built without foundation: he ought first to have determined the nature of weight ...' [27].

Thus, we are led to think that Darwin used the methodologies of Galileo, as well as those of Francis Bacon during his research; they were also based on the dialectic of the classical Greeks and Descartes. Darwin considered geology and biology to be fully historical and physical sciences. First he would quickly formulate a working hypothesis, and then check it using a vast amount of accumulated data. Only after that did he formulate his hypothesis as a theory. Biblical literalists objected that the theory of natural selection was cold and mechanical. However, Darwin thought that his theory showed the lifeblood of the natural world. In the final sentence of '**The Origin of Species**', he exclaimed that ..... *from the war of nature, from famine and death, the most exalted object which we are capable of conceiving, namely, the production of the higher animals, directly follows. There is grandeur in this view of life, with its several powers, having been originally breathed into a few forms or into one; and that, whilst this planet has gone cycling on according to the fixed law of gravity, from so simple a beginning endless forms most beautiful and most wonderful have been, and are being, evolved...* [2, p. 490].

1.3 Darwinism, Its Criticism. Simplicity and Beauty in Science

It is known that for many years Darwin did not intend to publish his theory of evolution, since he did not want to clash with the official ideology [29]. However, in **The Voyage of the *Beagle*** Darwin presented a few thoughts concerning the evolution of species. In particular, he formulated the following important ideas *... Seeing this gradation and diversity of structure in one small, intimately related group of birds, one might really fancy that, from an original paucity of birds in this archipelago, one species had been taken and modified for different ends ...* [13, p. 456], *.... some check is constantly preventing the too rapid increase of every organized being left in a state of nature. The supply of food, on an average, remains constant; yet the tendency in every animal to increase by propagation is geometrical, ...* [13, p. 215]. A few other statements about evolution [11, 13, 30–32] also appeared.

Darwin intended to present his theory in a book to be published posthumously. (In this regard he proved to be more courageous than the great mathematician Karl Gauss, who did not dare to publish his non-Euclidean geometry, because he feared the 'clamour of the Boeotians'.) Fortunately, however, he heeded the opinion of his friends, who urgently recommended that he publish his work and indeed helped with the publication. This situation is similar to the history of the publication of Newton's major work.

However it is remarkable that not only in the first edition of **The Voyage of the *Beagle***, but also in its second edition which was published in 1845 [13], Darwin concealed his ideas about evolution of species. Indeed he continued to do so up until to 1858–1859. If his theory had not been published then Darwin's name, at best, would have been placed close to some geniuses who could only anticipate the results of the development of science. Democritus of Abdera, for example, formulated even more far-reaching ideas: 'There are incalculable worlds of the various sizes. In some is not present either the sun, or the moon, in others they more than ours and them more than one. These worlds are located on various distances. One of them arises and prospers, others die. They can collide. Some worlds have no animals and plants, and even water'. As we know, however, the theory of evolution of species did get published, and Darwin subsequently became the most prominent and widely-known representative of science of his age [33, 34]. Darwin saw natural selection as the mechanism by which advantageous variations passed on to succeeding generations, in the process of which the less competitive traits of individuals gradually disappeared from populations. There is no other theory which has, for more than 150 years, been exposed to the most intense checks and criticisms and which has, each time, been found to be in conformity with the latest achievements of science. All objections have been extremely well answered by subsequent researchers in geology, physics, genetics and biology.

Lord Kelvin (William Thomson), a major physicist of the nineteenth century, launched one of the most powerful attacks against the theory of evolution. His point was that Darwin's theory is based on an explicit assumption of very great age of the Earth. Darwin had come to the conclusion that about 300 million years had passed since the Cretaceous period.[1] This was based on his analysis of the rate of erosion of the surface of a large valley in England (the Weald, in Kent). A comparably lengthy time period is an indispensable condition for biological evolution.

Indeed, it is only possible to discuss theories about biological science after the age of the Earth (measured in billions of years) has been decided. If it were definitely proven that geological history is much shorter than Darwin's estimate, then the results of the theory of evolution become mere speculations which have no basis in reality. With regard to this point, Kelvin emphasized that the Sun's energy is necessary for biological evolution, and in his opinion the Sun could not shine for more than 3,000 years if its energy derived from chemical reactions. On the assumption that gravitation is the source of the Sun's energy, then, Kelvin reached a result that was more favourable to Darwin, that the period of evolution may last for about 20 million years – but that is still far too short for the slow process of natural selection. This was one of the strongest blows dealt to the theory of evolution. Kelvin was one of the best-known scientists in Great Britain, and he was operating with the 'infallible' data of physics and mathematics. Darwin respected mathematical knowledge. He regretted that, in his own words, he was not able ... *to understand something of the great leading principles of mathematics; for men thus endowed seem to have an extra sense. But I do not believe that I should ever have succeeded beyond a very low grade* ... [18, p. 30]. Having taken Kelvin's opinion into consideration, in the later editions of **The Origin of Species** Darwin excluded his estimation of the age of the Earth. But while he showed respect for his opponent, he did not accept Kelvin's view.... *namely, that though our continents and oceans have endured for an enormous period in nearly their present relative positions, we have no reason to assume that this has always been the case; consequently formations much older than any now known may lie buried beneath the great oceans. With respect to the lapse of time not having been sufficient since our planet was consolidated for the assumed amount of organic change, and this objection, as urged by Sir William Thomson, is probably one of the gravest as yet advanced, I can only say, firstly, that we do not know at what rate species change, as measured by years, and secondly, that many philosophers are not as yet willing to admit that we know enough of the constitution of the universe and of the interior of our globe to speculate with safety on its past duration. That the geological record is imperfect all will admit; but that it is imperfect to the degree required by our theory, few will be inclined to admit. If we look to long enough intervals of time, geology plainly declares that species have all changed; and they have changed in the manner required by the theory, for they have changed slowly and in a graduated manner....* [2, Chapter XV]. Many of Kelvin's contemporaries accepted his opinion – but Darwin was right about this most important matter. Modern estimates give the age of the Earth as almost 4.7 billion years, which is more than enough for

[1] It is now known to have ended about 66 million years ago.

evolution theory. For all of that time, the vivifying rays of the Sun have warmed the Earth! In the late 1920s the English astronomer Sir Arthur Stanley Eddington theorized that the Sun contains a large quantity of energy, which is released during nuclear processes. Like Darwin's estimate, Eddington's theory met with much criticism. Its opponents postulated that the temperatures inside stars were not high enough for thermonuclear reactions to occur. Eddington responded to these objections saying: 'We do not argue with the critic who urges that the stars are not hot enough for this process; we tell him to go and find a hotter place'.[2]

At the same time, the Russian physicist George Gamow formulated the concept of quantum tunnelling to explain the radioactive disintegration of some atomic nuclei, in which enormous energy is released. On the basis of this idea and Eddington's hypothesis, Hans Bethe constructed (in 1938) a theory of thermonuclear processes in stars, which completely swept aside Kelvin's objections. A few years later Gamow constructed the first adequate theory of the evolution of stars by this thermonuclear energy source, which led in time to the Big Bang theory and the theory of the origin of matter in the Universe.

In June 1867 Darwin's theory again came under attack when Fleeming Jenkin, Professor of Engineering at the University of Edinburgh, published an article in which he criticized natural selection as a source of evolution. In Darwin's time, the standard model of inheritance was 'blood' or some fluid. Inheritance was considered to be a result of the mixture of the 'bloodlines' of the father and the mother. For example, the colour of a child's eyes was some 'blend' of the father's eye colour with the mother's eye colour. Darwin's theory obviously used this idea. According to Jenkin, however, the useful variations of species could persist only if they occurred at once in a great number of individuals and during a short space of time (in one generation), which contradicts Darwin's basic idea.[3] After examining Jenkin's objections, Darwin wrote to Alfred Russell Wallace (January 22nd, 1869): *...I have been interrupted in my regular work in preparing a new edition of the 'Origin,' which has cost me much labour, and which I hope I have considerably improved in two or three important points. I always thought individual differences more important than single variations, but now I have come to the conclusion that they are of paramount importance, and in this I believe I agree with you. Fleeming Jenkin's arguments have convinced me.* ... He called these arguments '*Jenkin's nightmare*'.

That 'nightmare' tormented Darwin till the end of his days. In response he developed a hypothesis about the inheritance of variations, which he called 'pangenesis'. Darwin understood that the variations might be inherited not like fluid, but like particles. He wrote ... *I give my well-abused hypothesis of Pangenesis. An unverified hypothesis is of little or no value. But if ... such hypothesis could be*

[2] A polite way to say 'Go to Hell!'?

[3] Jenkin assumed that features inherited from each parent got averaged at each generation; hence an isolated feature would get halved each time, and so it would effectively disappear after several generations.

established, I shall have done good service, as an astonishing number of isolated facts can thus be connected together and rendered intelligible . . . [18, p. 79].

According to his hypothesis, any cell of any organism contains special microparticles (gemmules), which are responsible for the inheritance of variations. Darwin emphasized that only cells could regenerate new tissue and new organisms. He theorized that the gemmules form cells which diffuse and then form the reproductive organs. It is important to remember that at this time, when Gregor Mendel had just formulated his laws of heredity (in 1865), the concept of atoms was far from being universally accepted. The electron was not discovered until 30 years later. Therefore, though Democritus of Abdera, Hippocrates of Cos and others had speculated that the transfer of parents' variations to their descendants provides a set of separate (discrete) hereditary units, we believe that Darwin was among the first to consider units of hereditary information. He noted that . . .*the existence of free gemmules is a gratuitous assumption.* (**The Variation of Animals and Plants under Domestication,** p. 452). Such was Darwin's authority that this idea disseminated through the scientific community. In the modern interpretation, gemmules may be considered as a mix of DNA, RNA, proteins, and other mobile elements that contain the heritable information.

We are not sure, but it is impossible to reject the possibility that Darwin's concept of gemmules influenced the formation of a revolutionary concept of the quantum of energy, and of the heredity gene. Thus, his hypothesis contains a representation of units for a material basis of heredity (genes and chromosomes) which was indeed later confirmed. The discovery of these units of heredity has allowed us to overcome 'Jenkin's nightmare'.

The outstanding Russian geneticist Sergey Chetverikov appears to have been the first to reject Jenkin's arguments, and took the first step towards connecting evolution theory and genetics. Later, another Russian geneticist, Theodosius Dobzhansky, wrote a book which was consonant with Darwin's [35]. Recent research by some French scientists (in 2010) allows us to consider so-called microvesicles, which are vials from 30 nm to 4 μm in size, which are universally encountered in the liquid of cells [36], as with Darwin's gemmules.

'Jenkin's nightmare' would probably not have haunted Darwin if he had known about the epochal discovery made by Gregor Mendel. On 8th March 1865, Mendel reported the results of experiments with pea plants to the Natural History Society of Brünn. These results were published in the Proceedings of the Society at the end of 1866, and that publication was distributed to 120 university libraries around the world. Mendel sent about 40 copies of his report to eminent researcher-botanists. Darwin also received the report, but it got buried under all his other papers and he never read it. (He did read German, but only with difficulty.) Although Mendel was initially convinced that he had made a great discovery, it did not arouse any interest among his contemporaries. Moreover, the results of his subsequent experiments did not support the laws he had posited, and as a result, he lost faith in it himself.

I have observed that similar events do sometimes occur in science. In 1917, Albert Einstein formulated his equations of the general theory of relativity, which when applied to the Universe as a whole predict that the Universe is unstable as a consequence of gravitation and energy. But the concept of an evolving universe, just like the theory of evolution of species, contradicted contemporary accepted ideas. Thus Einstein rejected this idea. He introduced a special parameter into his theory compensating the gravitation in cosmological scales and providing the stability of the Universe according to his equations. That parameter was called the cosmological constant, and it describes very weak repulsion between any two masses, which increases with distance. Later, after Aleksander Friedmann's calculations[4] and Edwin Hubble's observations (1929), the expansion of the Universe was accepted by everyone, including Einstein. He said that the introduction of the cosmological constant into his equations had been the greatest blunder of his life, and he regretted it until the day he died. However, in 1998 and the following years some observations produced unexpected results, which show convincingly that the Universe expands in a complex manner. As a first approximation, the expansion is determined by Einstein's cosmological constant. Now that once unacceptable theory of the expansion of the Universe has become a well-documented fact.

Darwin was bold enough to construct his brilliant theory 'on sand'; the gemmule idea was indeed a weak base for it. On the other hand, he understood clearly the incidental character of the process of natural selection. Though Darwin did not explicitly use probability methods, he did introduce probability into biology before Mendel. It should be remembered that in those years causal processes in science were practically never considered – Laplacean determinism reigned supreme. Thus, probability was not then a good base for a theory. Despite all of that, Darwin's theory was robust because it was based ... *on all sorts of facts* ... [2, Introduction]. .... *I cannot believe that a false theory would explain, as it seems to me that the theory of natural selection does explain, the several large classes of facts above specified. It is no valid objection that science as yet throws no light on the far higher problem of the essence or origin of life. Who can explain what is the essence of the attraction of gravity? No one now objects to following out the results consequent on this unknown element of attraction; notwithstanding that Leibnitz formerly accused Newton of introducing "occult qualities and miracles into philosophy."* [the 1861 3rd edition of **The Origin of Species**].

Recently, modern geneticists have formed a basis for that theory. Natural selection probably works universally, and not only among living (biological) systems.

The preceding discussion leads to the idea that the greatest discoveries of geniuses such as Nicolaus Copernicus, Galileo Galilei, Charles Darwin, Dmitri Mendeleev, James Maxwell, Albert Einstein and others were brilliant because they contained fundamental predictions that extended beyond current scientific knowledge. Galileo recalled that during his student years Copernicus' concept – that in

---

[4] Alexander Friedmann, a professor of Petrograd University, learned from Einstein's paper that without the cosmological constant, the theory had no static solution. Friedmann took a radical step that would immortalize his name; he broke away from the static paradigm and found the dynamical solutions of Einstein's equations.

representations of the Universe it was necessary to interchange the positions of the Sun and the Earth – seemed utterly wild. Even the most educated people of the time believed that Copernicus' model had no advantages except simplicity and beauty. Indeed, everyone could see its apparent discrepancy by mere observation – the Earth was not seen to move. Similarly, the planetary model of an atom constructed by Niels Bohr was contradictory to Maxwell's equations. Certainly, all of these researchers saw the defects of their models. However, the bewitching beauty and simplicity of the theoretical constructions confirmed them in the belief that future scientists would be able to polish and refine those new theories. The basic idea that sustained them through doubts and uncertainty was expressed well by Werner Heisenberg 'If nature leads us to mathematical forms of great simplicity and beauty ..., we cannot help thinking that they are "true" ...'.

Henry Poincaré wrote: '... this special aesthetic sensibility that plays the part of the delicate sieve .... Sufficiently clear why the man who has it not will never be a real discoverer ...'. Truth is always beauty, but one can say more — that sometimes beauty is truth. Of course, beauty and simplicity are rather subjective concepts, and simplicity should not be transformed into a vulgarization. Some great minds have understood this very well. Einstein, who was a great admirer of beauty and simplicity, formulated the maxim that 'Everything should be as simple as possible, but not simpler'.

The words 'beauty', 'fine', and 'beautiful' occur more than 20 times in Darwin's book **The Origin of Species**. In certain cases Darwin used the word 'fine' in the sense of 'good' or 'excellent'. As with **The Voyage of the *Beagle***, Darwin included both scientific facts and personal impressions in **The Origin of Species.** His son Francis described how he always aspired to simple and emotional presentation, even where difficult elements of his theories were concerned [1]. In particular, **The Origin of Species** is filled with the high aesthetics that Poincaré speaks of. The emotional and expressive literary style of the author, the power of his thoughts and the persuasiveness of his arguments, make reading **The Origin of Species** a deeply aesthetic experience [37]. In his younger days Darwin had been very fond of poetry, art and music, but after the age of 30 these interests waned as he became increasingly involved in his research. He would come to regret this very much. Near the end of his life he wrote ...*This curious and lamentable loss of the higher aesthetic tastes ... is a loss of happiness* .... [18, p. 85].

Before Darwin, each species was perceived as being in its own eternal category; each had been made, individually, by God. But evolution by natural selection means that all species have a common ancestor; they are unified into one great family. In this sense Darwin has greatly simplified our understanding of Nature.

No other book has affected the development of society so quickly and extensively as **The Origin of Species**. It is almost as if human society had needed and waited for such a theory to emerge. Many cultures and countries were ready for the so-called 'Darwin revolution', although Darwin himself was certainly never a revolutionary or a militant atheist. He rejected any violence, which usually accompanied social and religious revolutions. As an example of the latter, take the encounter of Christian and Muslim religions, during which some great civilizations collapsed.

Darwin respected the religious beliefs of others – those of his wife, who was a deeply religious woman, and also those of some of his friends and teachers. However his book would lay a scientific base for atheism, and serve to promote and distribute it around the globe. In this way, indirectly, it would also serve to promote the rise of modern developed countries; the less influence that religion has in a country, the less influence of corruption and criminality, and the more stable and rich is the life of its people.

We want to emphasize that the influence of Darwin's researches expanded far beyond the realm of science. Thanks to Darwin, the idea of evolution is now central in the social and spiritual life of human society. The idea of progress which is implicit in this concept has become the basis for many social, political and technological doctrines. It seems that Darwin anticipated these things might happen; this may explain why he delayed publishing his research on evolution for almost 20 years. In **The Origin of Species** Darwin demonstrated, with mathematical precision, the uselessness of God as an explanation of the laws of Nature. He seemed to experience some guilt about this, a feeling that he had shattered mankind's dreams about God. As he wrote to his close friend Joseph Hooker: ... *It is as if one were confessing to a murder* ... But even Darwin could not have foreseen just how dangerous and outrageous his book would appear to some people. It would change the way that people thought about humanity, nature and indeed the Universe itself [1, 38].

The publication of **The Origin of Species** had many repercussions for its author. The uproar that it caused put him under psychological pressure that effectively ruined his health. By 1864 he had started to feel extremely unwell. Within the space of 2 years he had become a sickly old man with a huge gray beard. This picture of Darwin, with which we are all familiar, was taken when he was just 57 years old. (Doctors continue to debate the source of his illness.) Notwithstanding, he found the strength to continue his epoch-making researches [1].

Darwin began to study the faculty of the emotions, and in 1872 he published **The Expression of the Emotions in Man and Animals**. Probably, what led Darwin to this type of research was the obvious connection between a person's creative abilities and their emotional life.

1.4 Darwin's Geological, Physical and Mechanical Interests

We also want to emphasize that Darwin was well-known in the scientific world as a naturalist and geologist before the publication of **The Origin of Species**. In particular, the majority of his publications up until 1859 are fully or at least partly devoted to geology. **The Voyage of the *Beagle*** is the first considerable publication that introduced readers to this field of Darwin's interest. His three most important geological discoveries on the voyage are as follows.

First, Darwin established that oceanic islands lying far from continents are either formed by corals or else consist of volcanic matter, and this result was developed further in his subsequent research. He showed that both continental and island volcanoes are connected with large cracks of the Earth's crust. Those cracks were formed by the lowering and raising of the seabed and the continents [11, 13, 30]. He wrote ... *we see that many scattered volcanic islands and small groups are related, not only by proximity, but in the direction of the fissures of eruption to the neighbouring continents-* .. [30, p. 101] and ... *it would appear that volcanoes burst forth into action and become extinguished on the same spots, according as elevatory or subsiding movements prevail there* ... [13, p. 575].

Darwin came to the conclusion that over many millions of years the continent of South America had been subjected to numerous movements of raising and lowering, which alternated with periods of stillness [11, 13, 31]. **The Voyage of the *Beagle*** described the newest geological discoveries and how the surface of the Earth had evolved over many millions of years. Prior to embarking on the voyage, and like many from a religious background, Darwin had no doubts about the Bible dogma that animal and plant species were invariant, and all had been created by God about 6,000 years ago. Perhaps, he knew that the Primate of All Ireland, Archbishop James Ussher, had used the Bible and other sources to pinpoint the data of creation to Sunday 23 October 4004 BC. Isaac Newton thought the year was 3988 BC. But in 1749 the French naturalist Comte de Buffon had found that the Earth could be up to 70,000 years old. Then the well-known philosopher Immanuel Kant had suggested that the Earth could have existed, not for thousands but for millions of years. However, neither Buffon nor Kant attempted to verify these assumptions. It was the Scottish natural philosopher James Hutton and the Scottish geologist Charles Lyell [26, 27] who formally presented the proof that the Earth was significantly older than 6,000 years; it could have existed for hundreds of millions of years, and maybe more. However, their results were known only to a few scholars, whereas **The Voyage of the *Beagle*** was read by many thousands of people. Thus, the results of Darwin's geological researches, as well as his theory of the origin of species, were revolutionary. He was apparently the first person to think about the possible correlation of the development of life on Earth and the evolution of geological formations. In particular, with regard to coral formation, he wrote ... *the nature of the reefs having been governed by the nature of the earth's movement* ... [13, p. 574].

Darwin's third important geological result is his theory of the origin of coral reefs [11, 13, 32], which was entirely determined by his hypothesis of the lowering and raising of continents ... *the great continents are for the most part rising areas; and from the nature of the coral-reefs, that the central parts of the great oceans are sinking areas.* ... [13, p. 574].

In the above publications, Darwin presented his remarkable ideas with strict logic, simplicity and universal coverage of all parts and features of the geological processes. We stress again that Darwin had already clearly stated his fundamental geological ideas in the first edition of **The Voyage of the *Beagle***.

In February 1859 the Geological Society of London awarded Darwin a great honour – the Wollaston Medal – for his considerable contribution to that field of knowledge. However it must be said that Darwin's scientific interests were much more wide-ranging than those ideas and discoveries which made his name immortal. Reading **The Voyage of the *Beagle***, one is impressed by the depth and scope of Darwin's knowledge. His knowledge of natural history is of course to be expected, but he also displays sound knowledge of chemistry and physics. In particular, Darwin confidently applied this knowledge to explain the results of his observations. For example, he impressed experts in the fields of hydrodynamics, waves and impact mechanics with his surprisingly deep understanding of mechanical phenomena. Reflecting on the flight of birds, he wrote:.. *In the case of any bird soaring, its motion must be sufficiently rapid, so that the action of the inclined surface of its body on the atmosphere, may counterbalance its gravity. The force to keep up the momentum of a body moving in a horizontal plane in that fluid (in which there is so little friction) cannot be great, and this force is all that is wanted*....[11, p. 224]. It would be difficult to explain more clearly the bases of aerodynamics! The Wright brothers probably didn't know more about aeromechanics than Darwin when they began to design their airplane.

Darwin observed the mechanical interaction of small sea animals and ocean waves: ... *Thus do we see the soft and gelatinous body of a polypus, through the agency of the vital laws, conquering the great mechanical power of the waves of an ocean, which neither the art of man, nor the inanimate works of nature could successfully resist*. ... [11, p. 548]. He was interested in the influence of small animals on the structure of the ocean surface. The stability of layers of fresh and salt water also interested him. .... *In the sea around Tierra del Fuego, and at no great distance from the land, I have seen narrow lines of water of a bright red colour, from the number of crustacea,.... how do the various bodies which form the bands with defined edges keep together?* ... [11, pp. 18 and 19] and ... *When within the mouth of the river, I was interested by observing how slowly the waters of the sea and river mixed. The latter, muddy and discoloured, from its less specific gravity, floated on the surface of the salt water*... [11, p. 44].

Darwin observed that wave movement is not directly connected with a displacement of particles of water... *this movement bears no relation to the actual displacement of the fluid from the bulk of the vessel*... [11, p. 378]. Considering vibrations of mountains caused by earthquakes, he wrote..., *that the surface of any body, when vibrating, is in a different condition from the central parts*... [11, p. 371].

With precision and vibrancy, he described an occurrence of catastrophic waves: ...*down came a mass with a roaring noise, and immediately we saw the smooth outline of a wave travelling towards us. The men ran down as quickly as they could to the boats; for the chance of their being dashed to pieces was evident. One of the seamen just caught hold of the bows as the curling breaker reached it. He was knocked over and over, but not hurt*;...[13, pp. 273–274].

The phenomenon he described, that is, the tsunami, is now well-known. The highest measured tsunami was caused by an earthquake in Alaska in 1958, when the impact of a mass of rocks falling into the water of a sea-gulf produced a wave that destroyed trees up to a height of 524 m above sea-level.

Darwin explained the occurrence of porous structure in volcanic bombs by non-stationary processes [11, p. 589], which may be called cavitation in modern terminology. Thus, he described for the first time the cavitation behavior of magma. He also described the huge bubbles and ebullition of sea-water during seaquakes.

Those pages of **The Voyage of the *Beagle*** which describe the 1835 Chilean earthquake are classic descriptions of natural non-stationary phenomena. Darwin wrote that the earthquake and its aftermath had probably affected him more than anything else during the voyage …. *I have not attempted to give any detailed description of the appearance of Concepcion, for I feel it is quite impossible to convey the mingled feeling with which one beholds such a spectacle …. It is a bitter and humiliating thing to see works, … .overthrown in one minute; yet compassion for the inhabitants is almost instantly forgotten, from the interest excited … In my opinion, we have scarcely beheld since leaving England, any other sight so deeply interesting…* [11, pp. 376–377].

Actually, Darwin's account of seismic and mechanical phenomena did not arouse anything like the interest in his well-known geological discoveries, and it was practically never discussed in the scientific literature [39–41]. In our opinion his thoughts related to the 1835 Chilean earthquake and the Earth's dynamics have yet to be fully appreciated.

1.5 The Earth as a Huge Seismically-Active System

Darwin asked … *Where on the face of the earth can we find a spot on which close investigation will not discover signs of that endless cycle of change to which this earth has been, is, and will be subjected?* … [13, p. 590]. In particular, earthquakes changed the surface of the Earth, as he had witnessed; he landed at Valdivia and then at Concepcion, Chile, just before, during, and after the great earthquake, which demolished hundreds of buildings, killing and injuring many people (Fig. 1.3).

It was a huge natural catastrophe. Darwin described the land rising, shaking and cracking before his eyes, volcanoes erupted and giant ocean waves attacked the coastline. Some of these events might seem unbelievable; a number of volcanoes erupted simultaneously, even though separated by a great distance, and islands were lifted up to 3 m. His description could easily be read as a report from Indonesia or Sri Lanka after the catastrophic tsunami of 26 December 2004. Darwin was the first geologist to observe and describe the effects of a great earthquake and its aftermath. Severe earthquakes sometimes have similar effects; but great earthquakes such as in Chile 1835, and giant earthquakes as in Chile 1960, are rare, and they remain completely unpredictable. This is one of the few areas of science where experts remain largely in the dark [42–47].

Let us consider some more of Darwin's geological ideas. He suggested that the Earth's motions were induced by *… the rending of strata, at a point not very deep*

Fig. 1.3 Remains of the cathedral at Concepcion after the 1835 Chilean earthquake (After J.C. Wickham)

below the surface of the earth ... [20, p. 70] and ... *The two kinds of movements may, possibly, be explained, by considering that when the crust yields to the tension, caused by its gradual elevation, there is a jar at the moment of rupture, and a greater movement may be produced by the tilting up of the edges of the strata and by the passage of the fluid rock between them. In breaking a long bar of steel, would not a jar be caused by the fracture, as well as a vibration of the two ends when separate?*... [20, p. 73]. According to Darwin, the global evolution of the Earth's surface may be explained by its vertical and horizontal movements.

We emphasize that Darwin wrote about crust tension, and he connected the earthquake and the eruptions with rupture of the crust, which might resemble the breaking of a long steel bar. The vibrations of the edges of crust elements were a source of this great movement and were an important component of his theory.

Darwin considered the results of the earthquake as important support for his theory of the lowering and raising of continents:... *within the same period one part of the continent has been elevated more than another,*... [20, p. 82]. Indeed, this idea allowed him to explain the formation of coral reefs. Usually Darwin took into account only very slow processes. However, he decided to use the idea of impact vertical motion to explain the shock-like results of the earthquake.

Darwin hoped to develop a universal law for the evolution of the Earth. In particular, he concentrated on an important geological issue: that of crust elevation and subsidence [39, 40]. As Sandra Herbert has observed '... Darwin was also ambitious. He believed that it would be possible to create a "simple" geology based on an understanding of the vertical motions of the earth's crust, ...' [39, p. 356]. At the same time he thought ... *this large portion of the earth's crust floats in a like manner on a sea of molten rock.* ... [20, p. 81]. There was his special focus ... *on the coast of South America,... or to similar ones,* ... [20, p. 74]. Darwin wrote about. . .*the principal phenomena generally accompanying the earthquakes on the west coast of South America; ... the intimate connexion between the volcanic and elevator force;* ... [20, p. 54]. This concurs with plate tectonic theory, according to which seismicity and volcanism on the Earth's surface occurs at plate boundaries.

Thus, Darwin formulated big ideas about the Earth's evolution and its dynamics, and some of his ideas paved the way for future tectonic plate theory. However, plate tectonics does not completely explain the global evolution of the Earth's surface. In particular, why do some earthquakes occur within the plates [42–45]? Darwin emphasized that there are different kinds of earthquakes ... *I confine the foregoing observations to the earthquakes on the coast of South America, or to similar ones, which seem generally to have been accompanied by elevation of the land. But, as we know that subsidence has gone on in other quarters of the world, fissures must there have been formed, and therefore earthquakes*... [20, p. 74]. For example, the magnitude 8.6 earthquake that occurred in the Indian Ocean on 11th April 2012 suggests a new way in which catastrophic earthquakes can come about. Events of this magnitude normally occur in a subduction zone, where one tectonic plate slides beneath another. But that earthquake struck hundreds of kilometres from the nearest subduction zone. It was surprising also that the rupture was shared between four distinct faults, three of which were oriented perpendicular to one another, creating a rough zigzag pattern [48, 49].

Darwin considered the Earth as a gigantic, seismically active system. According to him ... *It has frequently happened, that during the same convulsion large areas of the globe have been agitated,* ... [20, p. 59] and ... *most earthquakes, though appearing sudden, are in truth parts of a prolonged action,*.... [20, p. 66]. Seismologists are convinced that a strong earthquake in one place can cause a similar event to occur at another location 1,000 km away.

Thus, some observations cannot be accounted for by plate tectonics. Indeed, Darwin emphasized that there is some balance of vertical motions; the uplift of mass at one place determines the down movement of mass at another place. He wrote.... *within the same period one part of the continent has been elevated more that another,... but how it can explain the slow elevation ... of great continents, I cannot understand. Within the same view, ... the configuration of the fluid surface of the earth's nucleus is subject to some change, – its cause completely unknown,- its action slow, intermittent, but irresistible*... [20, p. 82].

Darwin explained global change of the Earth's surface by the movements of its continents. However, he almost always included the movement of melted materials in his analysis as well. In a notebook entry written in late 1838 or early 1839, Darwin observed that ... *Assuming from Sir W. Herschel's views earth originally fluid, then cooling process must go from surface toward the interior, – who knows how far that may have penetrated. lower down the temperature may be kept up far higher from circulation of heated fluid or gases under pressure* ... [39, p. 204]. On the other hand ..., *there must have been great horizontal extension, and this, if sudden, would have caused as many continuous outbursts of volcanic matter* ... and ... *if the force had acted suddenly, these portions of the earth's crust would have been absolutely blown off*, ... [20, p. 79]. Darwin did not completely exclude vertical movement even under the Earth's surface. His ideas do not totally agree with the theory of plate tectonics.

Darwin believed [20, p. 81] that the crust sank into molten rock in some places, and in other places the molten rock could erupt from the fractured areas (see Figs. 3.3). A plume of hot mantle material could actively push it up from below and break through to the surface. In particular, this process could partly explain the occurrence of volcanoes far from tectonic plate boundaries.

Plate tectonics cannot explain everything that occurs on the surface of our planet [50]. According to some researchers, evidence of vertical movements within the Earth, reshaping its surface, can be found almost everywhere. Dietmar Müller of the University of Sydney is one who thinks so: 'Geology is on the cusp of another revolution like plate tectonics' [51]. He is referring to the hypothesis of mantle plumes; the plumes are rooted in the lower mantle, below the level of vigorous convection associated with plate tectonics. It is seen then that some of these revolutionary ideas coincide with Darwin's theory.

In his interesting book on this subject, G. F. Davies wrote [52, p. 23]: 'There is another story that developed in the shadow of plate tectonics: the story of mantle plumes. Plumes are a distinct component of mantle convection. They have played a significant role in the history of the continents, and possibly had a large role early in earth history, though that is still quite uncertain. They may have had decisive effects on the history of life. The story of the idea of plumes is just as long but not as complex as the story of plates. Perhaps four names can be identified as principals: Darwin, Dana, Wilson and Morgan, with a footnote for Holms.'

In 1971 J. W. Morgan developed the idea by proposing that there are plumes of hot material rising from the lower mantle [53–55]. As shown above, Charles Darwin had pioneered this approach a full 130 years earlier. In particular, he wrote ... *On this view we are led to conclude, that the unstratified mass forming the axis of any mountain, has been pumped in when in a fluid state* ... [11, p. 381].

It is postulated that plumes rise through the mantle and begin to partially melt on reaching shallow depths in the asthenosphere. This would create large volumes of magma. The plume theory posits that this melt rises towards the surface and forms the 'hot spots'.

Even plume 'tails' can generate enormous amounts of volcanism (see Fig. 1.4). For instance, Mauna Kea on the island of Hawaii is only about one million years old, but is actually the tallest mountain on Earth (measured from the seafloor). Many observations qualitatively support Darwin's thoughts ... *I believe that the frequent quakings of the earth on this line of coast are caused by the rending of the strata, ... and their injection by fluidified rock. This rending and injection would... form a chain of hills; ... I believe that the solid axis of a mountain differs in its manner of formation from a volcanic hill only in the molten stone having been*

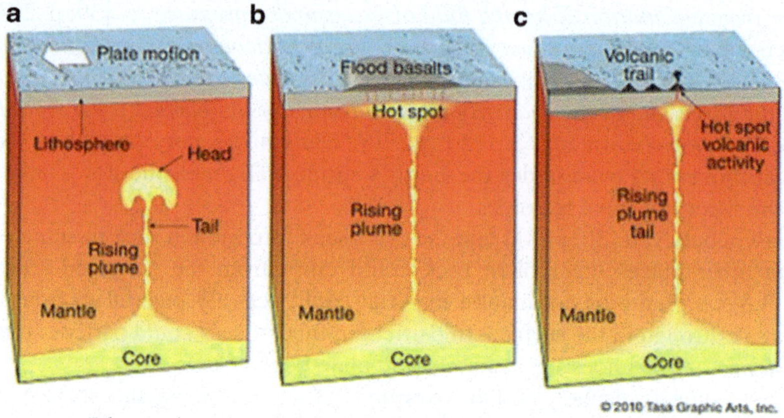

Fig. 1.4 Formation, evolution and interaction with the land surface of a heat plume. Due to Rayleigh-Taylor instability a vortex may form in the plume head. http://georneys.blogspot.co.nz/2010/12/geology-word-of-week-h-is-for-hotspot.html. Plume head and tail (Image from Tasa Graphics)

repeatedly injected, instead of having been repeatedly ejected...[13, p. 377]. He wrote that molten rock erupted from the fractured areas; a plume of hot mantle material could get actively pushed up from below and break through to the surface.

Thus, the issues that Darwin broached about 200 years ago have only recently begun to be discussed. He postulated, for instance that the crust is a system where fractured zones, and zones of seismic and volcanic activities interact: Darwin formulated the task of considering together the processes which are now studied as seismology and vulcanology [11, 56–59]. However, the study of the interactions between earthquakes and volcanoes presents many difficulties and as such is still in its early stages.

In particular, analysing the movement of the Earth's surface during an earthquake, Darwin wrote ..., *the displacement at first appears to be owing to a vorticose movement beneath each point thus affected;* ...[11, p. 376]. He reflected on the possibility of vortex formation under the surface. Indeed, occurrence of vortex motion within the Earth is quite probable, because of heat convection (Fig. 1.4).

It has been seen that Darwin was close to a qualitative explanation of different phenomena of the Earth's dynamics. In regard to this, he sometimes spoke about the effects of greater movements and of sudden processes; in other words, he spoke about catastrophic geological processes determining the motion of Earth's surface. On the whole he agreed with the opinions of Lyell, who had advanced the idea of the effects of local catastrophes [58, 59]. Therefore, although in biology Darwin might be considered an advocate of uniformitarianism, he did not exclude catastrophic events in geology. His geological observations showed the possibility of the rapid (sudden) evolution of the Earth's surface. Smooth development may be interrupted for a short time by a fast development. Development may also result from many events, for example, it may be induced by earthquakes.

This book is devoted to wave processes, which do not relate to the classical principle of catastrophism. However, Darwin's interest in unusual, even catastrophic, wave phenomena has significant bearing on this subject. We will discuss how very recent data on seismic and volcanic events support Darwin's observations and ideas about the 1835 Chilean earthquake. In addition, we will show how modern mechanical impact tests and simple experiments with weakly-cohesive materials also support his reports. The most recent data facilitates a much fuller exploration and analysis of this topic than we have been able to achieve before [60–73].

Research into different aspects of the Charles Darwin's scientific activity has been ongoing at the University of Auckland since the 1970s [74–78]. We consider this book to be a development of this long-time interest. Moreover, simultaneous tsunamis, volcano eruptions and earthquakes such as those described by Darwin are a constant threat in New Zealand [79]; for this reason too, Darwin's discoveries are of great importance to us.

1.6 Vertically-Induced Destructive Seismic Waves, and Content of the Book

Darwin's theory posited the elevation (rising movement) and the subduction of crust plates. In particular, Darwin experienced strong vertical ground motion induced by the Chile earthquake. It is now known that severe earthquakes generate large-amplitude waves in the Earth and its surface [80, 81]. The generation and interaction of the waves can explain the elevation up to 15 m and vertical accelerations of the ground surface up to 3.8 g (g being gravitational acceleration) [80–84] as well as simultaneous volcanic eruptions [11, 85–87]. Understanding these unfamiliar phenomena will require more detailed investigations of physical processes that control the strong ground motion, and of the laws governing the propagation of large-amplitude seismic waves near the land surface [46, 47, 88]. The results of such a study may be very important since many cities are built on sedimentary basins (for example, Los Angeles, Tokyo, New York, Shanghai, Calcutta, Jakarta, Mexico City, Delhi), and some are built near the coastline or on the tops of ridges. From the standpoint of an earthquake risk, however, these locations are the least-desirable places to build, because of the topographic effect. It is now recognized that hills, sedimentary basins, lakes, and continental shelves are natural resonators, where vertical seismic waves may be amplified very strongly. The geological topographies can increase the damaging effects of earthquakes by trapping and amplifying seismic waves.

A theory of vertical seismic waves excited in natural resonators was presented in [89]. The theoretical results were illustrated by the extreme oscillations at Tarzana Hill during the Northbridge 1994 Southern California earthquake. This hypothesis was developed in [60–70]. Generally speaking, earthquakes cause horizontal or vertical movement. The earthquake described by Darwin was primarily vertical. According to Darwin the vertical acceleration of the Earth surface exceeded g. The margins of safety against gravity-induced static vertical forces in constructed buildings usually provide adequate resistance to dynamic forces induced by the vertical acceleration during an earthquake. However, the earthquakes in Port-au-Prince (2010, Haiti; the depth of the earthquake – 13 km) [90], Christchurch (2011, New Zealand; the depth of the earthquake – 5 km) [91] and Ludian County (2014, China; the depth of the earthquake – 10 km) [92] are examples of vertical seismic shocks which caused the collapse of cities. The earthquake magnitudes were not very large – about 7, 6.3 and 6.1. However, the results were catastrophic (Fig. 1.5 left).

In Christchurch (2011) the vertical acceleration reached 2.2 g (Fig. 1.5 right). Land movement was varied around the area horizontally—in both east and west directions—and vertically; the Port Hills were raised by 40 cm. The upwards acceleration was greater than the downwards, which had a maximum recording of 0.9 g; the maximum recorded horizontal acceleration was 1.7 g.

The vertical acceleration measured during the 14th June 2008 Iwate-Miyagi earthquake in Japan was 3.8 g [82]. This acceleration was 5.5 times the horizontal acceleration. The depth of the earthquake was 8 km. A magnitude-9.0 earthquake struck off the coast of the Tohoku region in north-eastern Japan on March 11th, 2011 [93, 94]. The vertical acceleration was 2.7 g, about 10 times the horizontal acceleration. The ocean floor above the earthquake epicentre rose by 10 m. We stress that large vertical acceleration usually takes

Fig. 1.5 Destruction caused by vertical shock in Christchurch, 22nd February, 2011 (*left*) (http://www.shutterstock.com/pic-71750008/stock-photo-christchurch-nz-feb-destruction-caused-by-earthquake-on-february-in-christchurch.html?src=vIQW8OCGcQm0BrtHWeo6bA-1-53). Vertical ground acceleration was 2.2 g. The *green star* indicates earthquake epicentre (*right*) [91]

Fig. 1.6 Results of the 19th September 1985 Michoacan earthquake. Some 300 buildings collapsed and more than 20,000 people died [96]

place during shallow earthquakes, in which the depth is more important than the earthquake intensity.

On the other hand, earthquake results can be determined by local topographical and resonant effects. In particular, strong resonant ground motion may be more dependent upon the ground properties at a site than on the proximity or intensity of the earthquake sources. Resonant amplification of about 75 was produced in the Mexico City sediments during the September 19th 1985 Michoacan earthquakes. That earthquake caused tremors along the coast of Mexico about 400 km west of Mexico City. Near this coast the ground tremors were mild and caused little damage. As the seismic waves raced inland, the ground shook even less, and by the time the waves were 100 km from Mexico City the shaking had nearly subsided. Nevertheless, the seismic waves induced severe shaking in the city (Fig. 1.6), and some areas continued to shake for several minutes after the seismic waves had passed. This resulted from trapped and resonant amplification of the seismic waves by sediment basins located within the city [95, 96].

In the Michoacan earthquake and others like it, earthquake damage depends not only on the seismic event itself but also upon local geology and site circumstances. This conclusion, drawn by observations over the last few decades, completely concurs with Darwin's results regarding the 1835 Chilean earthquake *...the effects of one great cause, modified only by local circumstances...* [20, p. 60] (Figs. 1.7 and 1.8 left).

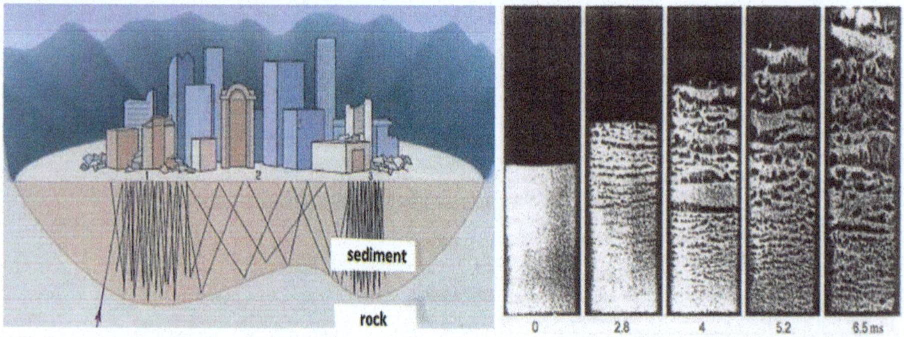

Fig. 1.7 The scheme of oscillations of vertical seismic waves within of a city foundation (Concepcion (Chile, 1835), Port-au-Prince (Haiti, 2010), Christchurch (New Zealand, 2011) and Ludian County (China, 2014)) (*left*) [96]. The scheme of a collapse of the foundation soil by vertical seismic waves (*right*) [97]

Fig. 1.8 Sketch of the coastal amplification of a tsunami (*left*) (Internet) (http://www.waikatoregion.govt.nz/Services/Regional-services/Regional-hazards-and-emergency-management/Coastal-hazards/Tsunami/); Sketch of the action of an extreme ocean wave (the Draupner wave) which struck an oil platform in the North Sea on New Year's Day, 1995 (*right*)

In this book we illustrate the 'shivering' of islands and ridges, volcanic eruptions and seaquake-induced waves, which support Darwin's theory that these events are parts of a single phenomenon (as in the case of the 1835 Chilean earthquake). We also draw the reader's attention to the existence of catastrophic ocean waves (Fig. 1.8 right), such as that which the *Beagle* encountered near Cape Horn. We show that similar extreme waves may appear in different media (gas, liquid, solid and scalar field) since nonlinear waves in different physical fields are described by similar wave equations. These waves can be excited by natural causes in 'natural resonators'- structures such as sediment layers, mountains, oceans and the Universe itself in which resonant effects can cause rapid and very large amplitude growth. The conception of resonances provides a common ground to reduce disciplinary boundaries and to unify scientific results.

This book contains eight chapters. Chapter 2 considers the behaviour of the Earth's surface layers, and volcanoes excited by vertical waves. Then (in Chap. 3) we discuss Darwin's thoughts about the behaviour of the Earth's surface and oceans during certain great earthquakes; specifically, extracts from [11, 13, 98, 99] and the **Autobiographies** [18] are presented. In particular, we consider a few days in a life of Darwin related to the 1835 Chilean earthquake. Those days were perhaps among the most important days of his life. He observed and analysed rapid change of the terrestrial surface as a result of seismic impact, which confirmed his assumptions about the existence of variability in Nature. So that the reader can trace Darwin's ideas, we group the material of the extracts and illustrate them by modern results.

Chapter 4 is devoted to ground vertical waves. It considers the loosening, rupture and lifting of sediment layers, soils and magma as results of rapid decompression after seismic impact. Results of modern experimental and theoretical researches are presented. The issues discussed in the Chap. 4 are illustrated in Fig. 1.7, which shows oscillations of seismic waves in a sediment basin (left) and the fracture and the elevation of initially-compressed granular material (glass balls) within a rarefaction wave (right) [97]). It is emphasized that Darwin's theory that earthquakes influence volcanoes has since been supported by different observations [85–87, 100–103]. The energy of vertical seismic waves concentrates near a volcano crater, as a result of which eruption begins. Darwin had stressed that earthquakes and volcano eruptions can destroy entire countries: and indeed, a recent study by Professors Yoshiyuki Tatsumi and Keiko Suzuki at Kobe University has warned that one major volcanic eruption could make Japan 'extinct'. They conclude that the chance of a big eruption that would disrupt the lives of everyone in Japan are about 1 % over the next 100 years (http://blogs.wsj.com/japanrealtime/2014/10/23/major-volcanic-eruption-could-make-japan-extinct-study-warns/).

Chapter 5 is devoted to tsunami (Fig. 1.8 left) and other vertically-induced ocean waves. Darwin's description of the evolution of a tsunami in coastal waters is modelled. It is shown that when a tsunami approaches a coastline, nonlinear effects become important. The tsunami amplitude grows up to 40 m [104]. Coastal resonance may be responsible for this amplification. An ebb forms ahead of the tsunami, whose form begins to be determined by the shape of the coast. Then the vertical-induced inner and surface waves in water and granular layers are studied. The surface waves are considered as strongly-nonlinear cases of Faraday waves. Bubbles, cavitation and highly-nonlinear waves appear within the layers if the forced acceleration is greater than the gravity acceleration g. A vast amount of accumulated experimental data is then discussed. The counterintuitive data and the atomization of the layers are described using elastica forms. It is stressed that the processes described take place during seaquakes in shallow seas.

Extreme (rogue, freak or catastrophic) waves are considered in the Chap. 6 (Fig. 1.8 right). From Darwin's book, we know that the *Beagle* encountered such a wave near Cape Horn. The height of these waves may be up to 70 m. The existence of catastrophic waves demands that the design of ocean-going vessels and oil platforms should be fundamentally reconsidered. In particular, the threat of these waves should be heeded by ship-building companies [105, 106]. It is not only the extreme height of catastrophic waves that increases the danger; these waves can also have a very steep front, a so-called 'water wall'. The steep front and the

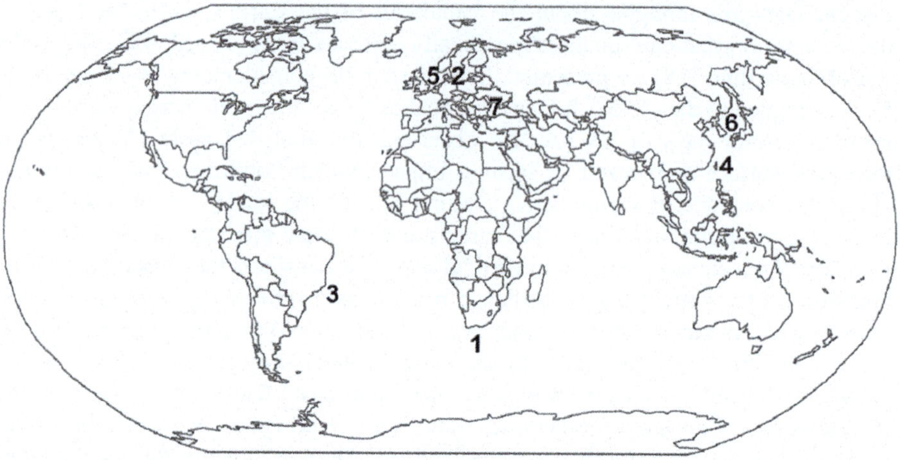

Fig. 1.9 Some instrumental registrations of extreme waves. (*1*) Offshore from Mossel Bay (1,563 events, 100 m depth, gas-drilling platform). (*2*) The Baltic Sea (414 events, 7–20 m depth, buoys). (*3*) Campos Basin near Rio de Janeiro (276 events, 1,050 m and 1,250 m depth, buoys). (*4*) Off the eastern coast of Taiwan (175 events, 43 m depth, buoys). (*5*) The North Sea (at least 107 events, 126 and 85 m depth, platforms). (*6*) Sea of Japan (14 events, 43 m depth, ultrasonic submerged gauges). (*7*) The Black Sea (3 events, 85 m depth, buoy) [109]

overturning process are, probably, the most important parameters defining the result of a wave action – more important than the height of the wave. The occurrence of these waves is explained by resonance effects. As an example of catastrophic amplification, the transresonant evolution of ocean waves going over underwater topography is analysed. Long and short extreme waves are considered. Nonlinear wave equations for ocean waves in Lagrangian coordinates are used [62, 64, 66, 70, 107, 108].

An extreme (rogue) ocean wave is a rare event, and may be recorded only by means of long-time regular measurements. At present, a couple of hundreds of extreme waves have been registered. Figure 1.9 shows the areas where they have been continuously measured over a period of years [109].

Extreme waves were usually recorded in shallow water (at a depth of around 100 m or less). These waves may be considered as weakly-dispersive long waves [66, 70]. The waves measured near Rio de Janeiro may be explained by a theory presented in this Chapter [66, 70]. Extreme waves with a breaking front are also considered, and an interaction of these waves with hull elements is then modelled. Experimental data on underwater explosions and hull cavitation are presented, and the effects of deformation and hull cavitation on the interaction are studied [110].

In the Chap. 7 we continue to study extreme waves in natural resonators. A theory of extreme waves excited in oceans, sediment layers, volcanoes and scalar fields is briefly described. Our goal is the development of a universal, interdisciplinary understanding of the appearance and evolution of extreme waves as a transresonant phenomenon. Results of the analysis are compared with many

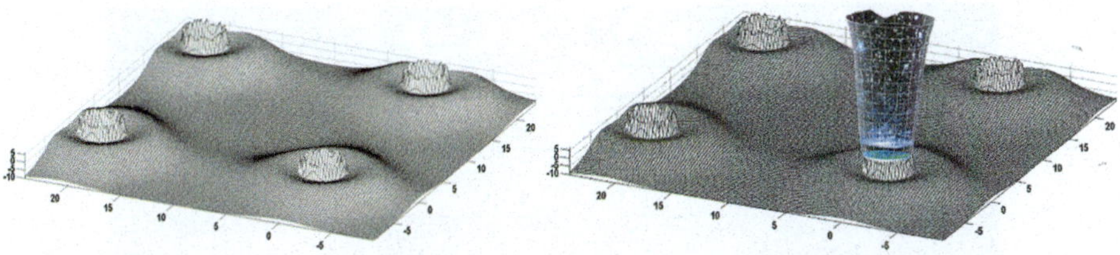

Fig. 1.10 The two-dimensional landscape of the pre-Universe (*left*) and the scheme of the origin of the Universe due to quantum perturbations, bifurcations and resonances (*right*)

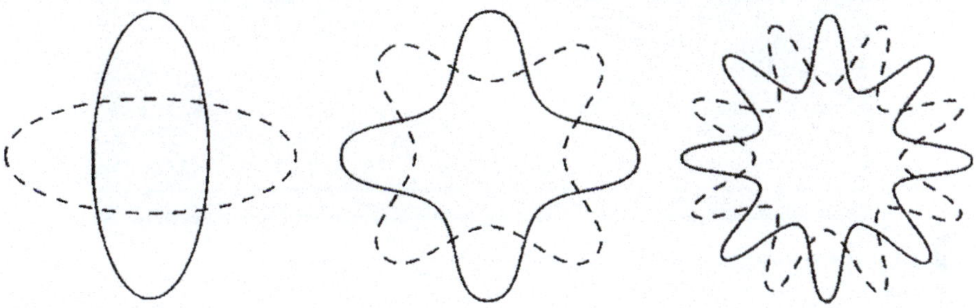

Fig. 1.11 Rough schemes of some modes of oscillations of multidimensional spheres at the tops of the hills [115]

experiments. The theory qualitatively describes the strongly-nonlinear wave processes considered in this book. In particular, different approximate solutions of the nonlinear Klein-Gordon equation (NKGE) which describe the origin of the Universe are illustrated by data of the Faraday-type water experiments and experiments with the Bose-Einstein condensate. We believe that scalar field equations and solutions of them might shed light on the emergence and initial evolution of the Universe [111–120].

Certain solutions can be considered as a description of a weakly oscillating pre-Universe. According to the theory, the scalar field potential describes a landscape which consists of hills and valleys (Fig. 1.10 left). At any moment this multidimensional structure gives birth to the billions of 'seeds' of rapidly evolving universes, one of which accidentally formed our Universe (Fig. 1.10 right). In the scenario thus developed, the explanation of the beginning of the Universe differs greatly from Big Bang and inflation theories.

The highest energy density is reached at the tops of the hills where multidimensional spheres (bubbles) oscillate. The pure radial oscillations of the spheres are studied, although more complex oscillations may also be considered (Fig. 1.11).

There is an extremely small chance that some spheres can grow into big universes (Figs. 1.10 and 1.12) [113–117]. The origin of the Universe is being studied as an evolution of an initial scalar field (sphere) into other fields (particles), energies, waves and vortices.

Fig. 1.12 A rough scheme of the landscape evolution into universes [117]

Linear and weakly-nonlinear models and perturbation methods are used by the researchers to describe the evolution of the early Universe. We developed the strongly-nonlinear approach to describe the origins of the Universe, dark matter and dark energy. Effect of the nonlinearity of quantum scalar fields becomes extremely important, if some resonant conditions take place. Modern cosmology is phenomenally successful, but its successes have also uncovered further deep and complex mysteries. In this Chapter, partly written together with Galiyev T. Sh., we attempt to shed light some upon them.

In the Chap. 8, the results of recent catastrophic earthquakes are presented and discussed. The importance of Darwin's geophysical research and geodynamic ideas for modern science is underlined.

Many books have been written about Darwin as the author of evolution theory. Here, however, we consider him primarily as a researcher in the field of dynamic geophysical processes. We illustrate Darwin's important contribution to our understanding of nonlinear dynamics of the Earth, using publications in high-prestige journals (about 300 publications) which have been published during the last 15–20 years. While there are many publications on the subject of extreme waves, problems of the catastrophic amplification of seismic and ocean waves now warrant our attention, having become especially important due to events of the last 5 years.

References

1. Browne J (2006) Darwin's "origin of species". Atlantic Books, London
2. Darwin C (1872) The origin of species, 6th edn. John Murray, London
3. Nature (2008) Nature 456 (7220)
4. Science (2009) Science 323 (5911)
5. New Scientist (2009) New Scientist (2692)
6. New Scientist (2009) New Scientist (2734)
7. Culotta E (2009) On the origin of religion. Science 326:784–787
8. Scientific American (2009) Scientific American 300 (1)
9. New Scientist (2012) Mr Darwin's book. New Scientist 2884:48
10. Galloway D, Timmins J (eds) (2010) Aspects of Darwin: a New Zealand celebration. Friends of the Knox College Library, Dunedin
11. Darwin C (1839) Journal of researches into the geology and natural history of the various countries visited by H.M.S. Beagle, under the command of Captain FitzRoy, R.N. from 1832 to 1836, 1st edn. Henry Colburn, London
12. Goddard J (ed) (2010) Concise history of science and invention. National Geographic, Washington
13. Darwin C (1890) Journal of researches into the natural history and geology of the countries visited during the voyage of H.M.S. Beagle around the world, under the command of Captain FitzRoy, R.N. T. Nelson and Sons, London
14. Albert E (1955) The story of English literature. Collins, London
15. Desmond A, Moore J (2010) Darwin's sacred cause: race, slavery and the quest for human origins. Penguin, London
16. Gopnik A (2009) Angels and ages: a short book about Darwin, Lincoln, and modern life. Knopf, New York
17. Thomson KS (2010) The young Charles Darwin. Yale University Press, New Haven
18. Neve M, Messenger S (eds) (2002) Darwin C. Autobiographies. Penguin, London
19. Darwin CR (1851) Section VI: Geology. In: Herschel JFW (ed) A manual of scientific enquiry; prepared for the use of her majesty's navy: and adapted for travellers in general. John Murray, London
20. Darwin C (1977) On the connexion of certain volcanic phenomena in South America; and on the formation of mountain chains and volcanoes, as the effect of the same power by which continents are elevated. (1840, Read March 7, 1838). In: Barrett PH (ed) The collected papers of Charles Darwin. The University of Chicago Press, Chicago
21. Brangwynne CP, Hyman AA (2012) The origin of life. Nature 491:524–525
22. Pereto J, Bada JL, Lazcano A (2009) Charles Darwin and the origin of life. Orig Life Evol Biosph 39(5):395–406
23. Zimmer C (2006) Evolution. The triumph of an idea. Harper Perennial, New York
24. Rose M, Travis J (eds) (2009) Evolution. The Belknap Press of Harvard University Press, Cambridge
25. Kline M (1985) Mathematics and the search for knowledge. Oxford University Press, London
26. Repcheck J (2003) The man who found time. Perseus Publishing, New York
27. Brake ML (2009) Revolution in science: how Galileo and Darwin changed our world. Palgrave Macmillan, New York
28. Verschuuren GM (2012) Darwin's philosophical legacy. Lexington Books, New York
29. Stone I (1980) The origin. A biographical novel of Charles Darwin. Doubleday & Company, New York
30. Barrett PH, Freeman RB (eds) (1986) The works of Charles Darwin. The geology of the voyage of H.M.S. Beagle. Part II: geological observations on the volcanic islands, vol 8. William Pickering, London

31. Barrett PH, Freeman RB (eds) (1986) The works of Charles Darwin. The geology of the voyage of H.M.S. Beagle. Part III: geological observations on South America, vol 9. William Pickering, London
32. Barrett PH, Freeman RB (eds) (1986) The works of Charles Darwin. The geology of the voyage of H.M.S. Beagle. Part I: structure and distribution of coral reefs, vol 7. William Pickering, London
33. Skiena SS, Ward CB (2013) Who's bigger?: Where historical figures really rank. Cambridge University Press, Cambridge
34. Bowler PJ (2013) Darwin deleted: imagining a world without Darwin. The University of Chicago Press, Chicago
35. Dobzhansky T (1937) Genetics and the origin of species. Columbia University Press, New York
36. Romanovskiya T (2010) Molecular biology. http://www.bio.bsu.by/genetics/romanovskayatv_ru.html
37. Volkenshtein M (1999) Beauty of a science. Internet. http://n-t.ru/tp/br/kn.htm1
38. McGrath AE (2010) Science and religion: a new introduction, 2nd edn. Wiley-Blackwell, London
39. Herbert S (2005) Charles Darwin, Geologist. Cornell University Press, Ithaca
40. Rudwick M (2005) Darwin's first love. Nature 436:330
41. Chiesura G (2002) Charles Darwin geologo. Hevelius Edizioni, Benevento (in Italian)
42. Chui G (2009) Shaking up earthquake theory. Nature 461:870–872
43. Taira T, Silver PG, Niu F, Nadeau RM (2009) Remote triggering of fault-strength changes on the San Andreas fault at Parkfield. Nature 461:636–639
44. Parsons T (2009) Last earthquake legacy. Nature 462:42–43
45. Stein S, Liu M (2009) Long aftershock sequences within continents and implications for earthquake hazard assessment. Nature 462:87–89
46. Anderson JG (2007) Physical processes that control strong ground motion. In: Schubert G (ed) Treatise on geophysics. Elsevier, Amsterdam/Boston
47. Rhoades DA, Zhao JX, McVerry GH (2008) A simple test for inhibition of very strong shaking in ground-motion models. Bull Seismol Soc Am 98(1):448–453
48. Meng L, Ampuero JP, Stock J et al (2012) Earthquake in a Maze: compressional rupture branching during the 2012 M_w 8.6 Sumatra Earthquake. Science 337:724–726
49. Yue H, Lay T, Koper KD (2012) En échelon and orthogonal fault rupture of the 11 April 2012 great intraplate earthquakes. Nature 490:245–249
50. Foulger GR, Natland JH (2003) Is "hotspot" volcanism a consequence of plate tectonics? Science 300:921–922
51. Ananthaswamy A (2012) Earthly powers. New Scientist 2878:38–41
52. Davies GF (1999) Dynamic earth: plates, plumes and mantle convection. Cambridge University Press, Cambridge
53. Morgan WJ (1971) Convection plumes in the lower mantle. Nature 230:42–43
54. Nature Geoscience (2011) Forty years of plume research. Nat Geosci 4:813
55. Sobolev SV, Sobolev AV, Kuzmin DV et al (2011) Linking mantle plumes, large igneous provinces and environmental catastrophes. Nature 477:312–316
56. Hill DP, Pollitz F, Newhall C (2002) Earthquake-volcano interaction. Physics Today 55(11):41–47
57. Kawakatsu H, Yamamoto M (2007) Volcano seismology. In: Schubert G (ed) Treatise on geophysics. Elsevier, Amsterdam/Boston
58. Maclennan J (2013) All rise for the case of the missing magma. Nature 494:182–183
59. Zhou H, Dick HJB (2013) Thin crust as evidence for depleted mantle supporting the Marion Rise. Nature 494:195–200
60. Galiev SU (1998) Topographic amplification of vertical-induced resonant waves in basins. Adv Hydrosci Eng Vol. 3, pp. 179–182

61. Galiev SU (1999) Topographic effect in a Faraday experiment. J Phys A: Math Gen 32:6963–7000
62. Galiev SU (2003) The theory of nonlinear trans-resonant wave phenomena and an examination of Charles Darwin's earthquake reports. Geophys J Int 154:300–354
63. Galiev ShU, Mallinson GD (2006) Resonant acoustic model of catastrophic tsunami. The 13th international congress on sound and vibration. http://www.iiav.org/archives_icsv/2006_icsv13/pdf/icsv13Final00761.pdf
64. Galiev SU (2008) Strongly-nonlinear wave phenomena in restricted liquids and their mathematical description. In: Perlidze T (ed) New nonlinear phenomena research. Nova Science Publishers, New York
65. Galiev ShU, Mallinson GD (2009) Modelling of Darwin's description of coastal evolution of catastrophic tsunami. The 16th international congress on sound and vibration. http://www.iiav.org/archives_icsv/2009_icsv16/content/technical_papers/863.pdf
66. Galiev ShU (2009) Modelling of Charles Darwin's earthquake reports as catastrophic wave phenomena. http://researchspace.auckland.ac.nz/handle/2292/4474
67. Galiev ShU (2010) Charles Darwin's earthquake reports. Geophysical research abstracts, EGU2010. Vienna, 3738. http://www.researchgate.net/publication/234540315_Charles_Darwin%27s_earthquake_reports
68. Galiev ShU (2010) Darwin's triggering mechanism of volcano eruptions. Geophysical research abstracts, EGU2010. Vienna, 3693. http://meetingorganizer.copernicus.org/EGU2010/EGU2010-3693.pdf
69. Galiev ShU (2010) Modelling of Charles Darwin's tsunami reports. Geophysical Research Abstracts, EGU2010. Vienna, 2588
70. Galiev SU (2011) Геофизические Сообщения Чарльза Дарвина как Модели Теории Катастрофических Волн (Charles Darwin's geophysical reports as models of the theory of catastrophic waves). Centre of Modern Education, Moscow (in Russian). http://books.google.co.nz/books/about/%D0%93%D0%B5%D0%BE%D1%84%D0%B8%D0%B7%D0%B8%D1%87%D0%B5%D1%81%D0%BA%D0%B8%D0%B5_%D1%81%D0%BE%D0%BE%D0%B1%D1%89%D0%B5.html?id=H0KekgEACAAJ&redir_esc=y
71. Ilgamov MA (2012) О столетии и двухсотлетии Чарльза Дарвина (On Charles Darwin's centennial and bicentennial). Herald ASRB 17(2):66–69 (in Russian)
72. Talipova TG, Pelinovsky EN (2012) Рецензия на книгу Галиев, Ш.У. Геофизические Сообщения Чарльза Дарвина как Модели Теории Катастрофических Волн. Центр современного образования, Москва, 2011, 655с. Fundam Appl Hydrophys 5(3):96 (in Russian). (Review of the book: Galiev, ShU. Charles Darwin's geophysical reports as models of the theory of catastrophic waves. Centre of Modern Education, Moscow, 2011, p. 655)
73. Troshchenko VT, Astanin VV (2013) Рецензия на монографию Галиева, Ш.У. Геофизические Сообщения Чарльза Дарвина как Модели Теории Катастрофических Волн. Центр современного образования, Москва, 2011, 655 с. (Review of the book: Galiev, ShU. Charles Darwin's geophysical reports as models of the theory of catastrophic waves. Centre of Modern Education, Moscow, 2011, 655 p.). Strength Mat 1:168–171 (in Russian)
74. Tee GJ (2014) Charles Darwin's New Zealand correspondents. J Hist Stud Group 48:35–60
75. Tee GJ (1990) Review of the book: Nicholas FW, Jan M (1989) Charles Darwin in Australia. Cambridge University Press. J Polyn Soc 101(2):201–203
76. Tee GJ (1979) Another link between Marx and Darwin. Ann Sci 36:176
77. Tee GJ (1985) Review of the book: a calendar of the correspondence of Charles Darwin, 1821–1882 (Burkhardt F, Smith S (eds) (1985) Garland Publishing Inc, New York). J R Soc N Z 15:341–343
78. Tee GJ (1985) Review of the book: Dempster WJ (1983) Patrick Matthew and natural selection. Paul Harris Publishing, Edinburgh. New Zeal J Hist 18(1):66–67
79. Morton J (2014) Living with disaster: warning for Kiwis – be ready for the big one. The NZ Herald, June 18

80. Coen DR (2012) The earthquake observers: disaster science from Lisbon to Richter. University of Chicago Press, Chicago
81. Valencius CB (2013) The lost history of the New Madrid earthquakes. University of Chicago Press, Chicago
82. Aoi S, Kunugi T, Fujiwara H (2008) Trampoline effect in extreme ground motion. Science 322:727–730
83. O'Connell DRH (2008) Assessing ground shaking. Science 322:686–687
84. Johnson JB, Lees JM, Gerst A et al (2008) Long-period earthquakes and co-eruptive dome inflation seen with particle image velocimetry. Nature 456:377–381
85. Donne DD (2010) Earthquake-induced thermal anomalies at active volcanoes. Geology 38(9):771 (see, also, Geoscience: volcanoes respond to earthquakes. 2010, Nature 467:887)
86. Pritchard ME, Jay JA, Aron F et al (2013) Subsidence at southern Andes volcanoes induced by the 2010 Maule, Chile earthquake. Nat Geosci. doi:10.1038/ngeo1855
87. Takada Y, Fukushima Y (2013) Volcanic subsidence triggered by the 2011 Tohoku earthquake in Japan. Nat Geosci. doi:10.1038/ngeo1857
88. Denolle MA et al (2014) Strong ground motion prediction using virtual earthquakes. Science 343:399–403
89. Galiev ShU (1997) Resonant oscillations governed by the Boussinesq equation with damping. In: Proceeding of 5th international congress on sound and vibration, Adelaide, pp 1785–1796. http://www.acoustics.asn.au/conference_proceedings/ICSVS-1997/pdf/scan/sv970428.pdf
90. Schiermeier Q (2010) Quake threat looms over Haiti. Nature 467:1018–1019
91. Summary and recommendations (2012) vol 1. Canterbury earthquakes. Royal Commission. http://canterbury.royalcommission.govt.nz/Final-Report
92. Qiu J (2014) Chinese data hint at trigger for fatal quake. Nature 513:154–155
93. Chester FM et al (2013) Structure and composition of the plate-boundary slip zone for the 2011 Tohoku-Oki earthquake. Science 342:1208–1211
94. Chung E (CBC News) 2013. Japan tsunami's huge size blamed on slimy, slimy fault. http://www.cbc.ca/m/touch/technology/story/1.2452553
95. Flores J, Novaro O, Seligman TH (1987) Possible resonance effect in the distribution of earthquake damage in Mexico City. Nature 326:783–785
96. Rial JA, Salzman NG, Ling H (1992) Earthquake-induced resonance in sedimentary basins. Am Sci 80:566–578
97. Anilkumar AV (1989) Experimental studies of high-speed dense dusty gases. Dissertation, California Institute of Technology. http://thesis.library.caltech.edu/3299/
98. FitzRoy R (1839) Narrative of the surveying voyages of His Majesty's Ships Adventure and Beagle, between the years 1826 and 1836, describing their examination of the southern shores of South America, and the Beagle's circumnavigation of the globe. In: Proceedings of the second expedition, 1831–1836, under the command of Captain Robert Fitz-Roy R.N., vol 2. Henry Colburn, London
99. Keynes RD (ed) (1979) The Beagle record. Cambridge University Press, London
100. Linde AT, Sacks IS (1998) Triggering of volcanic eruptions. Nature 395:888–890
101. Brodsky EE, Sturtevant B, Kanamori H (1998) Earthquakes, volcanoes, and rectified diffusion. J Geophys Res 103(23):827–838
102. Prejean SG, Haney MM (2014) Shaking up volcanoes. Science 345:39
103. Brenguier F, Campillo M, Nakeda T et al (2014) Mapping pressurized volcanic fluids from induced crustal seismic velocity drops. Science 345:80–82
104. Pease R (2014) Double-whammy tsunami? Science 346:18
105. Proceedings of the 15th international ship and offshore structures congress, vol 1, 20–25 Aug 2003, San Diego. Dynamic Response, Elsevier Science, pp 193–264, (September 15, 2005)
106. Proceedings of the 16th international ship and offshore structures congress, vol 2, 20–25 Aug 2006. Naval Ship Design. University of Southampton, School of Engineering Sciences, Southampton, pp 213–263 (June 2006)
107. Galiev SU (2005) Strongly-nonlinear two-speed wave equations for coastal waves and their application. Physica D 208:147–171

108. Galiev ShU, Mace BR (2014) Lagrangian description of extreme ocean waves. Известия Уфимского Научного Центра Российской Академии Наук (Herald of Ufa Scientific Center, Russian Academy of Sciences (RAS)) 4:6–13
109. Kharif C, Pelinovsky E, Slunyaev A (2009) Rogue waves in the ocean. Springer, Berlin
110. Galiev SU, Flay RGJ (2014) Interaction of breaking waves with plates: the effect of hull cavitation. Ocean Eng 88:27–33
111. Galiev SU, Galiyev TS (2001) Nonlinear transresonant waves, vortices and patterns: from microresonators to the early Universe. Chaos 11:686–704
112. Galiev ShU (2003) Transresonant evolution of waves and the Richtmyer-Meshkov instability: from microstructures to the Universe. In: Khovanskii AG, Nourgaliev DK, Petrova NK (eds) Proceedings of international conference on 'new geometry of nature' Kazan, vol I, Aug 25–Sept 5, 2003, pp 88–91. https://istina.msu.ru/collections/7729472/
113. Galiev ShU, Galiyev TSh (2013) Scalar field as a model describing the birth of the Universe. InterNet. https://researchspace.auckland.ac.nz/bitstream/handle/2292/21292/Galiev%20-%20Galiyev%20%2018-12-2013.pdf?sequence=2
114. Galiev ShU, Galiyev TSh (2014) Nonlinear scalar field as a model describing the birth of the Universe. Известия Уфимского Научного Центра Российской Академии Наук (Herald of Ufa Scientific Center, Russian Academy of Sciences (RAS)) 2:7–33
115. Greene B (1999) The elegant universe. Vintage Books, New York
116. Vilenkin A (2006) Many worlds in one. Hill and Wang, New York
117. Linde A (2014) Теория инфляционной Вселенной, или теория Мультивселенной (Мультиверса) (Theory of the inflationary Universe, or theory of the Multiuniverse (Multiver)). InterNet. http://scisne.net/a1075
118. Underwood B, Zhai Y (2014) Non-linear resonance in relativistic preheating. JCAP 04 (2014) 002. doi:10.1088/1475-7516/2014/04/002
119. McKernan B, Ford KES, Kocsis B, Haiman Z (2014) Stars as resonant absorbers of gravitational waves. Mon Nat R Astron Soc Lett 445:L74–L78
120. Galiev ShU, Galiyev TSh (2014) Coherent model of the births of the Universe, the dark matter and the dark energy. https://researchspace.auckland.ac.nz/handle/2292/23783

Chapter 2
Introduction

*...The voyage of the Beagle has been by far the most important event in my life... (**Darwin**)*

During the voyage of the *Beagle*, Darwin landed at Valdivia and Concepcion in Chile just before, during, and after the 20 February 1835 Chilean earthquake [1]. For several days his attention was devoted to results of the earthquake. He observed and described the 'shivering' of islands and ridges, fissured ground, volcanic eruptions, and a tsunami. In particular, Darwin wrote ... *Among the other most remarkable spectacles which we have beheld may be ranked ... an active volcano; and the overwhelming effects of a violent earthquake. These latter phenomena, perhaps, possess for me a particular interest, from their intimate connection with the geological structure of the world. The earthquake, however, must be to every one a most impressive event: the earth, considered from our earliest childhood as the type of solidity, has oscillated like a thin crust beneath our feet; and in seeing the laboured works of man in a moment overthrown, we feel the insignificance of his boasted power* ... [2, pp. 601–602].

Darwin suggested that the earthquake was a result of ... *the rending of strata, at a point not very deep below the surface of the earth* ... [3, p. 70] and ... *From the intimate and complicated manner in which the elevatory and eruptive forces were shown to be connected during this train of phenomena, we may confidently come to the conclusion that the forces which slowly and by little starts uplift continents, and those which at successive periods pour forth volcanic matter from open orifices, are identical. From many reasons, I believe that the frequent quakings of the earth on this line of coast are caused by the rending of the strata, necessarily consequent on the tension of the land when upraised, and their injection by fluidified rock. This rending and injection would, if repeated often enough (and we know that earthquakes repeatedly affect the same areas in the same manner), form a chain of hills; and the linear island of S. Mary, which was upraised thrice the height of the neighbouring country,* ... [2, pp. 376–377].

Thus, Darwin wrote about crust tension and we emphasize that he connected the earthquake and the eruptions with the crust rupture. Darwin thought that the earthquake-induced vertical motion of the land surface could explain his

Sh.U. Galiev, *Darwin, Geodynamics and Extreme Waves*,
DOI 10.1007/978-3-319-16994-1_2

observations. Perhaps that is true if the accelerations of the land are large enough. However, usually the earthquake-induced acceleration of the land surface is not large (smaller than g). And the vertical ground motion is usually much smaller than the horizontal motion. Therefore, earthquake-hazard assessment studies have usually been focussed on the horizontal shocks, and the effects of strong vertical shocks were practically never discussed. The margins of safety against gravity-induced static vertical forces in constructing buildings usually provide adequate resistance to dynamic forces induced by the vertical acceleration during an earthquake.

But what does modern science know about the amplitude of strong wave motions of the Earth's surface? Can the strong vertical motion of the land be responsible for the results of the 20 February 1835 Chilean earthquake? Can modern experimental and theoretical data explain Darwin's observations? We try to reply to these questions here and later in this book.

2.1 Extreme Vertical Dynamics of the Land Surface During Earthquakes

Recent history of observational seismology has been marked by frequent reassessment of the maximum possible ground acceleration which an earthquake can make. Normally, ground motions, that can cause damage of structures, exceed about 0.1 g. The first strong-motion accelerogram was recorded at Long Beach, California, during the 10 March 1933 Long Beach earthquake – it had peak horizontal and vertical accelerations of 0.19 and 0.28 g, respectively [4]. It was assumed that no earthquake could generate ground acceleration as high as g.

However, recent data shows [4–8] that there are about 30 records with extreme peak accelerations greater than g and a few peaks were near 2.5 g [4]. Is there any evidence for substantially more severe motions? An unprecedented vertical surface acceleration of nearly 4 gravities (3.8 g), more than twice its horizontal component, was measured in Japan [9]. The vertical acceleration was distinctly asymmetric – the upward pulses were larger in amplitude than the downward ones. They tended also to be narrow and sharp, whereas, the downward pulses were broader and longer-lasting. On a dome of the volcano Santiaguito, Guatemala, the local peak vertical acceleration of 2–3 g was measured recently [10]. This peak was associated with an abrupt vertical surface displacement of the volcano dome, in which 20–50 cm uplift originates at the central vent and propagates at ~50 m/s toward the 200 m diameter periphery.

Thus, earthquake-induced vertical accelerations can be large. On the whole, this agrees with Darwin's remarks about the convulsions and great movements of the ground surface.

Porosity, surface geology, free surface and its relief may amplify the earthquake-induced surface shock. In particular, the seismic wave amplitude is greater in a low-density, low-velocity soil, than in a high-density, high-velocity rock. Robert

FitzRoy emphasized this phenomenon. Describing the ruins of Talcahuano he wrote 'three houses only, upon a rocky foundation, escaped the fate of all those standing upon the loose sandy soil' [11, p. 257]. Now, it is well-known that sediments amplify the ground motion relatively to bedrock [4, 12].

If the peak acceleration down is greater than g, then surface objects may be thrown up in the air by an earthquake. Darwin reported . . . *I was much interested by finding on the highest peak of one range (about 700 ft above the sea) a great arched fragment, lying on its convex or upper surface. Must we believe that it was fairly pitched up in the air, and thus turned?* ... [1, p. 255] and ... *during the earthquake... which in 1835 overthrew Concepcion, in Chile, it was thought wonderful that small bodies should have been pitched a few inches from the ground,* ... [1, p. 256]. Thus, Darwin described a severe earthquake in which the ground surface acceleration had been greater than g. Because of the topographic effect[1] the earthquake-induced acceleration of tops of mountains and volcanoes might be much greater than g.

A famous example was documented by Oldham [4], who reported displaced boulders or other stone objects during the 1897 earthquake in Assam, India. They were thrown from their original locations up to 2.5 m. Other examples, including eight cases where objects have been thrown from 2 to 4 m, are given by Midorikawa [4].

During the earthquake in Christchurch (February, 2011, New Zealand) cases were registered of people being thrown up in the air.

Even more surprising phenomena occur during seaquakes, excited by strong vertical perturbations of a seabed. Huge water columns, waves and gas bubbles appeared on the sea surface, caused by transient cavitation processes near the sea bottom. If the earthquake epicentre is directly under a vessel, then it starts to vibrate strongly and in certain cases can separate for an instant from the water [13].

2.2 Features of Destruction of Weakly-Cohesive Materials by Extreme Vertical Dynamics

During strong vertical earthquake-induced vibrations an upper-lying weakly-cohesive sediment can separate from a rock base [14–17]. As a result, a gap is formed between the materials. The annihilation of the gap and the following collision of the rock and the sediment generate strongly-nonlinear waves, which can have spiky or discontinuous profiles.

Vertical-induced bouncing motion There is a very interesting series of experiments with sand which was excited by vertical shaking [18].

[1] Or 'whiplash effect.'

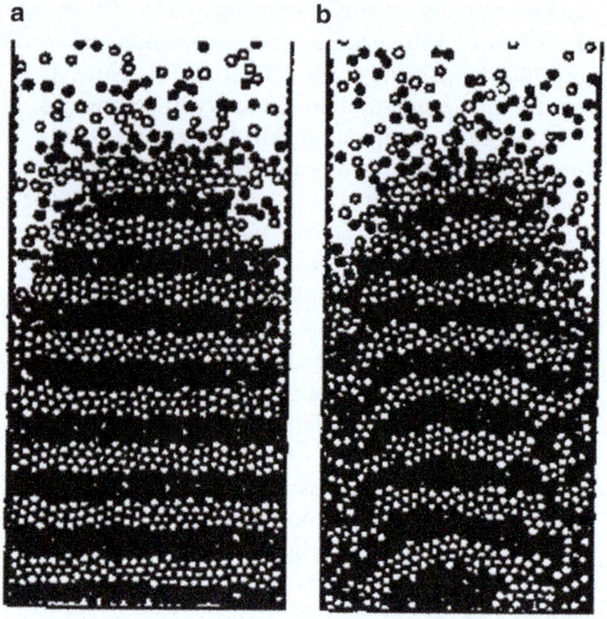

Fig. 2.1 Positions of grains of sand at the same phase in the motion of two successive shakes [18]

Figure 2.1 shows grains of sand after two successive shakes. The grains are coloured black and white, so that one can follow the net motion integrated over one full shake. The upper grains are flying free, and there is vertical motion in the central bulk of the material. It is important that this motion is largest near the bottom. The motion near the lateral boundaries is determined by boundary effects. Similar effects of weakly-cohesive material/rigid base interaction may be very interesting for seismology. Darwin wrote . . . *a vibratory movement of overwhelming force. . . ., the fragments have been levelled into one continuous sheet . . . what must we say to a movement which has caused fragments, many tons in weight (like so much sand on a vibrating board), to move onward and find their level?* . . . [1, pp. 255–256].

Michael Faraday [19] began to study the effect of vertical vibrations for a layer of weakly-cohesive materials, a few months before Darwin's voyage. He found that small heaps were being formed and a slow convection of the particles took place due to the vibrations. Recent experimental studies of vertically-vibrated granular media have demonstrated a rich variety of nonlinear wave phenomena. In particular, for acceleration down larger than g, the effective gravity becomes negative and the grain layer loses contact with the base. A small gap is formed between the base and the layer bottom and then the gap decreases. Examples of the heaps (surface waves) and the gap are given in Fig. 2.2. After such a collision the layer has the velocity of the base, and it remains in contact with the base until the next cycle, where the process repeats. This probably relates to effects of the earthquake and seaquake [13] described by Darwin [1] (cf. Chap. 5 of this book).

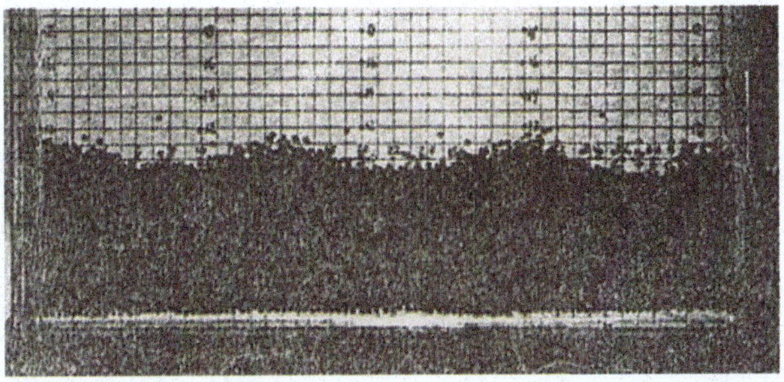

Fig. 2.2 An example of a Faraday-type experiment. A photograph of surface waves excited by vertical vibrations of granular material in a container [20]. The gap between the flying compact mass and the base is seen clearly. The vertical acceleration was 3.3 *g*, less than the vertical acceleration (3.8 *g*) which was measured during the 14 June 2008 Iwate-Miyagi earthquake in Japan [9]

When the down acceleration of the base exceeds *g*, the material can exhibit surprisingly complex behaviour. This includes states with the particles forming surface waves, or a 'cloud' with little or no structure (see Fig. 2.1). On the other hand, the particles can vibrate together like a compact mass (the bouncing motion). We think that Figs. 2.1 and 2.2 illustrate phenomena, which can take place during severe earthquakes.

The scheme of the bouncing motion has been used [16] to explain extreme oscillations of Tarzana hill, during the Northridge 1994 Southern California earthquake. The peak horizontal acceleration of the hill was 1.78 *g*, the peak vertical acceleration was 1.14 *g*. It was assumed that there was some threshold level of the earthquake-induced acceleration. Very strong amplification of the Tarzana hill oscillations can occur if the acceleration increases above this level. I have suggested that the earthquake-induced vertical oscillations of Tarzana hill resembled the bouncing motion of a ball on a vibrating base [16].

To illustrate this new idea I assumed [16] that the base rock may be simulated by an oscillating metal piston and the upper-lying sediment may be considered as liquid (water). In this way the high impedance contrast between the Tarzana hill soil and the underlying rock was modelled. Then results of Natanzon's experiments [21] with a water column were used for analysis of the Tarzana hill oscillations [14–17].

Oscillograms of the pressure, which is measured near the piston, are presented in Fig. 2.3 [16, 21]. The sinusoidal curves show the piston position. During the experiments the piston acceleration was slowly increased from 0.2 to 2 *g* and then was slowly reduced to 0.37 *g*. One can see that in the frequency range extending from 7.7 to 17 Hz there are large variations of the pressure waves. When the down acceleration increases above *g*, then the harmonic oscillations transform into peaks which correspond to collisions of the material and the base. The smooth parts of the pressure curves correspond to free flight of the material. During these moments above the base the rarefaction (cavitation) zone is formed. When the acceleration reduces, then a strong-hysteretic effect takes place. For this case, discontinuous waves exist in the system up to 0.5 *g*, and then instantly transform into harmonic

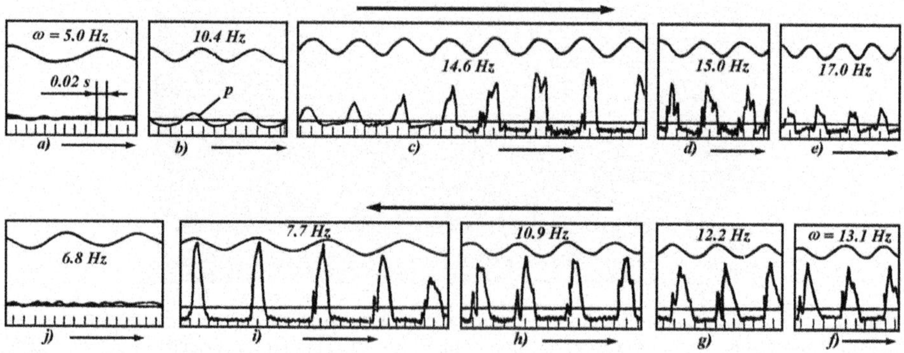

Fig. 2.3 Hysteretic dynamics of strongly-nonlinear oscillations of the water column (the exciting amplitude is 0.002 m and the column length is 7 m). The vertical acceleration is slowly increased from 0.2 to 2 g (forced frequency from 5 to 17 Hz) and then is slowly reduced to 0.37 g (forced frequency from 13.1 to 6.8 Hz) [16, 21]

waves of very small amplitude. Thus, according to the experiments (see Fig. 2.3) the strongly-nonlinear discontinuous waves may be generated in the upper-lying material, when the down acceleration of the base exceeds the gravity acceleration g. The linear continuous waves are excited if the vertical acceleration is smaller than 0.5 g.

Thus, if the forced acceleration increases above some threshold level, then the linear prediction breaks down. That nonlinearity changes the amplitude and form of the forced waves. In particular, the rarefaction (cavitation) wave is periodically generated above the vibrating base. However, according to Fig. 2.3 the amplitude of the forced waves is bounded and depends on the hysteretic effect. Because of this effect the maximal amplification takes place when the forced acceleration approximately equals 0.5 g. On the whole, this fact agrees with field observations. The amplification of strong seismic motions is often smaller than that observed for moderate seismic motions [4, 8].

The analogy noted above between oscillations of Tarzana hill and the water column in Natanzon's experiments agrees with the data of the Iwate-Miyagi 2008 earthquake in Japan [9]. An accelerogram measured during this earthquake is presented in Fig. 2.4. A very strong vertical acceleration of the surface of a sedimentary layer, located on a rigid volcanic tuff, is shown. The measured maximum positive acceleration was 3.8 g, while the negative acceleration did not fall lower than $-1.7\,g$. The last phenomenon is determined by the separation of the deposit layer from the tuff. Using the analogy noted above, I concluded [16] that the discontinuous oscillations of the water column (Fig. 2.3) can qualitatively describe discontinuous oscillations of the surface layers of the Earth during catastrophic earthquakes. A physical understanding of the anomalous acceleration 3.8 g [9] requires a detailed investigation of strong motion in upper layers during the ground shaking. We illustrated (see Figs. 2.1, 2.2, 2.3, 2.4 and 2.5 and Fig. 1.7 right) that

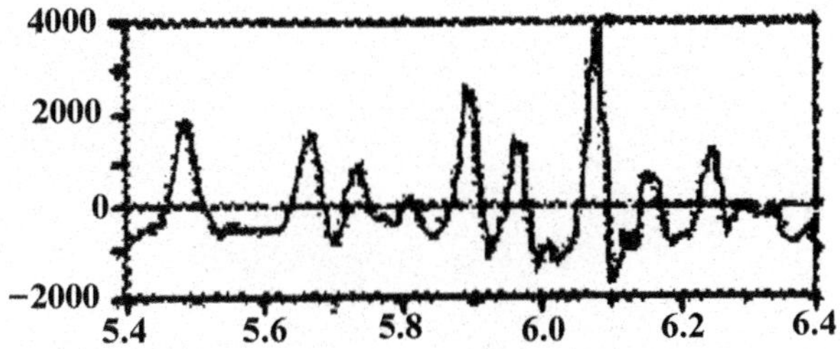

Fig. 2.4 Accelerogram of ground surface during of the Iwate-Miyagi 2008 earthquake in Japan. The acceleration is given in 0.001 *g*, time is given in seconds [9]

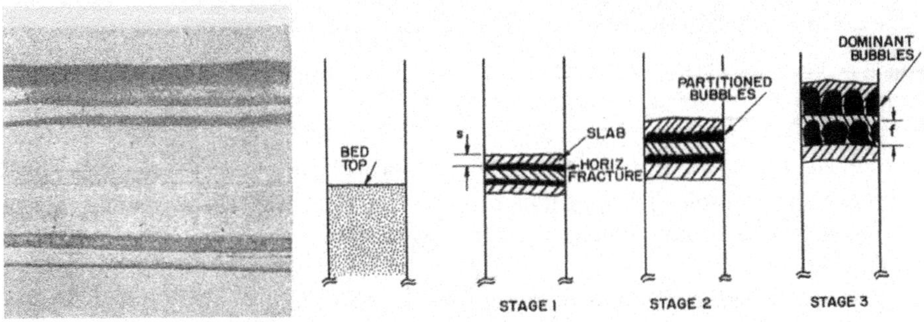

Fig. 2.5 The cavitation-layered rupture of water within the rarefaction wave (*left*) [22] and a scheme of the layered rupture of gassy material because of the growth of bubbles (*right*) [23]

this phenomenon may be connected with separations, collisions, and fractures of surface-ground layers.

Ruptures and cavitation inside vertical rarefaction waves As another illustration, we consider weakly-cohesive gassy materials, whose mechanical properties are very complex. However, the behaviour of these materials is much simpler within the vertical rarefaction waves (Figs. 2.5 and 2.6). This behaviour may resemble the behaviour of water or liquid-gas mixture. Compression and rarefaction waves may be generated in upper-lying weakly-cohesive material during great earthquakes. Within the compression zones the pore volume is reducing, and the gas pressure is increasing. In contrast, the pore volume is increasing within the rarefaction wave.

Thus, a collision of the surface layer with the rigid basis can generate a wave of compression with steep front, which is similar to a shock wave in gas. The reflection of this wave from the free surface of the layer generates a depression wave (expansion). The volume of the gas is increasing within the expansion, and the material transforms into gas-like state. As a result, the free surface of the layer is uplifted. Some material particles can begin to fly, practically independently (see Figs. 2.1 and 2.6). These processes are illustrated by Fig. 1.7 (right).

We will construct a few mechanical models for gassy deposits in this book. According to them, the rupture and loosening of these materials can resemble cavitation (boiling of liquid within rarefaction waves) (Fig. 2.5, left). The loosening

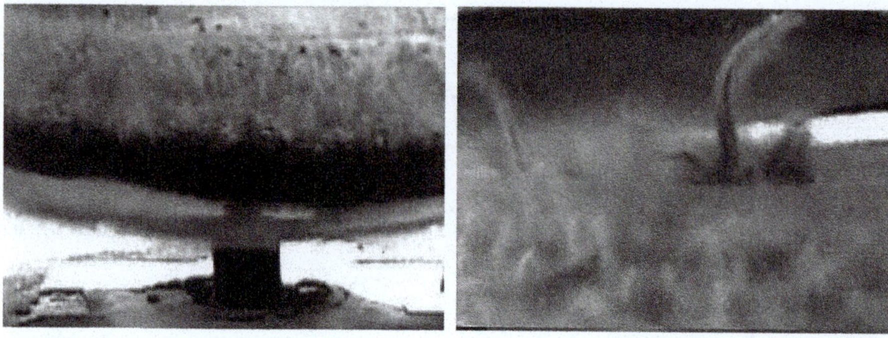

Fig. 2.6 Vertically-induced boiling of fine material (*left*). Vertically-induced local waves (jets) and craters on the surface of fine material (*right*) [24]

of high layers of the Earth can arise as a result of oscillations of the rigid base. This process can resemble the appearance of bubbles because of the boiling. These analogies do, to some extent, explain fast raising of loose deposits during earthquakes.

If there are high-amplitude vibrating fine materials, for example fine loose sand, then bubbles can occur near the bottom layer during the free-flight time. Then the 'bubbles' can move to the surface (Fig. 2.6, left), where they can form splashes, craters and waves. Another interesting phenomenon observed is rogue oscillons which are spewing the material [24, 25]. A few rogue oscillons may be oscillating in-phase with each other during resonance at the forced frequency (Fig. 2.6, right).

Different waves on an interface of two liquids may be observed in experiments developing Faraday's research. Recent researches show a huge variety of wave phenomena arising during similar experiments (cf. Chap. 5 of this book). In particular, formation and evolution of mushroom-like standing waves on an interface of two liquids were described [26].

We have considered vertically-excited waves. Similar waves may be formed by shallow earthquakes in upper sediments of the Earth. Different type of waves and bubbles may appear on the contact surfaces, inside and on the free surface of the layers from weak sediments. We can consider these waves as strongly-nonlinear Faraday-type waves, as a rough approximation.

On the whole, processes of fractures and loosening of upper-lying material and uplift of the free surface, which are shown in Figs. 2.1, 2.2 and 2.7, agree with data from many experiments (see, for example, [27]). These Figures also illustrate qualitatively one reason for ground elevation during severe earthquakes.

Thus, high-amplitude vertical accelerations may be excited by earthquakes [9]. If the down-going acceleration of the base is large enough, the upper-lying sediment material might lose contact with the base. As a result, the deposit layer can be fragmented and begin free-fall motion. Darwin wrote ... *the unsupported masses might be precipitated with the violence of an explosion* ... [3, p. 74]). Because of the following layer-base collision, strongly-nonlinear waves may be generated in the material. Of course, similar fractures of the upper-lying material can happen

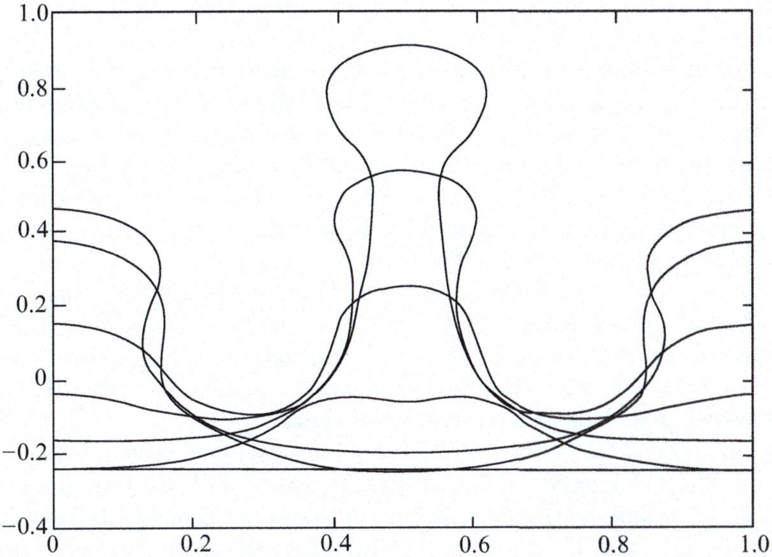

Fig. 2.7 Results of numerical studies of the formation and the evolution of mushroom-like standing waves on interface of two liquids during Faraday-type experiments [26]

when the vertical earthquake-induced acceleration is large enough. Apparently, these accelerations may be accompanied by the decomposition of surface weakly-cohesive geomaterials and the generation of many cracks (bubbles) near the ground surface. On the whole, this agrees with Darwin's note ... *That this is followed by many minor fractures, which, though extending upwards nearly to the surface, do not (excepting in the comparatively rare case of a submarine eruption) actually reach it* ... [3, p. 71].

2.3 Instability of the Land Surface and the Connection of Seismic and Volcanic Phenomena

Darwin attempted to develop a general theory of earthquakes. According to him, ... *It has frequently happened, that during the same convulsion large areas of the globe have been agitated, and strange noises propagated to countries many hundred miles apart* ... [3, p. 59] and ...*in other quarters of the world, fissures must there have been formed, and therefore earthquakes*. ... [3, p. 74]. Darwin began to think about instability of portions of the Earth crust which are formed by fissures. Discussing different kinds of earthquakes Darwin wrote ... *If the fluid matter, on which I suppose the crust to rest, should gradually sink instead of rising, there would be a tendency to leave hollows, and therefore a suction exerted downwards; or hollows would be actually left, into which the unsupported masses might be precipitated with the violence of an explosion. Such earthquakes, . . . would seldom be accompanied by eruptions, . . .* [3, p. 74].

On the other hand, earthquakes often precede volcanic eruptions. Interpretation of these relationships is a long-standing puzzle [28–31]. Is this eruption related to that earthquake? That is the question commonly asked by volcanologists and seismologists. Traditional answers to the question are of the form, 'Not likely' [29]. According to Darwin [1, pp. 380–381], the earthquake-induced vertical shock may be a common mechanism of simultaneous eruptions of volcanoes separated by long distances. In particular, Darwin wrote that . . . *the elevation of many hundred square miles of territory near Concepcion is part of the same phenomenon, with that splashing up, if I may so call it, of volcanic matter through the orifices in the Cordillera at the moment of the shock;. . .* [3, p. 60]. Indeed, according to experiments (see Fig. 1.7, right), weakly-cohesive initially-compressed material can erupt from the tube as a result of the fast decompression.

This analogy is fair, of course, not only for weakly-cohesive materials. It can explain boiling of magma within rarefaction waves. The last can appear due to collapse of volcanic craters and opening of magmatic conduits as a result of an earthquake (Fig. 2.8). Thus, this analogy explains a sequence explicitly existing in certain cases. Darwin noted – the earthquake first, the volcanic eruption next. Certainly, Fig. 2.8 is only a rough scheme. Magma properties are much more complex than properties of liquid [32]. However, it is known that fragmentation of magma resembles fragmentation of extremely viscous liquid [32–34].

The specified analogy with liquid allows us to explain the raising of magma as a result of vertical reflection of seismic waves of compression from its surface.

An earthquake-induced vertical acceleration of a volcanic base can induce these waves in a volcano conduit. Because of the topographic effect fractures may be maximal on mountain tops. Vertical accelerations can be amplified on tops of volcanoes and might initiate simultaneous eruptions of many volcanoes. According to Darwin . . . *a train of volcanoes situated in the Andes, . . . instantaneously spouted out a dark column of smoke. . .*.[1, p. 380] and . . . *If the earthquake . . . had acted in no other way, than in merely breaking the crust over the lava within the craters, a few jets of smoke might have been emitted, . . .* [3, p. 61] (see, also, Sects. 3.3.5, 4.2 and 7.4).

Darwin considered the crust as a system where fractured zones and zones of seismic and volcanic activities interact. He formulated the task of considering together systems which are being studied now in seismology and volcanology [4, 28–34]. If $g > 1$, then rarefaction waves may be generated near the ground surface. If an intensity of these waves is large enough, then the surface sediment layers may be ruptured during the earthquake. In particular, the surface material of volcano craters might collapse. Thus, the vertical acceleration can awake volcanoes. The vertical movement (acceleration) is an important characteristic of earthquake/volcano interaction.

We emphasize that Darwin explained these seismic effects by strong vertical vibrations of the surface layers of the ground. For us it is important that the possibility of such extreme vibrations has become accepted during recent years. It

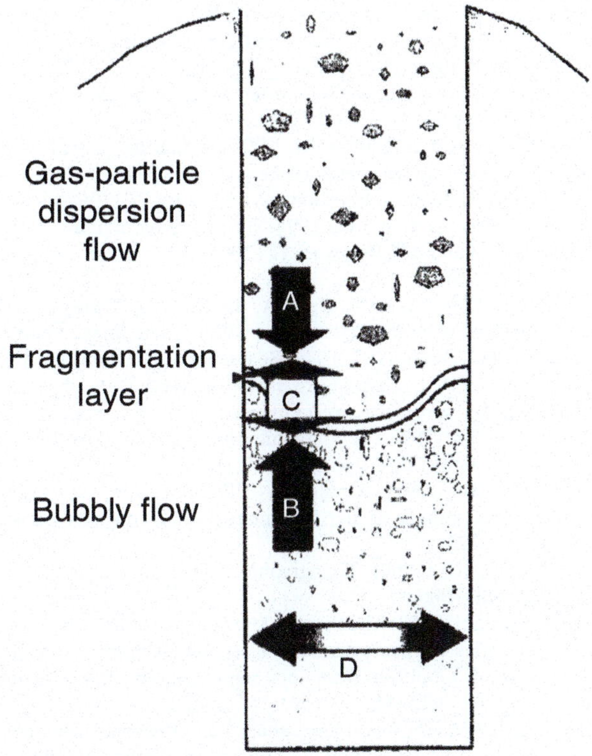

Fig. 2.8 A sketch of bubbly magma fragmentation [33]

has removed all doubts and has opened a way to our research. Remarkably, Darwin's interest in vertical vibrations of sediment layers and seas almost coincided with Faraday's experimental researches on vertically excited waves on a surface of liquid and granulated media.

References

1. Darwin C (1839) Journal of researches into the geology and natural history of the various countries visited by H.M.S. Beagle, under the command of Captain FitzRoy, R.N. from 1832 to 1836. Henry Colburn, London
2. Darwin C (1890) Journal of researches into the natural history and geology of the countries visited during the voyage of H.M.S. Beagle around the world, under the command of Captain Fitz Roy, R. N. T. Nelson and Sons, London
3. Darwin C (1840) On the connexion of certain volcanic phenomena in South America; and on the formation of mountain chains and volcanoes, as the effect of the same power by which continents are elevated. (Read March 7, 1838) In: Barrett PH (ed) (1977) The Collected Papers of Charles Darwin. The University of Chicago Press, Chicago
4. Anderson JG (2007) Treatise on geophysics. In: Schubert G (ed) Treatise on geophysics. Elsevier, Amsterdam/Boston

5. Singh SK, Mena E, Castro R (1988) Some aspects of source characteristics of the 19 September 1985 Michoacan earthquake and ground motion amplification in and near Mexico City from strong motion data. Bull Seism Soc Am 78:451–477
6. Aki K (1993) Local site effects on weak and strong ground motion. Tectonophysics 218:93–111
7. Oglesby DD, Archuleta RJ (1997) A faulting model for the 1992 Petrolia earthquake: can extreme ground acceleration be a source effect? J Geophys Res 102(B6):11877–11897
8. Field EH, Johnson PA, Beresnev IA, Zeng Y (1997) Nonlinear ground-motion amplification by sediments during the 1994 Northridge earthquake. Nature 390:599–604
9. Aoi S, Kunugi T, Fujiwara H (2008) Trampoline effect in extreme ground motion. Science 322:727–730
10. Johnson JB, Lees JM, Gerst A, Sahagian D, Varley N (2008) Long-period earthquakes and co-eruptive dome inflation seen with particle image velocimetry. Nature 456:377–381
11. Keynes RD (ed) (1979) The Beagle record. Cambridge University Press, Cambridge
12. Reiter L (1990) Earthquake hazard analysis. Columbia University Press, New York
13. Levin B, Nosov M (2009) Physics of tsunamis. Springer, Dordrecht/London
14. Galiev SU (1998) Topographic amplification of vertical-induced resonant waves in basins. Adv Hydrosci Eng Vol. 3, pp.179–182
15. Galiev ShU (1997) Resonant oscillations governed by the Boussinesq equation with damping. In: Proceedings of the 5th international congress on sound and vibration, Adelaide, pp 1785–1796
16. Galiev SU (1999) Topographic effect in a Faraday experiment. J Phys A Math Gen 32:6963–7000
17. Galiev SU (2003) The theory of non-linear transresonant wave phenomena and an examination of Charles Darwin's earthquake reports. Geophys J Inter 154:300–354
18. Kadanoff LP (1999) Built upon sand: theoretical ideas inspired by granular flows. Rev Modern Phys 71(1):435–444
19. Faraday M (1831) On a peculiar class of acoustical figures and on certain forms assumed by groups of particles upon vibrating elastic surface. Phil Trans R Soc Lond 52:299–340
20. Wassgren CR, Brennen CE, Hunt ML (1996) Vertical vibration of a deep bed of granular material in a container. Trans ASME J Appl Mech 63:712–719
21. Natanzon MS (1977) Продольные Автоколебания Жидкостной Ракеты (Longitudinal self-excited oscillations of a liquid-fuel rocket). Mashinostroenie, Moscow
22. Zamislaev BV, Iakovlev YS (1967) Динамические Нагрузки при Подводных Взрывах (Dynamical loads during underwater explosions). Sudostroenie, Leningrad
23. Anilkumar AV (1989) Experimental studies of high-speed dense dusty gases. Dissertation, California Institute of Technology
24. Thrasher ME (1997) Still playing in a sandbox. Internet
25. Xia H et al (2012) Oscillon dynamics and rogue wave generation in Faraday surface ripples. Phys Rev Lett 109:114502
26. Wring J, Yon S, Pozrikidis C (2007) Numerical studies of two-dimensional Faraday oscillations of inviscid fluids. J Fluid Mech 588:279–308
27. Mujica N, Melo F (1998) Solid-liquid transition and hydrodynamic surface waves in vibrated granular layers. Phys Rev Lett 80(2):5121–5124
28. Linde AT, Sacks IS (1998) Triggering of volcanic eruptions. Nature 395:888–890
29. Hill DP, Pollitz F, Newhall C (2002) Earthquake-volcano interaction. Phys Today 55 (11):41–47
30. Kawakatsu H, Yamamoto M (2007) Volcano seismology. In: Schubert G (ed) Treatise on geophysics. Elsevier, Amsterdam/Boston
31. Watt SFL, Pyle DM, Mather TA (2009) The influence of great earthquakes on volcanic eruption rate along the Chilean subduction zone. Earth Plan Sci Lett 277:399–407

32. Alidibirov M, Dingwell DB (1996) Magma fragmentation by rapid decompression. Nature 380:146–148
33. Spieler O, Dingwell DB, Alidibrov M (2004) Magma fragmentation speed: an experimental determination. J Volcanol Geotherm Res 129:109–123
34. Kameda M, Kuribara H, Ichihara M (2008) Dominant time scale for brittle fragmentation of vesicular magma by decompression. Geophys Res Let 35:L14302

Chapter 3
Extracts from Darwin's Publications, and His Basic Geophysical Ideas

...I a geologist... ***(Darwin.*** *Notebook M, p. 528)*

The 3-volume **Narrative of the Surveying Voyages of His Majesty's ships *Adventure* and *Beagle*** was published in London by Henry Colburn in 1839. Volume 1 contains FitzRoy's **Proceedings of the first expedition**, Volume 2 contains FitzRoy's **Proceedings of the second expedition** and Volume 3 contains Darwin's **Journal and Remarks**. Darwin's Volume 3 was reprinted a few weeks later as a separate book **Journal of Researches into the Geology and Natural History of the various countries visited by H. M. S. Beagle under the command of Captain FitzRoy, R.N. from 1832 to 1836**, which got published in German translation. Darwin revised that book significantly for the second edition published in 1845 by John Murray, and that 2nd edition has been reprinted very many times (with many variations in the title) and translated into many languages. The texts of the 1839 and 1845 editions of **Journal of Researches** can conveniently be compared at the website http://www.rockvillepress.com/tierra/texts/Journal-2.PHP. FitzRoy's Volumes 1 and 2 have twice been reprinted: in the facsimile edition of the 3 volumes (including Darwin's **Journal and Remarks**) by AMS Press, New York 1966; and in a 4-volume version (edited by Katharine Anderson), Pickering & Chatto, London 2012.

On February 20, 1835 Darwin experienced a great earthquake at Valdivia. The devastation was horrifying – nearly every building in the area was destroyed. Then the *Beagle* went to the city of Concepcion, and while the *Beagle* tried to make anchorage there, Darwin landed on the island of Quiriquina. There he found areas of the land that had risen a few feet due to the earthquake, and he was very excited by that discovery. It was direct evidence that the Andes Mountains, and indeed all of the South America, were very slowly rising above the ocean. That confirmed Charles Lyell's theory that land masses were rising in tiny increments during an extremely long period of time. Charles Lyell (1797–1875) was a great geologist, and the first edition of his famous **Principles of Geology** was Charles Darwin's 'vade mecum' during the voyage of the *Beagle* [1]. Lyell argued that the face of the Earth had changed gradually over long periods of time through continuing, cumulative effects of local disturbances, such as eruptions, earthquakes, erosion and deposition. The

Sh.U. Galiev, *Darwin, Geodynamics and Extreme Waves*,
DOI 10.1007/978-3-319-16994-1_3

planet was changing locally, eroding here, erupting there, in a state of perpetual directional flux during an unimaginable span of time. Darwin was greatly impressed by Lyell's ideas, and he considered them to be an important influence on his theory of natural selection. Given this fact of uplift, Darwin accepted the idea that the Earth must be extremely old. The next day he sailed to the town of Talcahuano and from there he rode a horse to Concepcion to meet up with the *Beagle*.

Darwin concentrated on one of the most important geological problems: elevation and subsidence of the Earth crust. This enhanced his interest in the effects of the great earthquake which he witnessed in Chile. He observed the earthquake-induced uplift of the ground surface, and that observation supported his opinion that the high Andes have been formed as a result of the ceaseless vertical movement of the crust. The modern theory explains this formation by movement of the tectonic plates.

In this Chapter we present some pages from the accounts by Darwin of that great earthquake. And there are some of Darwin's memories of the voyage of the *Beagle*, and of that period when he thought of himself as a geologist. At the end of the Chapter we shortly discuss Darwin's thoughts about the earthquake-induced catastrophic phenomena [2–5], the Earth's dynamics and extreme waves.

3.1 Extracts from Darwin's Journal of Researches

The emphasis on Darwin's priority in the description and the analysis of the results of the severe earthquakes is one from main goals of this book. The first time that Darwin reported his results to the Geological Society of London was on 7 March 1838. His report was printed in Transactions of the Geological Society of London, 2d ser., pt. 3, 5 (1840): 601–31 [5], where Darwin wrote . . .*It might at first be thought that, at Concepcion, the uplifting of the ground, which accompanied the first and great shock, would by itself have accounted for the whole phenomenon of the earthquake. The great shock, however, during the few succeeding days, was followed by some hundred minor ones (though of no inconsiderable violence), which seemed to come from the same quarter from which the first had proceeded: whilst, on the other hand, the level of the ground certainly was not raised by them; but on the contrary, after an interval of some weeks, it stood rather lower than it did immediately after the great convulsion, – a consequence, perhaps, of the settling down of the shaken ground. . . .*

From these considerations, we may, I think, fairly conclude, with regard to the earthquakes on the west coast of South America,

1st. That the primary shock is caused by a violent rending of the strata, which seems generally to occur at the bottom of the neighbouring sea.

2nd. That this is followed by many minor fractures, which, though extending upwards nearly to the surface, do not (excepting in the comparatively rare case of a submarine eruption) actually reach it.

3rd. That the area thus fissured extends parallel, or approximately so, to the neighbouring coast mountains.

4th. That when the earthquake is accompanied by an elevation of the land in mass, there is some additional cause of disturbance.

An lastly, That an earthquake, or rather the action indicated by it, relieves the subterranean force, in the same manner as an eruption through an ordinary volcano.

Darwin's descriptions of the great shock, the great waves and the volcanoes activity from **Journal of Researches** are presented in this section. First the title page of this book is shown here.

JOURNAL OF RESEARCHES

INTO THE

GEOLOGY

AND

NATURAL HISTORY

OF THE

VARIOUS COUNTRIES

VISITED BY H. M. S. BEAGLE,

UNDER THE COMMAND OF CAPTAIN FITZROY, R.N.

FROM 1832 TO 1836.

BY

CHARLES DARWIN. Esq., M.A. F.R.S.

SECRETARY TO THE GEOLOGICAL SOCIETY.

LONDON:

HENRY COLBURN, GREAT MARLBOROUGH STREET.

1839.

Extracts from this book (Chapter XVI) are presented in Sects. 3.1.1, 3.1.2 and 3.1.3.

3.1.1 The Great Shock

February 20TH. – The day has been memorable in the annals of Valdivia, for the most severe earthquake experienced by the oldest inhabitant. I happened to be on shore, and was lying down in the wood to rest myself. It came on suddenly, and lasted two minutes; but the time appeared much longer. The rocking of the ground was most sensible. The undulations appeared to my companion and myself to come from due east; whilst others thought they proceeded from south-west; which shows how difficult it is in all cases to perceive the direction of these vibrations. There was no difficulty in standing upright, but the motion made me almost giddy. It was something like the movement of a vessel in a little cross ripple, or still more like that felt by a person skating over thin ice, which bends under the weight of his body.

A bad earthquake at once destroys the oldest associations: the world, the very emblem of all that is solid, has moved beneath our feet like a crust over a fluid; – one second of time has conveyed to the mind a strange idea of insecurity, which hours of reflection would never have created. In the forest, as a breeze moved the trees, I only felt the earth tremble, but saw no consequences from it. Captain FitzRoy and the officers were at the town during the shock, and there the scene was more awful; for although the houses, from being built of wood, did not fall, yet they were so violently shaken that the boards creaked and rattled. The people rushed out of doors in the greatest alarm. I feel little doubt that it is these accompaniments which cause that horror of earthquakes, experienced by all those who have thus seen as well as felt their effects. Within the forest it was a deeply interesting, but by no means an awe-exciting phenomenon. The tides were very curiously affected. The great shock took place at the time of low water; and an old woman who was on the beach told me, that the water flowed very quickly, but not in big waves, to high-water mark, and then as quickly returned to its proper level; this was also evident by the line of wet sand. This same kind of quick but quiet movement in the tide happened a few years since at Chiloe, during a slight earthquake, and created much causeless alarm. In the course of the evening there were other weaker shocks, all of which seemed to produce in the harbour the most complicated currents, and some of great strength.

22D. – We sailed from Valdivia, and on the 4th of March, entered the harbour of Concepcion. While the ship was beating up to the anchorage, which is distant several miles, I was landed on the island of Quiriquina. The mayor-domo of the estate quickly rode down to tell us the terrible news of the great earthquake of the 20th; – "that not a house in Concepcion, or Talcuhano, (the port) was standing; that seventy villages were destroyed; and that a great wave had almost washed away the ruins of Talcuhano." Of this latter fact I soon saw abundant proof; the whole coast being strewed over with timber and furniture, as if a thousand great ships had been wrecked. Besides chairs, tables, book-shelves, &c., in great numbers, there were several roofs of cottages, which had been drifted in an almost entire state. The storehouses at Talcuhano had burst open, and great bags of cotton, yerba, and other valuable merchandise, were scattered about on the shore. During my walk round the island, I observed that numerous fragments of rock, which, from the marine productions adhering to them, must recently have been lying in deep water, had been cast up high on the beach. One of these was a slab six feet by three, and about two feet thick.

The island itself as plainly showed the overwhelming power of the earthquake, as the beach did that of the consequent great wave. The ground was fissured in many parts, in north and south lines; which direction perhaps was caused by the yielding of the parallel and steep sides of the narrow island. Some of the fissures near the cliffs were a yard wide: many enormous masses had already fallen on the beach; and the inhabitants thought, that when the rains commenced, even much greater slips would happen. The effect of the vibration on the hard primary slate, which composes the foundation of the island, was still more curious: the superficial parts of some narrow ridges were as completely shivered, as if they had been blasted by gunpowder. This effect, which was rendered very evident by the fresh fractures and displaced soil, must, during earthquakes, be confined to near the surface, for otherwise there would not exist a block of solid rock throughout Chile. This limited action is not improbable, as it is certain, that the surface of any body, when vibrating, is in a different condition from the central parts. It is, perhaps, owing to this same reason, that earthquakes do not cause quite such terrific havoc within deep mines, as would at first have been expected. I believe this convulsion has been more effectual in lessening the size of the island of Quiriquina, than the ordinary wear and tear of the weather and the sea during the course of an entire century......

3.1.2 The Great Waves

........In almost every severe earthquake which has been described, the neighbouring waters of the sea are said to have been greatly agitated. The disturbance seems generally, as in the case of Concepcion, to have been of two kinds: first, at the instant of the shock, the water swells high up on the beach, with a gentle motion, and then as quietly retreats; secondly, some little time afterwards, the whole body of the sea retires from the coast, and then returns in great waves of overwhelming force......

...........The whole phenomenon, it appears to me, is due to a common undulation in the water, proceeding from a line or point of disturbance, some little way distant. If the waves sent off from the paddles of a steam-vessel be watched breaking on the sloping shore of a still river, the water will be seen first to retire two or three feet, and then to return in little breakers, precisely analogous to those consequent on an earthquake. From the oblique direction in which the waves are sent off from the paddles, the vessel has proceeded a long way ahead, before the undulation reaches the shore; and hence it is at once manifest, that this movement bears no relation to the actual displacement of the fluid from the bulk of the vessel. Indeed, it seems a general circumstance, that in all cases where the equilibrium of an undulation is thus destroyed, the water is drawn from the resisting surface to form the advancing breaker.[1] Considering then a wave produced by an earthquake as an ordinary undulation proceeding from some point or line in the offing, we can see the cause, first of its occurrence some time after the shock; secondly, of its affecting the shores of the mainland and of outlying islets in a uniform manner – namely, the water retiring first, and then returning in a mountainous breaker; and lastly, of its size being modified (as appears to be the case) by the form of the neighbouring coast. For instance Talcuhano and Callao are situated at the head of great shoaling bays, and they have always suffered from this phenomenon; whereas, the town of Valparaiso, which is seated close on the border of a profound ocean, though shaken by the severest earthquakes, has never been overwhelmed by one of these terrific deluges. On this view, we have only to imagine, in the case of Concepcion, a point of disturbance in the bottom of the sea in a south-west direction, whence the wave was seen to travel, and where the land was elevated to a greater height than any other part, – and the whole phenomenon will be explained.

It is probable that near every coast, the chief line of disturbance would be situated at that distance in the offing, where the fluid which was most agitated, from overlying the shallow bottom near the land, joined on to that part which covered the depths (but slightly moved) of the ocean. In all distant parts of the coast the small oscillations of the sea, both at the moment of the great shock, and during the lesser following ones, would be confounded with the undulation propagated from the focus of disturbance, and hence the series of movements would be undistinguishable.

3.1.3 The Paroxysmal Vertical Movement and Volcanoes

The most remarkable effect (or perhaps speaking more correctly, cause) of this earthquake was the permanent elevation of the land. Captain FitzRoy having twice visited the island of Santa Maria, for the purpose of examining every circumstance with extreme accuracy, has brought a mass of evidence in proof of such elevation, far more conclusive than that on which geologists on most other occasions place implicit faith. The phenomenon possesses an uncommon degree of interest, from this particular part of the coast of Chile having previously been the theatre of several earthquakes of the worst class. . . .

. . .Some of the consequences which may be deduced from the phenomena connected with this earthquake are most important in a geological point of view; but in the present work I cannot do more than simply allude to the results. Although it is known that earthquakes have been felt over enormous spaces, and strange

[1] I am indebted to Mr. Whewell for explaining to me the probable movements on the shore, of an undulation of which the equilibrium has been destroyed.

subterranean noises likewise heard over nearly equal areas, yet few cases are on record of volcanoes, very far distant from each other, bursting out at the same moment of time. In this instance, however, at the same hour when the whole country around Concepcion was permanently elevated, a train of volcanoes situated in the Andes, in front of Chiloe, instantaneously spouted out a dark column of smoke, and during the subsequent year continued in uncommon activity. It is, moreover, a very interesting circumstance, that, in the immediate neighbourhood, these eruptions entirely relieved the trembling ground, although at a little distance, and in sight of the volcanoes, the island of Chiloe was strongly affected. To the northward, a volcano burst out at the bottom of the sea adjoining the island of Juan Fernandez, and several of the great chimneys in the Cordillera of central Chile commenced a fresh period of activity. We thus see a permanent elevation of the land, renewed activity through habitual vents, and a submarine outburst, forming parts of one great phenomenon. The extent of country throughout which the subterranean forces were thus unequivocally displayed, measures 700 by 400 geographical miles. From several considerations, which I have not space here to enter on, and especially from the number of intermediate points whence liquefied matter was ejected, we can scarcely avoid the conclusion, however fearful it may be, that a vast lake of melted matter, of an area nearly doubling in extent that of the Black Sea, is spread out beneath a mere crust of solid land.

The elevation of the land to the amount of some feet during these earthquakes, appears to be a paroxysmal movement, in a series of lesser and even insensible steps, by which the whole west coast of South America has been raised above the level of the sea. In the same manner, the most violent explosion from any volcano is merely one in a series of lesser eruptions: and we have seen that both these phenomena, which are in so many ways related, are parts of one common action, only modified by local circumstances.....

3.2 Extracts from Darwin's Autobiographies

Voyage of the *Beagle*:
from Dec. 27, 1831 to Oct. 2, 1836

On returning home from my short geological tour in N. Wales, I found a letter from Henslow, informing me that Captain FitzRoy was willing to give up part of his own cabin to any young man who would volunteer to go with him without pay as naturalist to the Voyage of the "Beagle" . . ., but my Father strongly objected, adding the words fortunate for me, – "If you can find any man of common sense, who advises you to go, I will give my consent." So I wrote that evening and refused the offer. On the next morning I went to Maer to be ready for September 1st, and whilst out shooting, my uncle sent for me, offering to drive me over to Shrewsbury and talk with my father. As my uncle thought it would be wise in me to accept the offer, and as my father always maintained that he was one of the most sensible men in the world, he at once consented in the kindest manner. . . .

. . ...The voyage of the "Beagle" has been by far the most important event in my life and has determined my whole career; yet it depended on so small a circumstance as my uncle offering to drive me 30 miles to Shrewsbury, which few uncles would have done, and on such a trifle as the shape of my nose. I have always felt that I owe to the voyage the first real training or education of my mind. I was led to attend closely to several branches of natural history, and thus my powers of observation were improved, though they were already fairly developed.

The investigation of the geology of all the places visited was far more important, as reasoning here comes into play. ......I had brought with me the first volume of Lyell's 'Principles of Geology,' which I studied attentively; and this book was of the highest service to me in many ways. The very first place which I examined, namely St. Jago in the Cape Verde islands, showed me clearly the wonderful superiority of Lyell's manner of treating geology, compared with that of any other author, whose works I had with me or ever afterwards read. . . .

.....As far as I can judge of myself I worked to the utmost during the voyage from the mere pleasure of investigation, and from my strong desire to add a few facts to the great mass of facts in natural science. But I was also ambitious to take a fair place among scientific men, – whether more ambitious or less so than most of my fellow-workers I can form no opinion.....

. . . I think that I can say with truth that in after years, though I cared in the highest degree for the approbation of such men as Lyell and Hooker, who were my friends, I did not care much about the general public. I do not mean to say that a favourable review or a large sale of my books did not please me greatly; but the pleasure was a fleeting one, and I am sure that I have never turned one inch out of my course to gain fame.

From my return to England

Oct. 2 1836 to my marriage Jan. 29 1839

.....On March 7th, 1837, I took lodgings in Great Marlborough Street in London and remained there for nearly two years until I was married. During these two years I finished my Journal, read several papers before the Geological Society, began preparing the MS for my 'Geological Observations,' and arranged for the publication of the 'Zoology of the Voyage of the "Beagle".' In July I opened my first note-book for facts in relation to the Origin of Species, about which I had long reflected, and never ceased working on for the next twenty years. .....

Religious belief

During these two years I was led to think much about religion. Whilst on board the Beagle I was quite orthodox, and I remember being heartily laughed at by several of the officers (though themselves orthodox) for quoting the Bible as an unanswerable authority on some point of morality. I suppose it was the novelty of the argument that amused them. But I had gradually come, by this time, to see that the Old Testament from its manifestly false history of the world, with the Tower of Babel, the rainbow as a sign, etc., etc., and from its attributing to God the feelings of a revengeful tyrant, was no more to be trusted than the sacred books of the Hindoos, or the beliefs of any barbarian. . . .

....Thus disbelief crept over me at a very slow rate, but was at last complete. The rate was so slow that I felt no distress, and have never since doubted even for a single second that my conclusion was correct. I can indeed hardly see how anyone ought to wish Christianity to be true; for if so the plain language of the text seems to show that the men who do not believe, and this would include my Father, Brother and almost all my best friends, will be everlastingly punished.

And this is a damnable doctrine. . . .

. . . A being so powerful and so full of knowledge as a God who could create the universe, is to our finite minds omnipotent and omniscient, and it revolts our understanding to suppose that his benevolence is not unbounded, for what advantage can there be in the sufferings of millions of the lower animals throughout almost endless time? This very old argument from the existence of suffering against the existence of an intelligent first cause seems to me a strong one; . . .

From my marriage, Jan. 29 1839, and residence in Upper Gower St. to our leaving London and settling at Down, Sep. 14 1842

... Besides my work on coral-reefs, during my residence in London, I read before the Geological Society papers on the Erratic Boulders of S. America, on Earthquakes, and on the Formation by the Agency of Earth-worms of Mould. I also continued to superintend the publication of the 'Zoology of the Voyage of the "Beagle".' Nor did I ever intermit collecting facts bearing on the origin of species; and I could sometimes do this when I could do nothing else from illness. ...

.....Whilst living in London, I attended as regularly as I could the meetings of several scientific societies, and acted as secretary to the Geological Society. But such attendance, and ordinary society, suited my health so badly that we resolved to live in the country, which we both preferred and have never repented of.

3.3 Darwin on Earthquakes, Land Elevation, Volcano Eruptions and Catastrophic Ocean Waves

On February 20, 1835 a violent earthquake affected the coastal area of Chile (for about 1,000 km) parallel to the Andes. The coincidence of the earthquake with the eruption of volcanoes in the Andes led Darwin to the conclusion that the earthquake and the volcanic activity were related.

3.3.1 The 1835 Chilean Earthquake Is a Part of One Great Phenomenon

The world's earthquakes are not distributed randomly over the Earth's surface. They tend to be concentrated in narrow zones. Why is this? And why are volcanoes also found in these zones, too?

Darwin connected the earthquake and the simultaneous eruptions of volcanoes with crust rupture (see Fig. 3.1). On the whole, this point of view agrees with modern data. Darwin reported ... *In a geological point of view, it is of the highest importance thus to find three great phenomena, – a submarine outburst, a period of renewed activity through many habitual vents, and a permanent elevation of the land, – forming parts of one action, and being the effects of one great cause, modified only by local circumstances. ...* [5, p. 60] and ... *The two kinds of movements may, possibly, be explained, by considering that when the crust yields to the tension, caused by its gradual elevation, there is a jar at the moment of rupture, and a greater movement may be produced by the tilting up of the edges of the strata and by the passage of the fluid rock between them. In breaking a long bar of steel, would not a jar be caused by the fracture, as well as a vibration of the two ends when separate? ...* [5, p. 73].

Apparently, Darwin implied that the gradual elevation and the slide of portions of Earth's crust were determined by the expansion of the Earth. Nowadays global

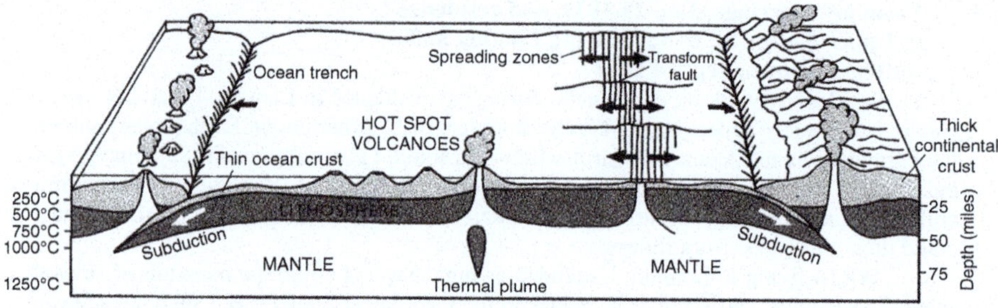

Fig. 3.1 Boundaries of tectonic plates as sources of earthquake and volcano activities. Diverging and converging zones of the lithosphere are shown. Strong vertical movement may be generated when the tectonic plates snap into a new position [6]

ruptures and elevation of the land are explained by slide motion, collisions and divergence of tectonic plates. However, Darwin formulated the big ideas about the rupture, the motion and the vibrations of the thin crust. The molten material rises through ruptures from the depth of the Earth.

It is well-known now, that the vast majority of strong earthquakes occur along fault zones and convergent boundaries of the tectonic plates, where one plate is pushed below another. A huge amount of energy is accumulating due to the friction and the compression of plate materials in the narrow band. As a result, the plate edge can melt and magma chambers are formed. The magma pressure in the chamber is, usually, very high. When the amount of energy becomes too much, the plates slide past each other into a new stable position. As a result, the energy accumulated within the plates gets released in the form of seismic waves – earthquakes. Seismic waves disturb the magma chambers, which are often connected with the land surface by volcano conduits. Usually the magma chamber is stable, but seismic waves can disturb the balance. The magma rises and erupts, and so the pressure decreases within the conduit-chamber system. As a result, gas bubbles begin to grow in the magma. This complex process is accompanied by high-amplitude pressure waves, travelling in the conduit. These waves can force vibrations of the land and the volcano.

Volcanoes are often located in subduction zones, where material of the tectonic plates is melted, or above a hot spot (a plume of molten material rising from the depth of the Earth) (Figs. 3.1 and 1.4). In particular, volcanoes are often located at oceanic ridges or rifts, where magma rises. When magma reaches the surface, it forms the ocean crust, which is layered plate of lava about 6 km thick.

We emphasize that the 1835 Chilean earthquake was great, but not the greatest Chilean earthquake. In May 1960, south–central Chile experienced a huge earthquake, the largest since instrument records began (9.5 magnitude). The energy released by the earthquake was accumulated for several centuries on the boundary of Nazca and South American tectonic plates. These plates converge at a rate of 8–9 cm per year.

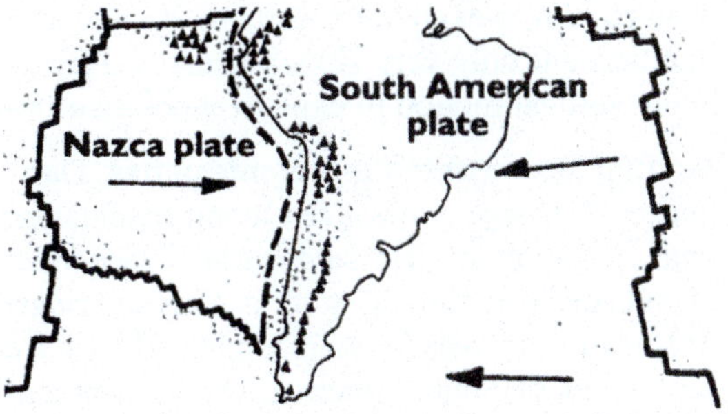

Fig. 3.2 Plate-tectonic schema of the west coast of South America. Arrows indicate the plate convergence [6]

Figure 3.2 shows the Nazca plate and the South American plate, with arrows showing the directions of displacements of the plates. The broken line determines a subduction zone (a fault) along the boundary of the Nazca and South American plates, the points show earthquake zones, and the triangles are volcanoes. The collision of the Nazca plate with the South America plate produces devastating earthquakes and it lifts up the Andes.

The earthquake of May 1960 resulted in ruptures within a zone 1,000 km long and 150 km wide. Large-amplitude ground shocks and vertical movements of the ground surface can occur, when there are such ruptures of tectonic plates and the impact interaction of them.

The subduction zone (fault) sharply separates the eastern side (continental shelf) from the deep ocean. There the depth changes rapidly from a few hundred metres to several thousand metres. The fault is located about 100 km offshore. Darwin reported ... *It is probable that near every coast, the chief line of disturbance would be situated at that distance in the offing, where the fluid which was most agitated, from overlying the shallow bottom near the land, joined on to that part which covered the depths (but slightly moved) of the ocean* ... [3, pp. 378–379].

Hence, Darwin believed that when big earthquakes occur under water, the seabed begins to vibrate. Movements of the seabed are the mechanism which forces tsunamis. Darwin wrote ... *On this view, we have only to imagine, in the case of Concepcion, a point of disturbance in the bottom of the sea in a south-west direction, whence the wave was seen to travel, and where the land was elevated to a greater height than any other part, – and the whole phenomenon will be explained* ... [3, p. 378].

Apparently, Darwin implied a simple one-dimensional model of the rupture. According to him, the crust breaks as a steel bar or as a spherical shell under internal pressure ... *in the bottom of the sea*...not far from Concepcion.

3.3.2 *The Earth as a Global Dynamic System*

Darwin considered results of that earthquake as important support for his theory of lowering and raising of continents. His idea of lowering and raising of continents enabled him to explain the formation of coral reefs. These theories of Darwin took into account only very slow processes. However, Darwin decided to use the idea about vertical motion to explain shock-like results of the earthquake.

Earthquakes within the tectonic plates Darwin distinguished two kinds of earthquakes. The first is localized along boundaries of the continents and is connected with elevation of the land surface (see Sect. 3.1). Thus, diverse catastrophic phenomena (the 'shivering' of islands and ridges, volcano eruptions, and generation of tsunami-like waves) were explained by the earthquake-induced elevation of the west coast of South America. On the whole, this idea agrees with the theory of tectonic plates. However, plate tectonics gives no insight into where and when earthquakes will occur within plates (continents) [7, 8].

According to Darwin ... *It has frequently happened, that during the same convulsion large areas of the globe have been agitated,* ... [5, p. 59]. Speaking about different kinds of earthquakes, Darwin emphasised ... *I confine the foregoing observations to the earthquakes on the coast of South America, or to similar ones, which seem generally to have been accompanied by elevation of the land. But, as we know that subsidence has gone on in other quarters of the world, fissures must there have been formed, and therefore earthquakes ... The earthquakes of Calabria, and perhaps of Syria, and of some other countries, have a very different character from those on the American coast. ... in Chile even the smaller shocks extend over the whole kingdom, and are propagated horizontally, whilst those which ... at Bologna, were of small extension, but instantaneous, and commonly explosive ... Such earthquakes, ... would seldom be accompanied by eruptions* ... [5, p. 74]. He thought that . . .*this large portion of the earth's crust floats in a like manner on a sea of molten rock.* ... [5, p. 81].

From Darwin's texts, it follows that his representations about a global structure of the Earth corresponded to the theory of the well-known French astronomer and mathematician Pierre Simon Laplace, who assumed that the land surface is a crust above melted material of the Earth. A similar representation about the structure of the Earth arose in ancient times. It returned as a scientific hypothesis through the authority of René Descartes and Gottfried Leibniz, and it remained in geology until the early 20th century. Geologists published imaginary pictures of an underground fiery ocean, with shakes in that ocean producing seismic waves in the Earth crust. Now that theory has been discarded, but it is known that there are separate regions of magma, melted materials and thermal plumes within the Earth (Figs. 1.4, 3.1 and 3.3). Thus, despite some specific historically-clear limitations of Darwin's views, his idea about dynamics of the earth crust and the causes of earthquakes are quite modern.

These predictions agree with results of recent observations and publications. In particular, Taira et al. [10] report that '... The fault-strength changes produced by the distant 2004 Sumatra-Andaman earthquake are especially important, as they suggest that the very largest earthquakes may have a global influence on the strength of the Earth's fault systems. As such a perturbation would bring many fault zones closer to failure, it should lead to temporal clustering of global seismicity ...'.

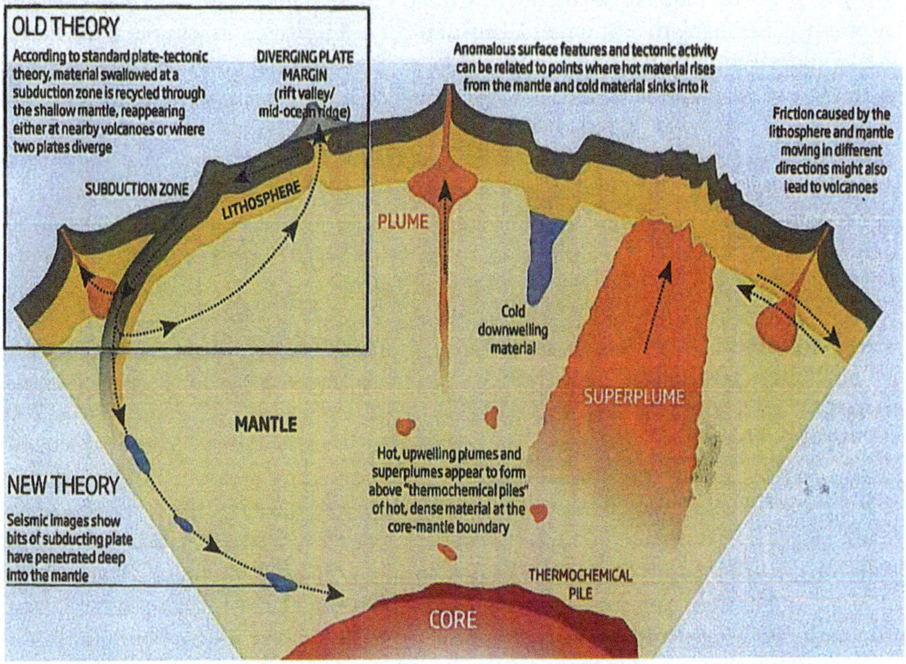

Fig. 3.3 Sketch of the new theory correcting the old theory (see Fig. 3.1). The deep Earth has an important effect on the surface features [9]

The magnitude M = 7.9 of the earthquake in Sichuan, China, in May 2008 was surprising, because it occurred on a fault that had had little recent seismicity. Such surprises occur often for earthquakes within continents. In contrast to the plate boundaries where large (magnitude M > 7) earthquakes occur at expected locations along boundary faults, the continental interiors like Sichuan or the central and east United States (for example, New Madrid and Charleston seismic areas) contain many old faults, most of which showed little seismicity over the past hundred years. Hence, we do not know the times and locations of future large earthquakes. Present earthquake hazard assessments typically assume that recent small earthquakes indicate the location of large future earthquakes. But what if some earthquakes are results of earlier large events [8]? According to Darwin... *most earthquakes, though appearing sudden, are in truth parts of a prolonged action, as evinced both the events which precede and those which follow it*... [5, p. 66]. Quite probably the earthquake in Sichuan (China, 2008) was provoked by the remote earthquake Sumatra-Andaman (2004), or by other similar earthquakes.

There is still no satisfactory explanation based on the theory of plate tectonics for some great earthquakes; for example, for the 1811–1812 earthquake at New Madrid, Missouri, USA, the 1886 earthquake at Charleston, South Carolina, USA, and the 2008 earthquake in Sichuan, China. Consequently, we will have to find alternative mechanisms. Stein and Liu write '... Improved assessments of earthquake hazard require treating the networks of faults within continents as complex systems and developing a better understanding of how such systems behave in time and space ...' [8]. An earthquake makes itself known in numerous ways. In some cases, earthquakes have come in clusters, such as the giant shock that hit off the northwest coast of Sumatra in 2004 and was followed by a series of nearby earthquakes. In other instances, they rupture large sections of faults, combining patches that had not been known to move together in previous seismic events. Earthquakes even hop from fault to fault, as seen in the earthquake that struck the town of Landers in California's Mojave Desert and the shock that hit central Alaska in 2002. Both of these earthquakes also had other claims to fame: they triggered tremors, geyser eruptions, and other seismic activity thousands of kilometres away [11].

The magnitude 8.6 earthquake that happened in the Indian Ocean on 11 April 2012 suggests a new way that catastrophic earthquakes can occur. It is important that the earthquake struck hundreds of kilometres from the nearest subduction zone. It was also surprising that the rupture had been shared between 4 distinct faults, 3 of which were oriented perpendicular to one another, creating a rough zigzag pattern.

Researchers now routinely talk about large earthquakes triggering others, both on nearby and distant faults. In fact, some say earthquakes are so complex that it may be impossible to pin them down, based on understanding of their physical nature alone [7, 8, 10–17].

Thus, large earthquakes can occur some distance away from a plate boundary. Some geologists believe that these earthquakes may be occurring along ancient faults that are buried deep in the continental crust. Recent seismic studies in the central United States have discovered such a fault in the Mississippi Valley.

It is now recognized that ruptures may be remarkably complex, with many undulations and intersecting branches. Prediction of severe earthquakes and strong ground motion requires understanding of the physical processes, such as 3-dimensional stress accumulation, which govern the initiation, propagation, and termination of an earthquake rupture. For example, the energy of the 3-dimensional tension state in the vicinity of the future seismic centre powers the initiation, distribution, and end of an earthquake.

Instability and vertical plumes Generally speaking, the systems of the faults may be unstable... *In a line of fracture, produced by subsidence, the distortion and overthrow of the strata would probably be even greater than in one of elevation, from the circumstance, that as soon as the weight of the mass overcame its cohesion, and it began to sink, there would be no counterbalancing power, like gravity during elevation, ...* [5, p. 75]. This does not agree completely with the theory of plate tectonics. Indeed, Darwin emphasized that there is some balance of vertical motions. The lift of mass at one place determines the down movement of mass at another place.

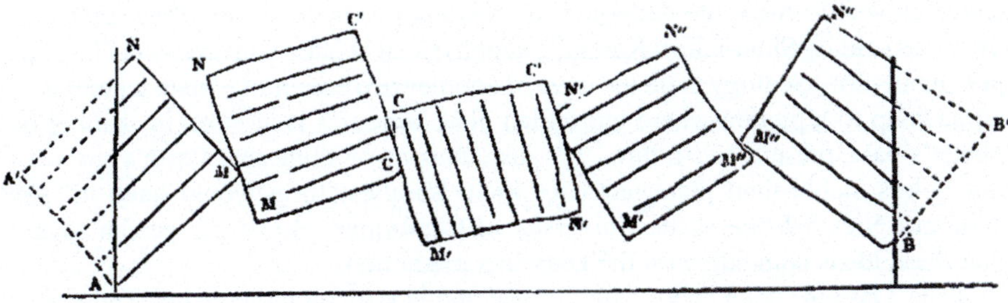

Overthrown crustal masses, from Diagram 3 in [5]

Darwin wrote that ... *for, in order to break up and throw over portions of very thick crust, as in Diagram 3, there must have been great horizontal extension, and this, if sudden, would have caused as many continuous outbursts of volcanic matter* ... and ... *if the force had acted suddenly, these portions of the earth's crust would have been absolutely blown off,* ... [5, p. 79]. According to Darwin, almost everywhere that we look, there is evidence of vertical movements within Earth reshaping its surface. Indeed, seismographic images suggest that the workings of the deep Earth have an important effect on surface features. A plume of hot mantle material can actively push up from below and break through to the surface (Figs. 1.4 and 3.3). In particular, this process explains partly the appearance of volcanoes.

Figure 3.3 agrees completely with Darwin's thoughts ... *–when we think of the increasing temperature of the strata, as we penetrate downwards in all parts of the world, ...as far as regards the cooling of the rock in the lowest abysses,... the extreme slowness with which heat can escape from such depths; – when we reflect how many and wide areas in all parts of the world are certainly known, some to have been rising and others sinking during the recent aera, even to the present day, and do not forget the intimate connexion which has been shown to exist between these movements and the propulsion of liquefied rock to the surface* ... [5, p. 81].

Darwin believed that the crust is sinking into molten rock in some places, and in other places molten rock can erupt from the fractured areas. A plume of hot mantle material can actively push its way up from below and break through to the surface. There are different kind of the plumes, corresponding to proposed origins at the core-mantle boundary (superplumes), at the base of the upper mantle (hot and cold spots), and in the lithosphere. The number the superplumes is about 6 to 9. In particular, Iceland, Hawaii, the southern Pacific Ocean, and the coast of Argentina are influenced by different superplumes. Southern Africa is being propped up by a superplume.

Small hot spots and small cold spots at the top of superplumes seem to correspond with local surface topography. The Congo basin, for instance, lies on a cold area and is hundreds of metres lower than its surroundings. The activity of Yellowstone is also determined by some local hot spot. There are submerged landscapes off the west coast of Scotland and undersea volcanoes in the southern Pacific: all of these observations cannot be explained only by plate tectonics. But these data agree with Darwin's theory of lowering and raising of land surface. 'At the time plate tectonics was formed, the deep interior was unknown, so people drew cartoons'. That was said by Shun-ichiro Karato, a geophysicist at Yale University [17]. Müller [17] thinks so: 'Geology is on the cusp of another revolution like plate tectonics'. It all adds up to a picture where more than plate tectonics is at work in shaping our planet's past, present and future. 'It's just amazing to think that Earth's surface is rather less stable than plate tectonics in its simplest form would have it,' says geologist Nicky White at the University of Cambridge. However, we can suggest that these ideas coincide with the Darwin's predictions.

Thus, Darwin spoke about both vertical and horizontal movements. These ideas do not correspond completely with the theory of plate tectonics, which usually takes into account only the horizontal motion of the plates. Darwin formulated large ideas about the earth evolution and its dynamics, and here we consider briefly these ideas. About 200 years ago Darwin considered very important problems, which have only recently begun to be discussed [7, 8, 10–14]. Therefore, on the whole, modern seismological results support Darwin's ideas concerning geodynamics and the nature of earthquakes, which he published almost 200 years ago.

Approximately from 1835 till 1850 Darwin did his excellent research in the field of geology and geodynamics (see Preface, Sects. 1.4 and 1.5). I agree with Sandra Herbert (see Sect. 1.5), that Darwin wanted to create a 'simple geology'. Using his scientific method (see Sect. 1.2) he collected and analysed very many factors connected with the dynamics and evolution of the Earth. However, he could not find the source of the motion for the global evolution of the Earth, which might be a rough analogue 'of struggle for existence' in the case of the evolution of species. He wrote *...The secular shrinking of the earth's crust has been considered by many geologists a sufficient cause to account for the primary motive power of these subterranean disturbances; but how it can explain the slow elevation, not only of linear spaces, but of great continents, I cannot understand....the configuration of the fluid surface of the earth's nucleus is subject to some change, – its cause completely unknown, – its action slow, intermittent, but irresistible...*

3.3.3 The Earthquake-Induced Elevation of the Land

Earthquakes are most awful natural catastrophes. The results depend on the earthquake's amplitude, distance from the epicentre, and geology of the ground. Soft (porous) upper-lying sediments have a tendency to increase the amplitude and duration of the seismic waves increasing the potential for damage of the sediment. As a result the pore volume increases and the volume of the sediment increases too. The free surface of the sediment is uplifted. In particular, ... *the sandy shores of the great Bay of Concepcion* ... were uplifted during the earthquake [18, p. 34].

Ocean islands which are formed by porous materials containing a lot of organically-generated gas bubbles may be elevated more strongly. Darwin wrote that ... *the island of Santa Maria, ... was found by Captain FitzRoy to have been elevated to nearly three times the height that the coast near Concepcion was upraised.* ... [5, p. 72] (see, also, Sect. 3.1.3). In [2, p. 259], Captain FitzRoy gave the following information 'Santa Maria was upheaved nine feet. It appeared that the southern extreme of the island was raised eight feet, the middle nine, and the northern end upward of ten feet.' Now earthquake-induced lifting of the ground is a well-known phenomenon. It is interesting, that an ocean coast and small islands were usually uplifted. The Yakutat Bay area of southeastern Alaska experienced one of the most notable earthquakes of the nineteenth century on September 10, 1899. Although, this shock was preceded by a magnitude 8.2 earthquake 1 week earlier, most of the effects were associated with the September 10 event, which is now rated as magnitude 8.6 on the Richter scale.

A field investigation in this area was undertaken in 1905 by a U.S. Geological Survey party. They reported that the largest uplifts in the land ranged from 30 ft (9 m) to about 47.5 ft (14.5 m) on the west coast of Disenchantment Bay. A large area was affected by changes up to 17 ft (5 m) or more. In a few cases depressions occurred – their depth was from 1 to 7 ft (0.3–2 m). The Great Alaskan earthquake of 1964 was the most powerful recorded earthquake in North American history. Ground fissures and failures occurred. Some areas near Kodiak were permanently raised by 30 ft (9.1 m), but some areas near Anchorage dropped by 8 ft (2.4 m). Some residents claimed the ground waves were over 3 ft (1 m).

The strongest recorded earthquake in the world occurred in Valdivia, Chile, on May 22 1960, with magnitude of 9.5 on the Richter scale. The observed changes in land level ranged from 6 m of uplift to 2 m of subsidence. A strong earthquake of magnitude of 8.1 hit the Solomon Islands and nearby areas at 7:40 a.m. Monday, April 2, 2007. The 245 km long rupture crossed the Solomon Sea, where Woodlark and Australian plates interacted. After the earthquake, Simbo Island (see Fig. 3.4), which covers an area of 12 sq km, subsided by 1 m. A clear ground lift was noted on Ranongga Island (32 km long and 8 km wide), ranging from 0.9 m to about 3 m from north to south.

Ranongga Island is located 10 km north of Simbo Island, and it spreads roughly from north to south. The earthquake-induced rupture along the tectonic plate boundary passes midway between these islands. Nowhere else on the Earth there is a pair of islands situated so close to one another across an active fault.

3.3.4 The Topographic Effect

Darwin reported (see Sect. 3.1.1) *...the superficial parts of some narrow ridges were as completely shivered, ...*. According to Darwin the height of these ridges was about 400 ft [18, p. 34]. He described the now well-known topographic effect, in which damages of a crest of a ridge may be much stronger than damages near the ridge base. Some recent recordings indicated, that crest amplifications with respect to the base reached a factor of 10 and, in one case, a factor 30 [19]. As an example, the amplification near the crest of a ridge is shown in Fig. 3.5. The average peak acceleration is about 2.5 times the average base acceleration.

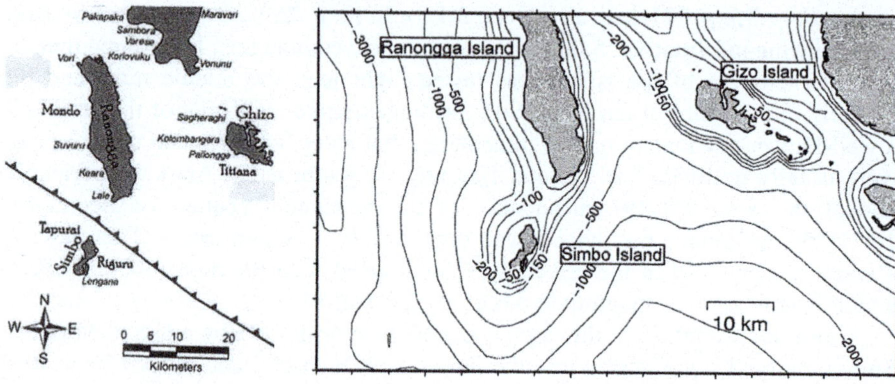

Fig. 3.4 Actuated fault and earthquake-excited islands during the 2007 Solomon Islands earthquake (*left*). The ocean depth around the islands (*right*). Internet. http://researchspace.auckland.ac.nz/handle/2292/4474

Fig. 3.5 Normalized accelerations (mean and error bars) recorded at different points along a mountain ridge at Matsuzaki, Japan [20]

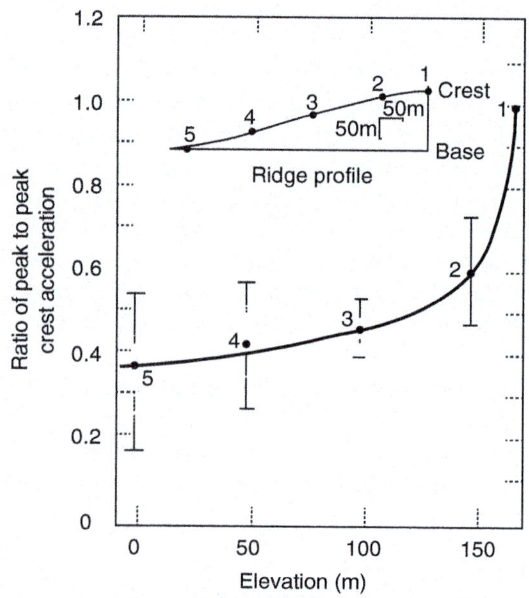

The reflection of an upward wave from the mountain slope may produce tensional stresses, which are sufficiently large to cause fractures of the surface. Focusing of the seismic waves near the mountain top can explain the violent crest shivers, which Darwin observed in Quiriquina.

He noted that … *the effects of one great cause, modified only by local circumstances* … [5, p. 60] may be explained by local geology and local topographic effects. In particular, the amplification of seismic waves at a volcano top can be a triggering mechanism for eruptions of volcanoes distant from each other.

The topographic effects can be caused by surface or vertical seismic waves. These waves can be trapped by topographies and amplified up to great amplitudes. In the case of surface waves the effect will be greatest if the wavelength is equal to one of the sizes of the topography base. For example, if the smallest size of a mountain base corresponds to the wavelength, then its top can start to oscillate like a pendulum. Resonance of horizontal oscillations of the topography takes place, if a wave frequency corresponds to a natural frequency of the topography. The amplitude of the oscillations will be greatest at the exact resonance.

A vertical seismic wave can be trapped by the surface sediment layers, and it can be amplified near resonances. The collapse of buildings at vicinities of the old dried lakes and the deaths of 20,000 people during the 1985 Mexican earthquake is connected with similar amplification [19–21]. The amplification can be so strong that the sediment layer starts to separate periodically from the rigid base. Probably, this effect explains the unexpected behaviour of Tarzana hill during the Californian earthquake of 1994 [22] (see, also, Sects. 1.6, 4.1.3 and 4.1.4).

3.3.5 Darwin's Triggering Mechanism of Volcano Eruptions

Darwin wrote that ... *the elevation of many hundred square miles of territory near Concepcion is part of the same phenomenon, with that splashing up, if I may so call it, of volcanic matter through the orifices in the Cordillera at the moment of the shock;* ... [5, p. 60] and ...- *a power, I may remark, which acts in paroxysmal upheavals like that of Concepcion, and in great volcanic eruptions,* ... [5, p. 61]. In particular, Darwin reported on simultaneous large eruptions of the four volcanoes: Robinson Crusoe, Minchinmavida, Cerro Yanteles and Peteroa. Thus, large earthquakes can stimulate large-scale volcano eruptions. Darwin suggested that there could be a common mechanism initiating eruptions of volcanoes, separated by long distances. This mechanism is ...*paroxysmal upheavals*... of the land (dynamic vertical displacement of the ground surface).

However, it is impossible to explain all documented events through a single mechanism [23–26]. The precise relationship between seismicity and magmatic activity remains enigmatic. In some cases, earthquakes induce magma movement and volcanic activity; but in other cases, the earthquake-volcano connection is not clear.

Charles Darwin reported that ... *If the earthquake or trembling of the ground (which, however, we have seen was less near these volcanos than elsewhere) had acted in no other way, than in merely breaking the crust over the lava within the craters, a few jets of smoke might have been emitted, but it could not have given rise to a prolonged and vigorous period of activity.* ... [5, p. 61]. Thus, the effect of vibrations and breaking of the crater surface may be important. On the whole, this agrees with information which has appeared in [27]. At the same time, according to Darwin, the effect of the break of a crater surface can be important, but is not

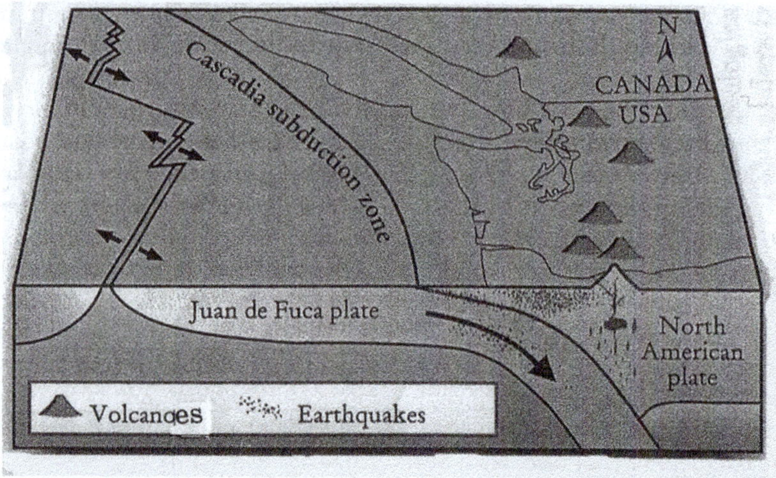

Fig. 3.6 Boundary of Juan de Fuca and North America plates where Mount St. Helens is located [24]

decisive for the beginning of the large eruption. Probably, he thought about other effects of trembling of the volcano base caused by the earthquake.

We shall compare Darwin's conclusions with data from the cataclysmic eruption of Mount St. Helens in Washington State (Fig. 3.6). The processes, effects, and products of this eruption were intensively studied and photographically documented (Figs. 3.7 and 3.8).

Geologists call Mount St. Helens a composite volcano (or stratovolcano), which is a term for a cone-like volcano having steep, often symmetrical sides. Composite volcanoes are constructed by alternating layers of lava flows, ash, and other volcanic deposits. Those volcanoes tend to erupt explosively; and Mount St. Helens exploded on May 18 1980, in the worst volcanic disaster in the recorded history of the United States. That was after 2 months of earthquakes and intermittent, relatively-weak eruptions,

Intense earthquake activity persisted at the volcano during and between visible eruptive activities. Seismographs began recording occasional spasms of volcanic tremors, a type of continuous, rhythmic ground shaking different from the discrete sharp jolts characteristic of earthquakes. The triggering earthquake was of slightly greater magnitude than any of the shocks recorded earlier at the volcano. For example, shocks of magnitude 3.2 or greater occurred at a slightly-increasing rate during April and May with 5 earthquakes of magnitude 4 or above per day in early April, and 8 per day in the week before May 18th. Initially, there was no direct sign of any eruption, but small earthquake-induced avalanches of snow and ice were reported by aerial observers.

About 8 o'clock in the morning there was an earthquake of magnitude 5.1, with epicentre at a depth of 2 km under the volcano. Edges of the crater of the volcano started to collapse because of a landslip. The pressure of the ground upon an exit of

Fig. 3.7 Dynamics of the volcano Mount St. Helens on May 18th 1980. It is clear that the earthquake of magnitude 5.1 caused the enormous avalanche which preceded the huge eruption (Photos by Gary Rosenquist)

the magmatic channel had fallen, gases and magma were then not constrained, and they began to erupt into the atmosphere. What happened next was described by the geologists Keith and Dorothy Stoffel, who were then in a small plane over the volcano's summit.

Among the events they witnessed, they 'noticed landsliding of rock and ice debris inward into the crater ... the south-facing wall of the north side of the main crater was especially active. Within a matter of seconds, perhaps 15 s, the whole north side of the summit crater began to move instantaneously. ... The nature of movement was eerie ... The entire mass began to ripple and churn up, without

Fig. 3.8 Mount St. Helens and the devastated area, 1 day after the devastating eruption. USGS Photograph taken on May 19, 1980, by Lyn Topinka

moving laterally. Then the entire north side of the summit began sliding to the north along a deep-seated slide plane. I (Keith Stoffel) was amazed and excited with the realization, that we were watching this landslide of unbelievable proportions. ... We took pictures of this slide sequence occurring, but before we could snap off more than a few pictures, a huge explosion blasted out of the detachment plane. We neither felt nor heard a thing, even though we were just east of the summit at this time.' Realizing their dangerous situation, the pilot put the plane into a steep dive to gain speed, and thus he was able to outrun the rapid eruption cloud that threatened to engulf them. The collapse of the north flank produced the largest landslide-debris avalanche ever recorded in historic time.

A volcano system (vent, conduit and magma chamber) can be quite stable. In this case, the deformation of the volcano crater and the volcano itself does not excite strongly the magma or gas eruptions, but stone/snow avalanches can be generated. The focusing of the stone/snow avalanches and the corresponding air-shock wave in the crater centre can form a vertical jet of a mixture of air, snow and solid particles, which may be sighted as a short-time eruption [28]. At the same time, the deformation can open the conduit vent. As a result, there may be a short-time gas emission which releases the pressure [5, p. 61], and then the magma chamber returns to the stable state. If the earthquake is large enough then dynamic processes can cause landslips and the crater can collapse. As a result the volcano vent unloads, producing conditions for a strong eruption to occur. Darwin wrote about these two possibilities (see the beginning of this subsection). Thus, earthquake-induced oscillations of the Earth surface can cause volcanoes to erupt.

We repeat this thought: land oscillations can be forced by earthquakes, those oscillations can be strongly amplified at the top of the volcano and it can collapse. As a result, a strong eruption can occur.

Darwin emphasized that an earthquake could be a trigger mechanism for simultaneous eruptions of volcanoes. In particular, a sharp peak in the number of eruptions on the same day as an earthquake was found [23]. The comparison of Darwin's conclusions with modern data shows that Darwin described some important mechanisms of earthquake-induced volcanic eruptions (see, also, Sects. 4.2, 7.4 and 8.2).

3.3.6 Tsunamis, Huge Ocean Waves and Resonant Amplification of Seismic Waves in Sediment Layers

Tsunamis Darwin [3, p. 377] wrote, that ... *the whole body of the sea retires from the coast, and then returns in great waves of overwhelming force.* ... and ...*The first wave was followed by two others,..* [29, p. 369]. Almost 200 years ago Darwin did not know that this wave was then named in Japanese as tsunami. Now his descriptions in his classic book **Voyage of the *Beagle*** could easily be read as a report from Sri Lanka after the tsunami of 26 December 2004. The sea water ebb during some earthquakes is now a well-known effect. For example, during the Great Alaskan Earthquake of 1964, at Kodiak Island the water in the harbour suddenly receded leaving fishing boats sitting on the seabed. When the waves struck, the boats were tossed into the town.

A tsunami may be considered as the tide-like wave which is excited by a earthquake-induced seabed uplift and submergence. Duration of the uplift may be of several seconds, or several tens of seconds. As a result, an initial local vertical elevation of the ocean surface occurs, and then the expanding ring of the tsunami begin to propagate. Far from shore the wavelength of a tsunami is very long and its amplitude is small (about 1 m). In this case the wave may be described by the linear theory. Nonlinear effects are important for tsunamis only in the vicinity of the shore, where the amplitude can rise up to 30 m [30, 31] (see Fig. 1.8 left).

Darwin emphasised the dependence of earthquake-induced waves upon the form of the coast and the coastal depth: ... *and lastly, of its size being modified (as appears to be the case) by the form of the neighbouring coast.* ... [3, p. 378]. This fundamental idea completely agrees with modern observations (see Fig. 3.9).

Captain Fitz-Roy gave a very interesting description of the Chilean 1835 tsunami [4, pp. 277–278] 'About half an hour after the shock, ...– the sea having retired so much, that all the vessels at anchor, even those which had been lying in seven fathoms water, were aground, and every rock and shoal in the bay was visible – an enormous wave was seen forcing its way through the western passage which separates Quiriquina Island from the mainland. This terrific swell passed rapidly along the western side of the Bay of Concepcion, sweeping the steep shores of every thing moveable within thirty feet (vertically) from high water-mark. It broke over,

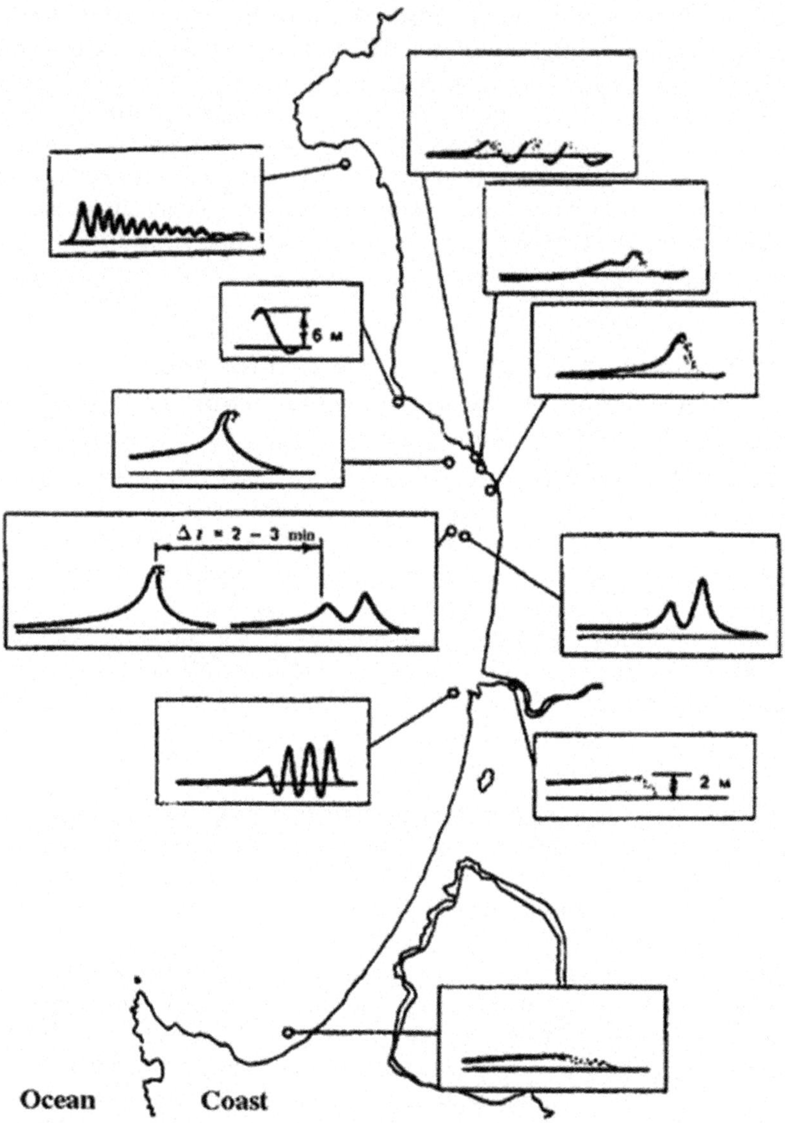

Fig. 3.9 Forms of tsunamis in different points of the North Akita Coast, Japan, during the Nihonkai-Chubu earthquake (26 May, 1983) [32]

dashed along, and whirled about the shipping as if they had been light boats; overflowed the greater part of the town, and then rushed back with such a torrent that every moveable which the earthquake had not buried under heaps of ruins was carried out to sea. In a few minutes, the vessels were again aground, and a second great wave was seen approaching, with more noise and impetuosity than the first; but though this was more powerful, its effects were not so considerable – simply because there was less to destroy. Again the sea fell, dragging away quantities of woodwork and the lighter materials of houses, and leaving the shipping aground.

After some minutes of awful suspense, a third enormous swell was seen between Quiriquina and the mainland, apparently larger than either of the two former. Roaring as it dashed against every obstacle with irresistible force, it rushed – destroying and overwhelming – along the shore. Quickly retiring, as if spurned by the foot of the hills, the retreating wave dragged away such quantities of household effects, fences, furniture, and other movables, that after the tumultuous rush was over, the sea appeared to be covered with wreck. Earth and water trembled: and exhaustion appeared to follow these mighty efforts.....

......Without explanation it appears astonishing how the shipping escaped destruction. There were three large whale-ships, a bark, two brigs, and a schooner, very near the town, in from four to seven fathoms water: they were lying at single anchor, with a good scope of cable: one only was well moored.

With the southerly breeze, which was rather fresh at the time of the earthquake, these vessels lay to seaward of their anchors, having their sterns towards the sea; and were left aground in this position. The captain of the port, D. Pablo Delano, was on board one of the whale ships at the time, with the hatches battened down, and dead lights shipped. All hands took to the rigging for safety. The first great wave came in an unbroken swell to the stern of the vessel, broke over and lifted her along without doing any material harm, more than sweeping her decks: and the slack chain dragging over the mud checked her gradually, as the first impetus of the wave diminished. Whirling her round, the water rushed out to seaward again, leaving the vessel stranded nearly in her former position. From two fathoms, then aground, the depth alongside increased to ten, as the water rose highest during the last swell. The two latter waves approached, and affected the shipping similarly to the former: all withstood their force, though the light anchors were dragged. Some of the vessels were thrown violently against others; and whirled around as if they had been in the vortex of a whirlpool. Previous to the rush of waters, the Paulina and Orion, two merchantmen, were lying a full cable's length apart; and after it had passed they were side by side, with three round turns in their cables. Each vessel had therefore gone round the other with each wave: the bow of one was stove in: to the other little damage was done'.

In the Chap. 5 of this book we will use the descriptions of the tsunami given above as tests of results of calculations.

Giant ocean waves The cultures of all seafaring nations contain reports of ships being swamped by gigantic waves, and of sightings of waves of unbelievable size. These waves appear suddenly, swallow a ship, and then vanish as quickly as they arise [33]. However, when Captain Dumont d'Urville, the French scientist and naval officer in command of an expedition to New Zealand in 1826, reported encountering waves up to 30 m height, he was openly ridiculed.

Perhaps the *Beagle* met such an extreme wave near Cape Horn. Darwin wrote ... *At noon a great sea broke over us,.... Had another sea followed the first, our fate would have been decided soon and for ever....* [29, pp. 264–265]. Extreme waves

Fig. 3.10 A huge wave strikes a cruise liner. Internet. http://www.usna.edu/Users/naome/phmiller/en358/EN358.htm

are not tsunamis caused by deep-sea geologic events, and are not tides caused by the gravitational pull of the sun and the moon. They often are wind waves generated by the movement of air over a water surface or the swell. In the ocean, swell generated by wind can travel thousands of kilometres beyond their origin – long after the wind has disappeared. The size and frequency of ocean waves are determined by the wind speed, by the distance travelled by them and by the water depth. Clashing currents can magnify the size of these waves, sometimes leading to the formation of extreme waves.

Modern wave formation theories (which work fairly successfully, in general) foretell very small probability of such waves occurring. The more extreme the wave, the less often it appears (if only it does not cause any geophysical accident). Even in a strong wind, such waves should appear so seldom that people never would face them (Fig. 3.10).

On April 16, 2005, the cruise liner Norwegian Dawn suffered heavy damage, although the wind-induced waves were not very large (between 14 and 28 ft high). However, the vessel suffered damage to the foredeck and to cabins on decks 9 and 10. The National Transportation Safety Board determined that the probable cause of the damage to the Norwegian Dawn, and of the injuries suffered by its passengers, was waves breaking over the bow during the ship's unavoidable encounter with severe weather and heavy seas. Perhaps, the liner met an attacking wave with height more than 20 m.

A similar extreme wave might have hit the *Beagle*. Indeed, such a wave is very dangerous for a small vessel. For example, the U.S. Coast Guard announced on April 22, 2013 that it was suspending the search for 4 fishermen whose boat is believed to have been destroyed by a extreme wave. The 50-ft vessel was tied to an oil rig in the Gulf of Mexico about 115 miles (185 km) southeast of Galveston. But in the early morning darkness, 'a rogue wave, a freak wave or something hit the side of the boat' John Reynolds, the sole survivor of the accident, told the Associated Press. Often ships do sink, but the cause is never studied as carefully as for an air crash: the tragedies are simply explained as 'bad' weather. At the same time these tragedies might be connected with the actions of extreme (catastrophic) waves (see Chap. 6 of this book).

According to [33], 22 supertankers have sunk or have received major damages as a result of encountering similar waves in the Pacific and Atlantic oceans during the period from 1969 to 1994, and 525 people were killed. And about 12 similar tragedies had occurred in the Indian Ocean in that same period. Sea oil platforms suffer from this phenomenon too. So, on February, 15th, 1982 an oil-drilling derrick of the company 'Mobil Oil' was wrecked by a rogue wave, killing 84 workers (see the sketch in Fig. 1.8 right) .

The resonant amplification of seismic waves The earthquake-induced ground waves may be amplified, like tsunami, in the surface sediment layers. The amplification of seismic waves is a complex process, which is connected with site geology and local topography. In particular, the concentration of seismic energy depends on the thickness variation of the layer, and it occurs near the edge of the layer (see Sect. 4.3).

That concentration can generate large and small cracks. FitzRoy reported in his Narrative [4, pp. 277–278] that '... The city of Concepcion stands upon a plain, very little higher that the level of the river Bio Bio. The soil is loose and alluvial. To the eastward and northward lie rocky irregular hills: from the foot of which the loose earth was every where parted by the great convulsion, large cracks being left, from an inch to more than a foot in width. It seemed as if the low land had been separated from the hills, having been more disturbed by the shock ....' This amplification of the convulsion depends also on the spatial variation of motion (the horizontal motion transforms partly into the vertical motion) and the impedance contrast between the layer and the down-lying rock. According to FitzRoy [2, p. 257] '....At Talcahuano the great earthquake was felt as severely on the 20th February as in the city of Concepcion. It took place at the same time, and in a precisely similar manner: three houses only, upon a rocky foundation, escaped the fate of all those standing upon the loose sandy soil, which lies between the sea-beach and the hills...'.

The waves considered in this subsection are generated by various causes, and they arose due to various physical conditions. However, the author believes that their amplification is connected with some or other resonant phenomenon. The amplification of the vertical movement is especially dangerous, when a frequency of vibrations forced by some earthquake coincides with some resonant frequency of sea water or of an upper sediment layer (see Fig. 1.7).

Thus, by examining some publications of Darwin we can formulate his ideas connected with the Chilean 1835 earthquake and its results. The earthquake source is at the coastal bottom of the ocean. An earthquake is a complex phenomenon which is defined by the previous development of the Earth crust and defines, appreciably, its subsequent dynamics. Because of oscillations caused by an earthquake, the raising of ocean islands and a land surface can take place. The earthquake-induced vertical movement can be amplified at tops of mountains and can awake volcanoes. Movement of the seabed can yield tsunamis and seaquakes. Similar phenomena can take place in sediment layers of variable thickness. Basically, these ideas completely agree with the data of modern science. We will consider of them attentively in the Chaps. 4, 5 and 6. At the same time, it is emphasized that some problems formulated by Darwin are still rather far from being resolved.

References

1. Sibson RH (2006) Charles Lyell and the 1855 Wairarapa, New Zealand earthquake: recognition of fault rupture accompanying an earthquake. Seismological Res Lett 77(3):358–363
2. Keynes RD (ed) (1979) The Beagle record. Cambridge University Press, Cambridge
3. Darwin C (1839) Journal of researches into the geology and natural history of the various countries visited by H.M.S. Beagle, under the command of Captain FitzRoy, R.N. from 1832 to 1836, 1st edn. Henry Colburn, London
4. FitzRoy R (1839) Narrative of the surveying voyages of His Majesty's Ships Adventure and Beagle, between the years 1826 and 1836, describing their examination of the southern shores of South America, and the Beagle's circumnavigation of the globe, vol 2. Henry Colburn, London. Proceedings of the second expedition, 1831–1836, under the command of Captain Robert FitzRoy R N
5. Darwin C (1840) On the connexion of certain volcanic phenomena in South America; and on the formation of mountain chains and volcanoes, as the effect of the same power by which continents are elevated. (Read March 7, 1838). In: Barrett PH (ed) (1977) The Collected Papers of Charles Darwin. The University of Chicago Press, Chicago
6. Robinson A (1993) Earth shock. Thames and Hudson, London
7. Parsons T (2009) Last earthquake legacy. Nature 462:42–43
8. Stein S, Liu M (2009) Long aftershock sequences within continents and implications for earthquake hazard assessment. Nature 462:87–89
9. Ananthaswamy A (2012) Earthly powers. New Sci 2878:38–41
10. Taira T, Silver PG, Niu F, Nadeau RM (2009) Remote triggering of fault-strength changes on the San Andreas fault at Parkfield. Nature 461:636–639
11. Chui G (2009) Shaking up earthquake theory. Nature 461:870–872
12. Field EH, Dawson TE, Felzer KR, Frankel AD, Gupta V, Jordan TH, Parsons T, Petersen MD, Stein RS, Weldon RJ II, Wills CJ (2009) Uniform California earthquake rupture forecast, Version 2 (UCERF 2). Bull Seis Soc Am 99(4):2053–2107
13. Akciz SO, Ludwig LG, Arrowsmith JR (2009) Revised dates of large earthquakes along the Carrizo section of the San Andreas Fault, California, since A. D. 1310±30. J Geophys Res 114: B01313
14. Chavarria JA, Malin P, Catchings RD, Shalev E (2003) A look inside the San Andreas fault at Parkfield through vertical seismic profiling. Science 302:1746–1748
15. Geller RJ (1995) The role of seismology. Nature 373:554
16. Geller RJ (1997) Earthquake prediction: a critical review. Geophys J Int 131:425–450
17. Müller DR (2011) Earth science: plate motion and mantle plumes. Nature 475:40–41
18. Barrett PH, Freeman RB (eds) (1986) The works of Charles Darwin. The geology of the voyage of H.M.S. Beagle. Part III: geological observations on South America, vol. 9. William Pickering, London

19. Reiter L (1990) Earthquake hazard analysis. Columbia University Press, New York
20. Kramer SL (1996) Geotechnical earthquake engineering. Prentice-Hall, Upper Saddle River
21. Singh SK, Mena E, Castro R (1988) Some aspects of source characteristics of the 19 September 1985 Michoacan earthquake and ground motion amplification in and near Mexico City from strong motion data. Bull Seism Soc Am 78:451–477
22. Galiev SU (1999) Topographic effect in a Faraday experiment. J Phys A Math Gen 32:6963–7000
23. Linde AT, Sacks IS (1998) Triggering of volcanic eruptions. Nature 395:888–890
24. Hill DP, Pollitz F, Newhall C (2002) Earthquake-volcano interactions. Phys Today 55 (11):41–47
25. Kawakatsu H, Yamamoto M (2007) Volcano seismology. In: Schubert G (ed) Treatise on geophysics. Elsevier, Amsterdam/Boston
26. Watt SFL, Pyle DM, Mather TA (2009) The influence of great earthquakes on volcanic eruption rate along the Chilean subduction zone. Earth Plan Sci Lett 277:399–407
27. Johnson JB, Lees JM, Gerst A, Sahagian D, Varley N (2008) Long-period earthquakes and co-eruptive dome inflation seen with particle image velocimetry. Nature 456:377–381
28. Galiev SU (2003) The theory of non-linear transresonant wave phenomena and an examination of Charles Darwin's earthquake reports. Geophys J Inter 154:300–354
29. Darwin C (1890) Journal of researches into the natural history and geology of the countries visited during the voyage of H.M.S. Beagle around the world, under the command of Captain Fitz Roy, R N. T Nelson and Sons, London
30. Galiev SU (2011) Геофизические Сообщения Чарльза Дарвина как Модели Теории Катастрофических Волн (Charles Darwin's geophysical reports as models of the theory of catastrophic waves). Centre of Modern Education, Moscow (in Russian)
31. Satake K (2007) Tsunamis. In: Schubert G (ed) Treatise on geophysics. Elsevier, Amsterdam/ Boston
32. Shuto N (1985) The Nihonkai-Chubu earthquake tsunami on the North Akita Coast. Coastal Eng Japan 28:255–264
33. Kharif C, Pelinovsky E, Slunyaev A (2009) Rogue waves in the ocean. Springer, Berlin

Chapter 4
Darwin's Reports on Catastrophic Natural Phenomena and Modern Science: Topographic Effect and Local Circumstances

> *...a vast majority of the volcanoes now in action stand either near sea-coasts or as islands in the midst of the sea ... the great continents are for the most part rising areas; and from the nature of the coral-reefs, that the central parts of the great oceans are sinking areas ... it would appear that volcanoes burst forth into action and become extinguished on the same spots, according as elevatory or subsiding movements prevail there ... (**Charles Darwin.** Journal of Researches. 1890, London, pp. 22, 574 and 575).*

Earthquakes, eruptions of volcanoes, connection of seismic and volcanic processes – all these phenomena stand among Darwin's most strong impressions during the round-world voyage of the *Beagle*. He saw the powerful forces which had been, for a long time, changing the surface of the Earth. Darwin thought over these processes and their connections, and that led to his reflection in the books [1, 2, 3] and his large article [4].

Local effects of great earthquakes can be determined by three main factors: distance from the earthquake source, travel path of seismic waves from the source to a particular location, and local site conditions [4–6]. Darwin saw personally only some results of the strong earthquake. Apparently, Darwin observed results of local amplification of the earthquake-induced ground motion.

4.1 Ground Elevation and Strongly-Nonlinear Topographic Effect

Darwin thought that there was: ... *one great cause, modified only by local circumstances* ... [4, p. 60]. The effects of local site conditions may be so great, that the propensity for earthquake damages at some locations may be more dependent upon these conditions, than from the proximity or the intensity of earthquake sources.

Large earthquakes are manifested at the Earth's surface by displaced deposits or by large cracks. Deep large cracks often do not reach the surface. In these cases, the surface seismic waves are responsible for rupture and elevation of the ground surface. Within the upper layers the earthquake-induced surface tensile stress

Sh.U. Galiev, *Darwin, Geodynamics and Extreme Waves*,
DOI 10.1007/978-3-319-16994-1_4

may substantially exceed the local compressive stress. As a result, within the surface layers, regions of compacted or loosened deposit can form.

When weakly-cohesive deposits are subjected to these local stresses, they can demonstrate strongly-nonlinear properties. Namely, within the deposits small cracks and pores appear, and the land surface can change. The phenomena described above, can explain the uplift of oceanic islands and of land. Here we ask, how the mechanical properties of deposits and nonlinear seismic waves are connected with the land elevation?

4.1.1 Some Physical Mechanisms for Ground Subsidence and Lift

During a large earthquake the density of sediments can change. The first seismic waves (for example, longitudinal waves) can transform the sediment into the liquefied state, and so its mechanical properties are changed. Therefore, the following waves (for example, shear waves) pass through regions having new mechanical properties. New configuration and geology of the sediment can strongly change the amplitude of the seismic wave.

Compaction and liquefaction of the surface material Deformation of weakly-cohesive materials is always accompanied by corresponding volume changes. In particular, the effect of compaction of granular materials under small-amplitude vibrations is well-known. The same phenomenon can take place as a result of moderate earthquakes [7, 8]. The soil particles disturbed by small-amplitude seismic waves can roll down into a new (lower) level of energy. In other words, under the action of seismic waves and gravity the loosely-packed individual soil particles drop down into a denser configuration (see Fig. 4.1) This phenomenon can partly explain the earthquake-induced sinking of a ground surface. In the case of a saturated soil the shear compaction can increase the water pressure, and in extreme cases the pore-water pressure may become much larger than the shear stresses. In such cases, the soil behaviour is determined by the pressure (as for water). This phenomenon is known as 'liquefaction', and it is often observed in low-lying areas near rivers, lakes, bays, and oceans.

Sometimes it is not necessary to add some water to granular materials, such as sand or dry soil, to make their properties like a liquid. These materials may simply become liquefied, if they are set in motion. For example, vibrations or impacts (see Figs. 2.1, 2.2 and Sect. 4.1.2), which separate individual grains from each other, transform the weakly-cohesive materials into the liquefied state. Of course, the amplitude of the vibrations or the impacts must be large enough.

Uplift of surface material Soil particles can uplift into a new (higher) level of energy, when they are forced by large-amplitude seismic waves. In other words,

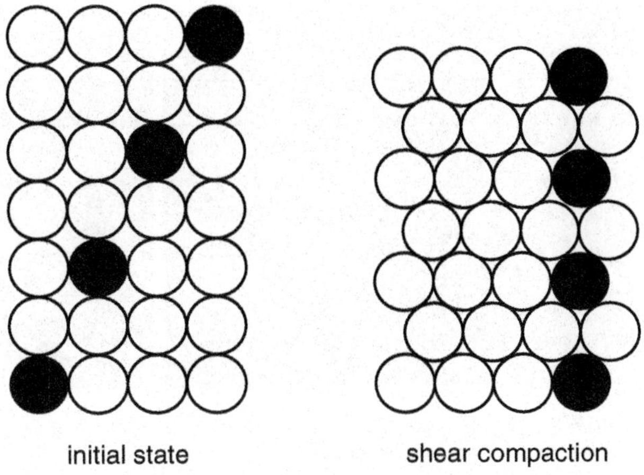

Fig. 4.1 Shearing of a packing of spheres introduces volume reduction [9]

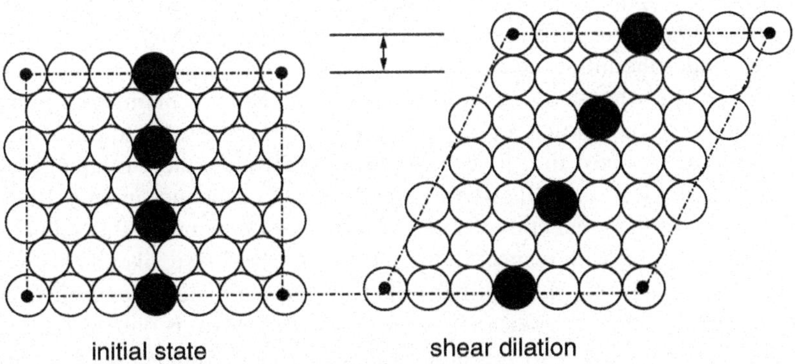

Fig. 4.2 Shearing of a close packing of spheres introduces volume expansion (dilatation) [9]

under the action of seismic waves, the loosely-packed individual soil particles can form a looser configuration (see Fig. 4.2).

Experimental evidence shows that mechanical and geometrical relationships exist between the stresses and the pores filled by a liquid or a gas. One of the most interesting and basic observations of such a relationship is the phenomenon of dilatancy. If an array of identical spherical grains (Fig. 4.2) is subjected by a shear deformation, then from pure geometrical considerations the particles must ride one over another. As a result, an increase in the volume of the bulk material will occur.

Osborne Reynolds – the expert in mechanics of liquid and granular materials- used the concept of dilatancy to explain the well-known phenomenon near the sea-shore, in which saturated sand dries next to the foot of a walker. The sand around the foot appears to be whiter. It indicates that the sand has dried instantly. Then upon removing the foot, one observes that a puddle of water has formed in the imprint in the sand. It is the result of the expulsion of water from the pores.

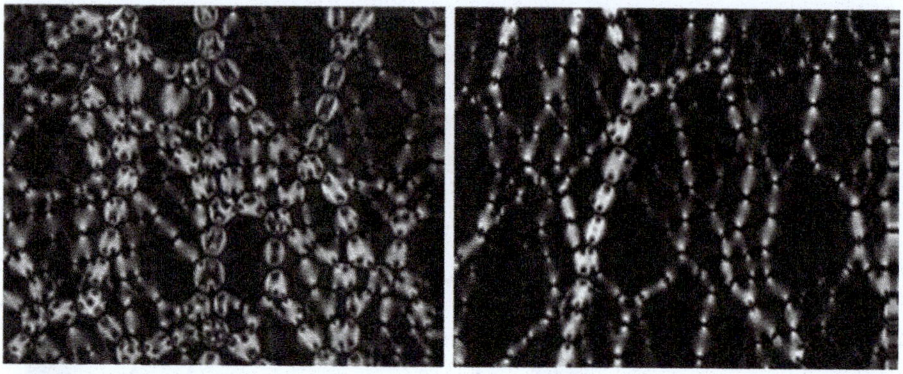

Fig. 4.3 Typical contact networks (skeletons) for isotropically-compressed (*left*) and sheared (*right*) states of granular material [12]

An example of a new more free structure of granulated material, which is generated by the shear displacement, is given in Fig. 4.3. Similar structures can be formed under the influence of seismic waves.

Of course, the wave-induced uplift of the land cannot remain for a long time. Indeed, it is well-known that an uplifted ground surface returns to the initial state during a relatively-short time (week or few weeks) after the earthquake. Robert FitzRoy reported [10, p. 260] 'when I was last at Talcahuano, in July 1835, only four months after the great convulsion, the shores of Concepcion Bay had regained their former position with respect to the level of the sea' (see, also, [11]).

It is emphasised, that this uplift mechanism is completely determined by the interaction of the land with seismic waves. The mechanism is quite different from the vertical elevation due to the deformation of the boundary of corresponding tectonic plates during an earthquake. That vertical elevation forms mountain chains on the boundary between plates. This rising, of course, does not come back to 'their former position'.

Uplift of a surface material within the rarefaction waves Darwin noted ... *mud containing organic matter in decay. In the bay of Callao, during a calm day, I noticed, that as the ship dragged her cable over the bottom, its course was marked by a line of bubbles...* [1, p. 374]. Thus, the soils around Concepcion had a high volume of gas. We can call such soils gassy soils. On the other hand, severe earthquakes can excite vertical accelerations which exceed g. In these cases the sharp-impedance contrast between the surface deposit layers and the base hard rock begins to play an important role. If the earthquake-induced negative pressure (tension) on the deposit/rock boundary exceeds the overburden pressure ρgh (ρ is the deposit density, h is the deposit thickness.), then several things may happen. First, a small gap could be generated (see Fig. 2.2). As a result of an annihilation of this gap (collision of the upper-deposit material with the rock bed) a strongly-nonlinear rupturing wave may be generated. An example of the similar wave is presented in Fig. 1.7 (right). This wave fragments the deposit material where cracks and pores are forming. Second, the resonance occurs as a seismic wave transmitted

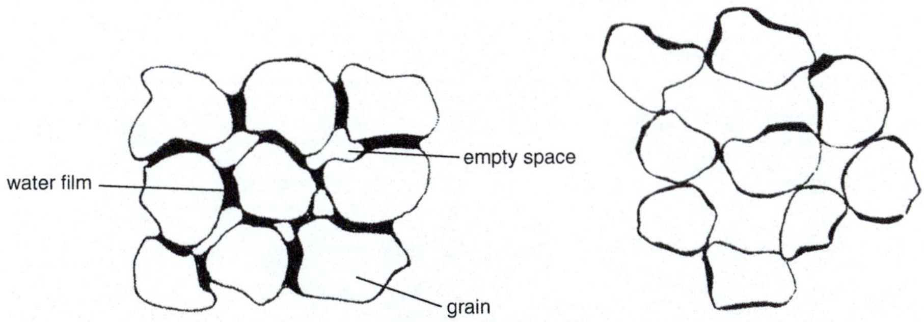

Fig. 4.4 Undisturbed (*left*) and disturbed (*right*) states of the material. The undisturbed material contains water, pores and cracks. The earthquake-induced strongly-nonlinear waves loosen the solid phase of the material, fragment it, and increase the pore volume [9]

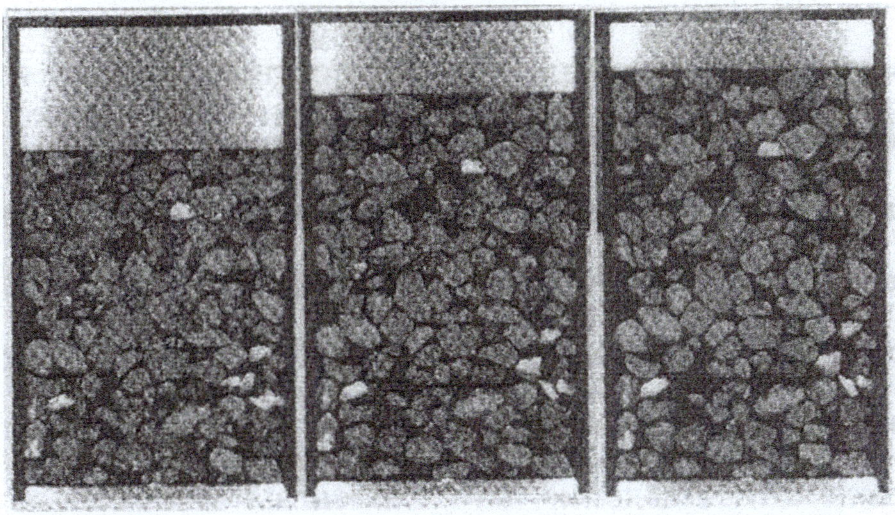

Fig. 4.5 Scheme of the uplift of the surface of initially compressed granular material during sharp decompression [9]

into the upper deposits themselves become trapped in these layers and begin to reverberate. If the seismic energy introduced into the deposits is high enough, then the layers can be fragmented. A result of such processes is shown in Fig. 4.4.

As a result, individual material fragments are formed. The porous volume and compressibility of the material increase strongly, and the strength of the material drops to zero.

The described process is illustrated by Fig. 4.5. A tube containing gas, solid particles and a piston is represented, where the gas gets strongly compressed by the piston. Three positions of the piston are shown. When the piston lifts very quickly the expanding gas moves the particles into new positions, forming new skeletons having a lesser and lesser strength. Similar phenomena can take place during a reflection of a seismic wave of compression from a free surface of a sediment layer.

Fig. 4.6 Sand can be sculpted into solid forms [13]

Thus, the strong earthquake-induced ground movement can transform weakly-cohesive materials into the liquefied state, in which the material volume increases strongly. As a result, the uplift of a land surface can occur. Perhaps, the uplift of islands and coastal regions during severe earthquakes is connected with the noted phenomenon. Apparently, this uplift of the land surface cannot take place during a long time.

However, the earthquake-induced vertical acceleration should be sufficiently large, greater then g.

4.1.2 Loosening of Sediments due to Vibrations

Granular materials, weakly-cohesive deposits and benthonic dirt, about which Darwin wrote, can be considered as some analogue of bubbly liquids. However, in rest these deposits resemble solid materials, since they can resist tension action (Fig. 4.6).

But if you shake this material vigorously, forcing particles to move relatively each other, the properties of the material change from 'solid' to 'granular liquid' states. In particular, standard experiments with granulated materials show an appearance of bubbles of gas in them under vibrations.

The vibrations transform the material of the surface layer into a mixture of solid particles, liquid, and gas (Figs. 2.1, 2.6 and 4.7). It is possible to talk about 'melting' a surface deposit, and the degree of shaking (or acceleration) might be related to an

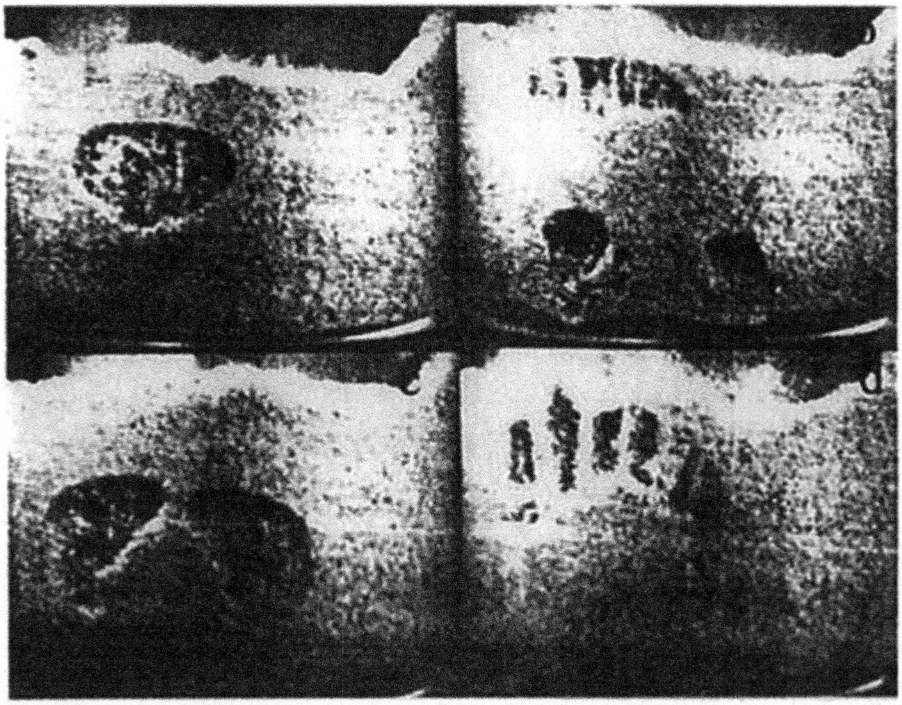

Fig. 4.7 In material consisting of small-diameter grains, that are intensively shaken bubbling of air voids can occur [14]

'effective temperature' of the deposit. In this state, the weakly-cohesive material is not in equilibrium – it needs a constant input of vibrational energy to retain its new state. If the amplitude of the vibrations is reduced, then the deposit (granular liquid) should reach a point at which it 'freezes' into a solid.

Thus, the behaviour of weakly-cohesive geomaterials depends on their history. In particular, properties of the materials before and after the earthquake may be quite different. Due to strong earthquake-induced vibrations, there are multiple relative displacements, cracks, tensional features between oscillating material blocks, and between the surface deposit and the solid base. The material is loosened and transforms into liquefied state, since the material strength and stiffness drop to zero [15].

The paper [15] explicitly describes these phenomena. A change of characteristics of the soil of Treasure Island during the 1989 Loma Prieta earthquake was described. The initial speed of a shear wave was about 160 m/s, but after the earthquake it became less than 10 m/s (Fig. 4.8). Figure 4.9 shows a change of soil properties caused by an earthquake. The variations of stress–strain diagram within an interval between 6 and 30 s after the beginning of the strong oscillations are shown.

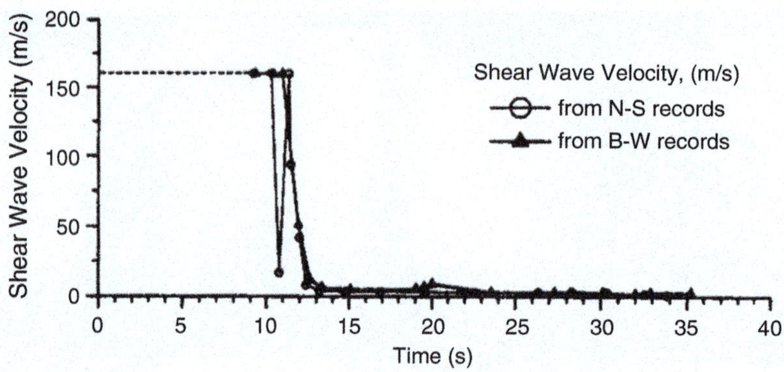

Fig. 4.8 Shear-wave velocity history of liquefied deposit at Treasure Island [15]

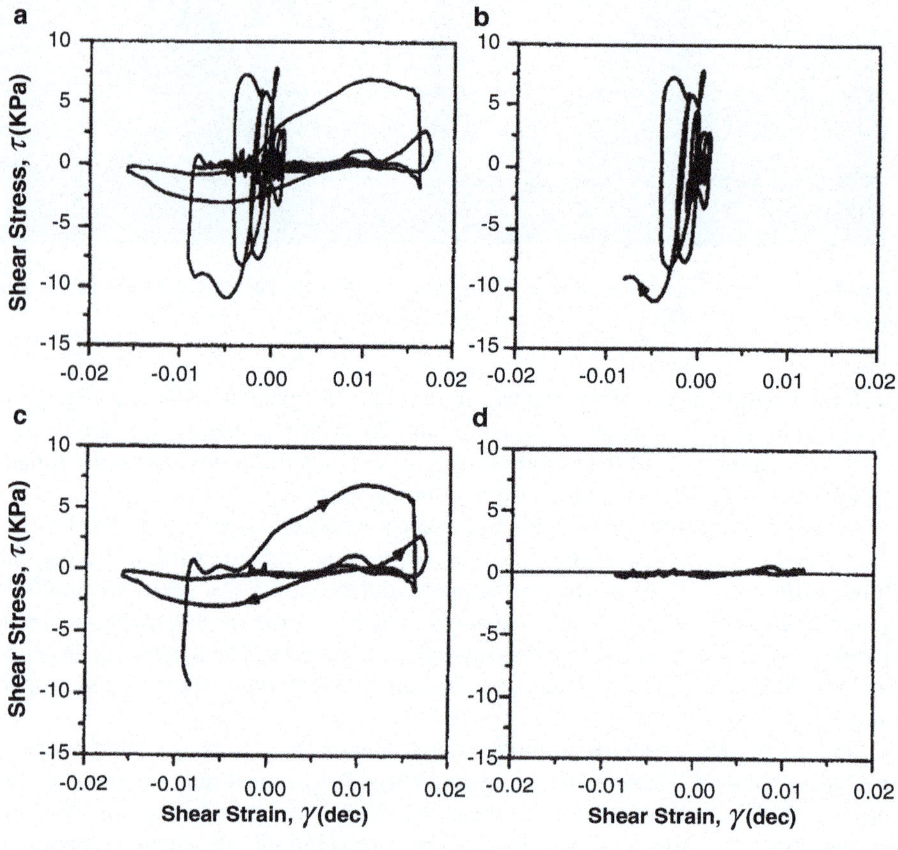

Fig. 4.9 East–west shear stress–strain history of liquefied deposit at Treasure Island: (**a**) Full record; (**b**) 8–14 s; (**c**) 14–22 s; (**d**) 22–32 s [15]

One can see progressive reductions in the shear modulus during the earthquake. The measured response of soil of Treasure Island may be considered as a reasonable analogue of the earthquake-induced change of properties of different deposits. Similar plots were recorded for soil of the site Wildlife [15], where the soil density is higher than for the sand of Treasure Island. This difference in the densities is reflected in Table 1, presented in [15]. The peak residual moduli show a reduction of 15–35 times, and the low-strain residual modulus shows a reduction of 75–500 times of the initial modulus.

The data presented are typical for many deposits. The particles of the surface layer can begin to lose their elastic properties, if the vertical acceleration is larger than the gravity acceleration *g* (see Sect. 4.1.3). If the surface slope is large enough, then the particles begin to move down and form wave-like avalanches. In the case of a volcano crater, the avalanche can focus in the lowest point of the crater and form a column (vertical jet) of air/particles mixture (see Sect. 4.2.3).

Offshore sediments frequently contain bubbles of gas, normally methane produced by biothermal and other processes. The presence of these gas bubbles can have a significant effect on the behaviour of the sea bed, the coastal materials, and small ocean islands during earthquakes. A quarter of the land surface of the northern hemisphere contains permafrost and about 100 billion tonnes of carbon. The properties of the permafrost depend strongly on the temperature. If it increases the carbon can yield carbon dioxide or methane, which bubbles out and transforms the land surface into gassy soil (see Fig. 4.10).

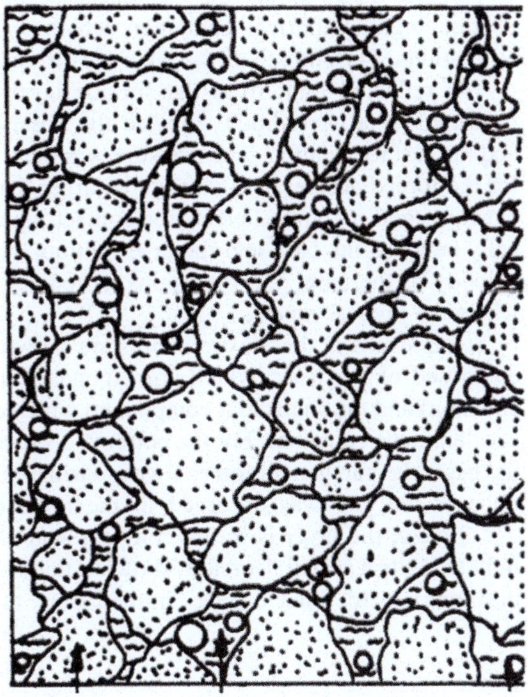

Fig. 4.10 Schematic presentation of gassy soil, deposits, containing gas hydrate and methane bubbles 'frozen' into soil [9]

A mixture of liquid (typically water) and gas (typically air or methane) often occupies pore spaces of many soils (gassy soils). Solid particles of the gassy soil form a sort of skeleton, and they can be considered as a continuum (the dense phase). Because of the high compressibility, the mechanical properties of gassy soils resemble more the properties of bubbly liquids or gas, rather than a solid body. In particular, the properties of weakly-cohesive materials resemble properties of gas within a depression zone. However, within a compression zones the properties resemble those of solid materials.

4.1.3 Loosening of Surface Layers due to Strongly-Nonlinear Wave Phenomena

Many weakly-cohesive materials are 3-dimensional aggregates of mineral grains, containing a complex assemblage of defects such as dislocations, grain boundaries, and fractures, which often contain impurities or fluids. These aggregates may often be modelled by granular materials. Their behaviour is like a solid, or a liquid, or a gas, depending on the energy injection and energy dissipation rates [14, 17]. A standard-granular material vibration experiment was used to study the effects of collision of a granular layer and a solid base, upon the pressure and surface waves [17]. In the experiment a thin layer of 0.106–0.125 mm diameter bronze particles, 15 particles along the depth, was placed at the bottom of a 40 mm diameter and 25 mm high cylindrical container, which was subjected to vertical sinusoidal oscillations. The surface of the layer was illuminated.

The reflected light from the surface layer was focused. The light intensity, measured by a photodiode, allowed the observer to examine the surface density and the surface dilation. The pressure was measured on the base. It was found that, for the forced acceleration amplitude larger than g, the effective gravity becomes negative, and the grain layer loses contact with the base. That gap forms periodically between the base and the layer bottom. After the collision, the layer has the velocity of the base, and it remains in contact with the base until the next cycle, where the process repeats.

Some results of these experiments are shown in Fig. 4.11. The pressure signal is composed of the peak, which corresponds to the gap (base/layer separation and collision) and the sinusoidal component, which corresponds to the force required to accelerate the container. For the bottom acceleration $G > g$ (G is the acceleration of the base) the gap generation is always visible in the pressure signal. In contrast, no trace of the gap was observed in the reflected light up to $G = 2g$. Thus, for $g < G < 2g$, the layer is in the compact state. However, for $G > 2g$ the reflected light exhibits a strong decrease, showing that the layer undergoes a transition from the compact to dilated state. Immediately after the pressure peak occurs the

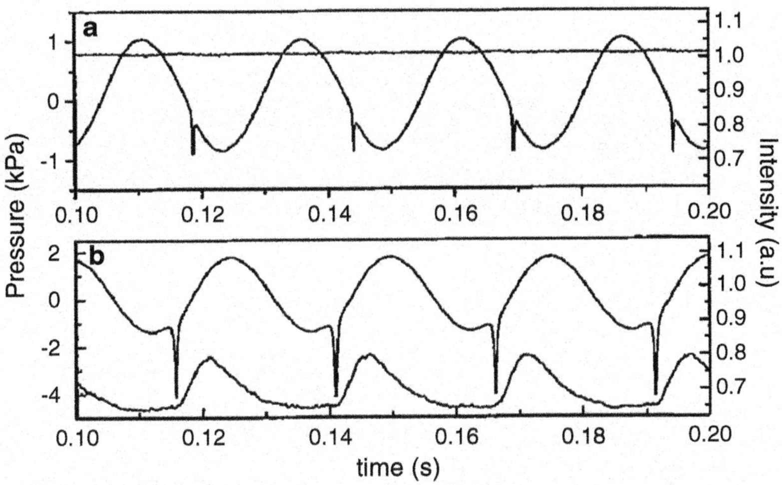

Fig. 4.11 Time series obtained from pressure and intensity sensors for a frequency of 40 Hz: (1) $G = 1.2\,g$ and (2) $G = 2.3\,g$ [17]

reflected light increases, indicating that the layer was dilated during the free flight, and that the small compression occurred due to the collision.

The author emphasizes that the strong-nonlinear effect of base/layer collision, just discussed, is not limited to small-scale experiments (see Fig. 2.3).

The undisturbed mechanical properties of many geomaterials may often be described by the theory of elasticity. However, due to the forced vibrations, the pore volume in geomaterials can increase, and then they begin to move like a liquid. Thus, the behaviour of a geomaterial depends on its history and it can change during earthquakes.

We think that during a severe earthquake, upper-lying material can separate from a hard rock base. As a result, a gap (see Figs. 2.2 and 4.12) is generated between these materials. The annihilation of a gap and the corresponding collision of the upper material with the rock may generate a strongly-nonlinear wave. Perhaps, the extreme acceleration more than 3.8 g, measured recently in Japan [19], is the result of a similar collision.

On a ground surface the acceleration peak was 3.866 g in the vertical direction (Fig. 4.13 left) and 1.435 g in the horizontal direction. Nothing like such data had been known previously – the accelerations and also their ratio are extremely abnormal. Indeed, the horizontal acceleration is usually much more than the vertical acceleration. The accelerations were measured also at a depth of 260 m, and there the horizontal and vertical accelerations had more familiar values namely, 1.036 g and 0.683 g. From this, it follows that the catastrophic amplification of the waves is connected with surface effects.

Figure 4.13 (left) shows forms of waves of the vertical surface acceleration which are essentially asymmetric about the zero line (see, also, Fig. 2.4). The positive peaks are higher and sharper, than the negative peaks. They strongly differ

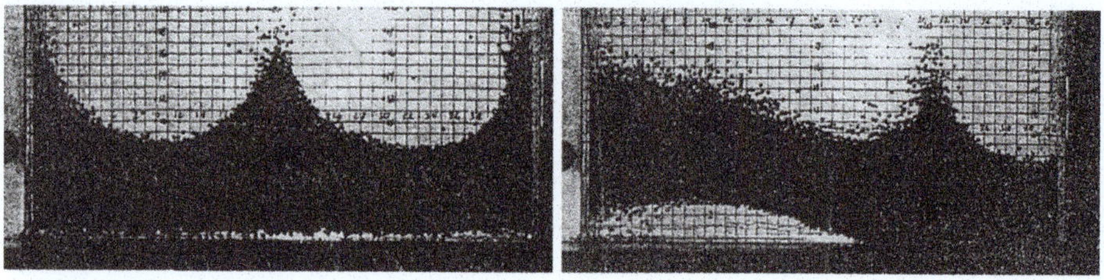

Fig. 4.12 Two photographs of large-amplitude surface waves excited in the container when the forced vertical acceleration $G = 6.2\,g$ [18]

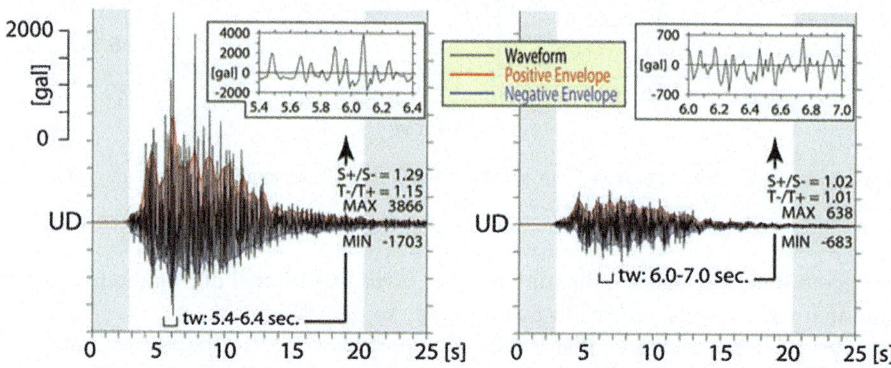

Fig. 4.13 Accelerograms of vertical (*left*) and horizontal (*right*) accelerations measured during Iwate-Miyagi earthquake (2008, Japan) [19]

from the horizontal surface components, and from the measurements on depth of 260 m. It is important to notice that the duration of the negative peaks, measured in the range from 5 is 7 s, is approximately twice the duration of the positive peaks. Therefore we can assume that during the Iwate-Miyagi 2008 earthquake the subharmonic strongly-nonlinear waves were excited within the upper-lying sediment layers. On the other hand, the duration of the negative peaks may be explained by the time which is need for fragmentation of the sediment within the tensile wave. Therefore, the duration of the negative peaks may be much larger than the duration of the positive peaks.

The experiments described above showed the variety of the nonlinear wave effects. In particular, they have shown that the behaviour of materials of surface layers in compression waves and depression waves can be rather diverse. Within the compression wave the materials conserve their strength properties. However, within the rarefaction wave the behaviour of the materials can resemble properties of gas or bubbly liquid. According to the data presented, the properties of some geomaterials can change during earthquakes. The earthquake-induced vertical oscillations of upper-lying sediment layers can resemble the hysteretic dynamics of bouncing motion of a water column on a vibrating base (see Figs. 2.3 and 5.11).

Fig. 4.14 View of Tarzana hill in Southern California [20]

4.1.4 Topographic Effect: Extreme Dynamics of Tarzana Hill

The analysis of Darwin's account an explosive behaviour of tops of mountains (see Sect. 3.4.4) is complicated, since there are only a few documented evidences of catastrophic behaviour of tops of mountains during earthquakes. Therefore, the results of measurements of vibrations of various points of a hill at Tarzana [20] are of great interest.

The Californian earthquake (on January, 17th, 1994; the depth 18 km) caused extensive destruction in Fernando's Dignity valley and adjoining areas. In many places the ground movement was amplified or strongly changed by local effects. However the most localized and extreme local effect has been fixed on Tarzana Hill. It is a small smooth hill, which is located 6 km to the south and 18.7 km over the earthquake epicentre. There the peak horizontal acceleration varied from 1.78 to 1.92 g, which is one of the greatest peaks ever measured. It is considerably higher than would be expected from empirical curves and numerical calculations, whereas the peak vertical acceleration was 1.14 g (Fig. 4.14).

The hill is 500 m in length, 130 m in width and about 15–20 m height above the surrounding district [20–22]. The extreme surface movement was measured in a radius of 50 m around the hilltop. Outside of that radius, the movement corresponded to a theoretically expected level of characteristics.

The thickness of soil on the hill is approximately 4 m. Measurements showed that the soil has low strength, with an average speed of shear waves about 200 m/s. A layer of soft clay is located at depth from 4–5 m to 12 m. Below that clay there is solid slate together with granite layers from 12 to 100 m, and there are also clay layers, and water appears at depth about 17–18 m – we emphasize that it appears at the interface of the hill material and the rigid basis. Probably, the presence of water explains the sharp change in speed of the seismic waves observed on this boundary. In particular, the average speed of compression waves changes there from 640 to 1,600 m/s. Hence, there is a very sharp impedance distinction of materials on that interface, and so the seismic wave may be trapped by the hill. All of the authors [20–22] thought that the amplification of seismic waves in the Tarzana was connected with topographic and resonant effects. I suggested [23] considering attentively a column of the hill material localized in the vicinity of the hill top

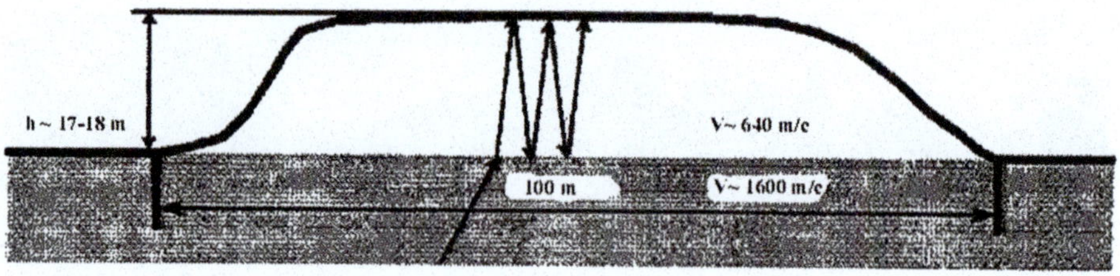

Fig. 4.15 Scheme of a section of Tarzana hill, and resonant oscillations of the vertical wave trapped by the hill [22]

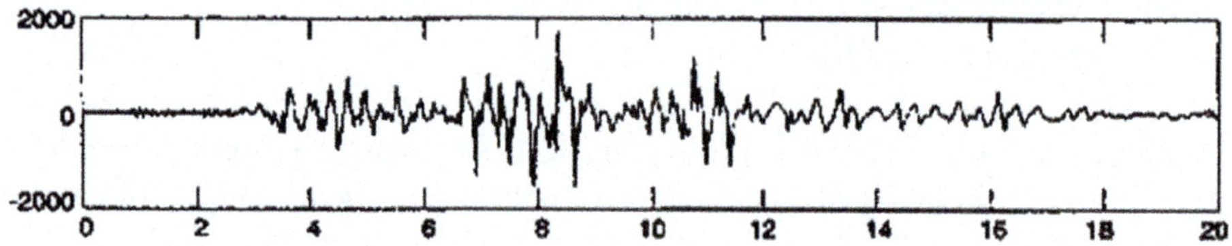

Fig. 4.16 The accelerogram of one point of Tarzana hilltop [21]. The acceleration is given in 0.001 g, the time is given in seconds

and having radius approximately 50 m (Fig. 4.15). Further, the earthquake centre was practically under the hill, hence the vertical seismic compression wave reached the base of the column specified above column at a very small angle (about 5°). Then the wave began periodical motions between the free surface and the basis (Fig. 4.15). As a result of resonance, accelerations larger than g could be generated, and some sequence of separations of the central part of the hill from the rigid base could take place.

The accelerogram of one point of Tarzana hilltop measured during the main shock is shown in Fig. 4.16. The graph of surface accelerations is non-symmetric about the zero line – the upward peak is larger than downward ones. Perhaps, that is a result of the separation of the central part of the hill from its base.

It is emphasized [23] that the remarks presented above explain the unexpected oscillations of the central part of Tarzana hill. We can also consider the aftershock accelerograms, which are presented in Fig. 4.17. They show that for a frequency 3.2 Hz the top/base amplification ratio was about 4.5 for horizontal ground motion orientated approximately perpendicular to the long axis of the hill (Fig. 4.17 left), and the ratio was about 2 for motions parallel to the long axis of the hill (Fig. 4.17 right).

Unfortunately, we have no accelerograms for other points of the top. Therefore, let us consider more attentively Fig. 4.17 and the seismograms W02 and S01. The accelerations at these points are rather asymmetric about the zero lines, with positive peaks larger and more sharp than the negative peaks. On the whole, in the central part, the curves W02 and S01 resemble the accelerations during of the Iwate-Miyagi 2008 earthquake in Japan (Figs. 2.4 and 4.13) [19] and the discontinuous oscillations of the water column in Natanzon's experiments (Fig. 2.3).

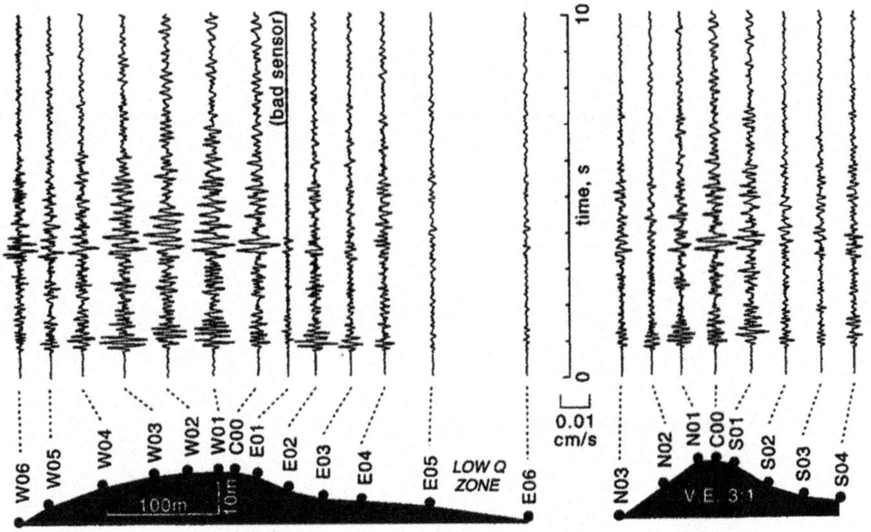

Fig. 4.17 North–south component seismograms measured along Tarzana hill [21]

I suggested [23] that oscillations of some topographies during earthquakes can resemble bouncing of solid balls on a vibrating base. As a matter of fact, this idea generalizes the remark of Darwin … *small bodies should have been pitched a few inches from the ground…* [1, p. 256].

On the whole, this remark corresponds well to Fig. 4.18. Recently, the Japanese researchers have measured a vertical acceleration of 3.8 g. They have explained this extraordinary result by using a model of a solid body on a vibrating base [19]. Results of the water column experiments (Fig. 2.3) also support our analysis of the oscillations of Tarzana hill [23–26].

Figure 4.18c (right) shows the complex motion and the convection of the ground particles during earthquakes. Similar effects were illustrated earlier by Fig. 2.1. We emphasized (Sect. 2.2) that Michael Faraday first studied similar effects. We will consider these effects additionally in subsections 5.2.1 and 5.2.2 where results of seaquakes are being discussed.

4.1.5 Topographic Effects: Uplift and Fissure of the Top of a Ridge

We have connected the uplift of the surface material with its loosening, which follows from results of the experiments presented above. The loosening occurs only if the acceleration of the upper layers is large enough. Therefore, we have to ask ourselves again about values of vertical accelerations, which occur during earthquakes. Can these accelerations be destructive for weakly-cohesive upper-lying sediments?

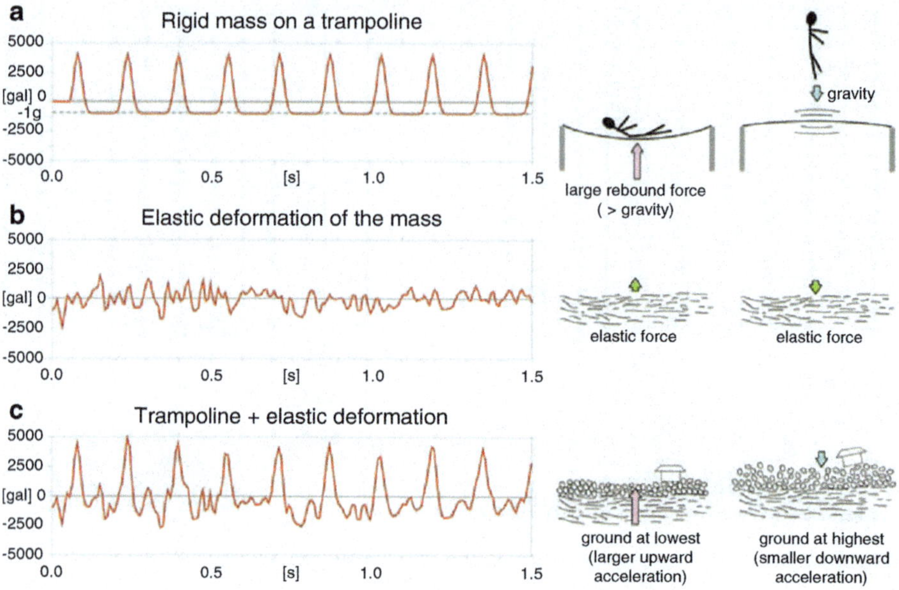

Fig. 4.18 (**a**) Simplified model of the motion of an undeformable mass bouncing on a trampoline. (**b**) Elastic deformation of a deformable mass, represented by a selected part of a downhole seismic record. (**c**) Simulated motion of a deformable mass bouncing on a trampoline, obtained as the sum of (**a**, **b**) [19]. To be compared with Figs. 2.1, 2.2, 2.3, 2.6 and 5.11, 5.12, 5.13

It was once assumed (see Sect. 2.1), that no earthquake could generate ground acceleration as high as g. However, some recent moderate earthquakes have caused peak ground accelerations over g, including the 1971 (magnitude 6.6) San Fernando earthquake (peak ground acceleration PGA = 1.25 g); the 1979 M = 6.4 Imperial Valley earthquake (PGA = 1.25 g); the 1984 M = 6.1 Morgan Hill earthquake (PGA = 1.29 g); the 1985 M = 6.8 Nahanni, Canada, earthquake (PGA > 2 g vertical; PGA = 1.25 g horizontal); the 1992 M = 7 Petrolia earthquake (PGA > 1.85 g); and the 1994 M = 6.7 Northridge (Californian) earthquake (PGA = 1.78 g) [16]. On 14 June 2008, the surface vertical peak acceleration 3.8 g was recorded for the Iwate-Miyagi earthquake in Japan, [19, 27]. This extreme vertical acceleration of the surface was more than twice the horizontal acceleration at the same place.

We do not know the accelerations during a great earthquake (like the Chilean 1835) or a giant earthquake (like the Chilean 1960). We noted above that, according to field observations [5], the crest amplification with respect of the base can reach a factor of 10 and more. According to these data, if the base acceleration is 0.5 g, then the crest acceleration could be 5 g. Thus, the vertical acceleration of the mountain tops during the 1835 Chilean earthquake might have been much more than g. In the literature, since ancient times, it is possible to find many descriptions of catastrophic earthquakes. In particular, Japanese sources contain data interesting to us. It is known that Japan is a zone of extreme earthquakes. In the book **Visions of a Torn World** (1212), which is a classic of Japanese literature Kamo no Chomei wrote 'In the second year Genrjaku (1185) there was a strong earthquake. Its kind was unusual: mountains broke up and buried under itself rivers... Rocks were dissected and rolled down in valleys'.

The Mino-Owari earthquake of 1891 was the greatest recorded inland earthquake which struck Japan. The seismologist John Milne reported 'the contortions produced along lines of railway, the fissuring of the ground, the sliding down of mountain sides and the toppling over of their peaks, the compression of valleys, and other bewildering phenomena' [6, p. 76]. Hartzell et al. [28] investigated severe damage on Robinwood Ridge in the earthquake at Loma Prieta, California, and they explained the phenomenon by the topographic effect.

To study this problem further we now consider some theoretical and experimental results. Firstly, we simulate analytically the topographic effects for narrow ridges and conical mountains.

It is known that elastic waves propagate along a cylindrical bar of uniform cross-section without change of the waveform. However, the shape of the wave as well as its amplitude changes if the cross-section is not uniform. To illustrate this fact, we will consider the propagation of a compression wave, described by the following expression for a displacement u,

$$u = r^{-n}F(r + ct) \tag{4.1}$$

Here $F(r + ct)$ is a travelling wave, r is the radial coordinate, t is time, c is the wave speed, and n is a constant. We may assume that $n = 0.5$ (cylindrical wave) is valid for narrow ridges, and that $n = 1$ (spherical wave) is valid for conical mountains. The acceleration is determined as follows:

$$\partial^2 u/\partial t^2 = r^{-n}c^2\partial^2 F/\partial(r + ct)^2 \tag{4.2}$$

Here $F = F(r + ct)$. It can be seen from (4.2) that the acceleration increases extremely near the top where $r \to 0$ (see Fig. 4.19). The acceleration depends also on the wave steepness – the steeper the wave front, the higher the acceleration.

This result agrees well with the strong amplification of earthquake effects near tops of mountains. Thus, we have shown that earthquake-induced accelerations of mountain tops may be very large. If the base acceleration is of the order of g, then the acceleration of the top may be many times g. In real situations along the mountain side complex weakening and amplification can occur. They lead to large and dangerous motions, especially near the upper parts. In particular, the reflection of the upward-seismic wave from the mountain slope may produce tension stresses. They may be sufficiently large to cause fractures, if the ridge is formed by the weakly-cohesive geomaterials. The same arguments are related to the propagation of elastic waves in a conical element with a small-apical angle, which is shown in Fig. 4.20. This Figure illustrates very approximately the earthquake-induced wave processes in ridges and mountains.

The generation of tensile stresses and ruptures is determined by the slope of the mountain. The compressive wave, radiated from the base, is reflected from the

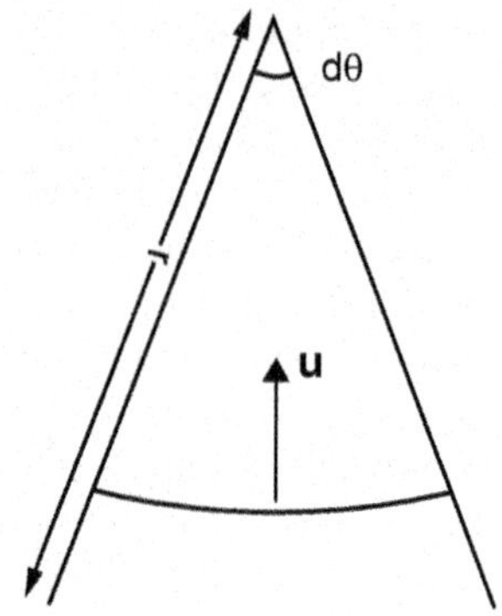

Fig. 4.19 Scheme of conical mountain having steep slope (dθ is small) and subjected to impact vertical displacement *u* [9]

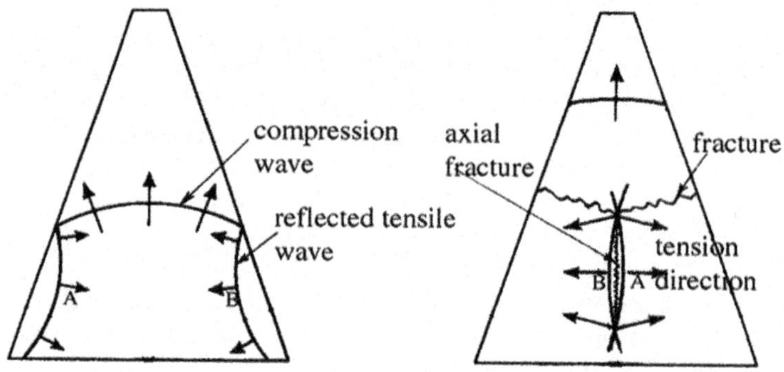

Fig. 4.20 Fracture pattern in a wedge-shape plate. Compressive and reflected waves (*left*), fracture generation (*right*) [9]

slope sides as a predominantly-tensile wave. At the same time the compressive wave moves towards the top. The waves reflected from the sides intersect on the axis of the mountain – this is the reason for axial fracture. It is not difficult to see, that if the reflected wave is of sufficient intensity, the axial fracture will start from the collision point of the reflected waves. Then the axial fracture spreads upwards and downwards from this point. Figure 4.20 does not depict a reflection of the compression wave from the top. We note only, that a short compression wave reflects from the top as a tensile wave, after which combined effects of the incident and reflected waves take place.

Now, we consider some experimental data. Figure 4.21 shows specimens in the form of a flat wedge-shaped plate, with charge detonated along the whole base of each plate during the experiments. The effects of the slope of lateral boundaries of the specimen with respect to the base were studied.

Multiple fractures (spalls) appeared near the apex after the detonation. This is true even for wedge slopes, which are not small, as Fig. 4.21 shows. Similar results have been observed at tests of other samples.

If a charge bows up in the centre of the basis of a cone of transparent material two effects were observed (Fig. 4.22). The first is defined by convergence of waves

Fig. 4.21 Fractures of wedge-shaped plates explosively loaded along the whole of the base [29]

reflected from the lateral surface. Thus, on the cone axis a zone of the destruction was generated, consisting of pores (bubbles) which resemble more a cavitation zone in a liquid, than a crack in a solid body. Secondly, the cone top collapses.

Thus, those experiments qualitatively confirm the results of the analytical research and graphic constructions of Fig. 4.20. They showed that fractures can gather together at tops and at axes of ridges and mountains. It is interesting that the axial fractures (pores) in Fig. 4.22 resemble cavitation bubbles in a liquid. Of course the vertical motion and the surface fractures may be very great. The localization of fractures in a cone-type mountain was observed, following a 1.7 kiloton underground nuclear blast [31]. Near the top the stress wave produced three slabs of equal thicknesses (about 35 m). The slabs moved upward with initial velocities of approximately, 2.4, 1.5 and 0.7 m/s, respectively. The surface slab of consolidated tuff, situated about 230 m above the blast, lifted on 225 mm.

A large volume of a rockfall as a result of an underground nuclear explosion is shown in Fig. 4.23. The ground fractures were generated by 2 m/s peak ground velocity and by 5–6 g peak ground acceleration, from two underground nuclear explosions along Rickey Cliffs on Pahute Mesa.

One can see, that the last observations, on the whole, support Darwin's notes ... *the superficial parts of some narrow ridges were as completely shivered, as if they had been blasted by gunpowder* ... [1, p. 370] and ... *I was much interested by finding on the highest peak of one range (about 700 feet above the sea) a great arched fragment, lying on its convex or upper surface. Must we believe that it was fairly pitched up in the air, and thus turned?* ... [1, p. 255].

Results presented above are very important for us, since they will allow us to explain many of Darwin's observations.

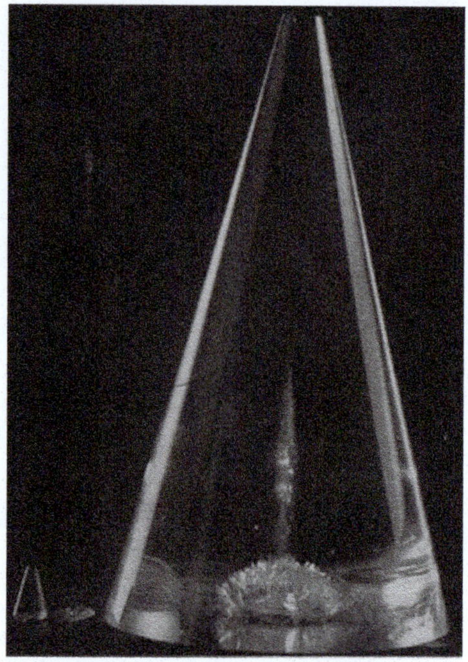

Fig. 4.22 Fracture in a Perspex cone as a result of the explosive load of the base [30]. The tip may break and fly off at high velocity. See the tip on the left

Fig. 4.23 The coarse rockfall as the results of underground nuclear explosions on Pahute Mesa [32]

4.1.6 *Topographic Effect: Uplift and Fissure of Island Surfaces*

The uplift of islands Islands are tops of ridges and mountains partly submerged in water (Fig. 4.24). Therefore, the earthquake-induced uplift of them may be explained by the concentration of seismic energy near the tops, by the high vertical acceleration of the top material, and by the loosening of this material.

Uplift of land was established by clear evidence after the 1835 Chilean earthquake. The greatest uplift occurred on the island of Santa Maria (Fig. 4.25), 35 miles southwest of Concepcion (up to 3 m). At Mocha Island (120 miles south-southwest of Concepcion), according to an observer whose accuracy Captain FitzRoy was inclined to trust, the uplift was about 2 ft (0.6 m).

We emphasise again, that uplift was not permanent during time. Captain FitzRoy stated that, for a few days after the earthquake, the sea did not rise to its usual marks by 4 or 5 ft, vertically. But, he added, this value was gradually diminished up to the middle of April. Darwin also noticed that, after a few weeks, the ground was lower than it had been immediately after the earthquake. He suggested that the later subsidence might be due to settling down of the uplifted crust.

Fig. 4.24 Ocean islands [9]

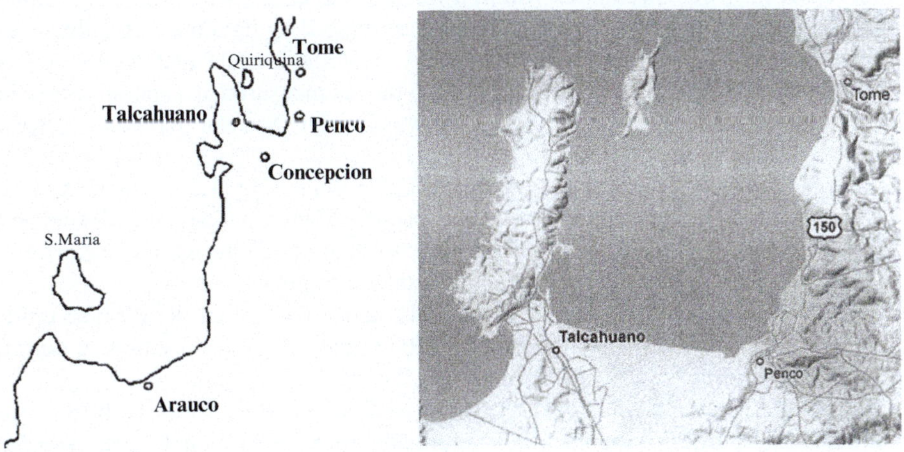

Fig. 4.25 Location of S. Maria and Quiriquina Islands (*left*). Location and the modern form of Quiriquina Island (*right*) [9]

Fig. 4.26 An example of earthquake-induced cracks in the ground near Lale village. Internet. researchspace.auckland.ac.nz/handle/2292/4474

The topographic effect can explain the uplift of Ranongga Island during the 1 April 2007 Solomons earthquake. We report that the seabed slope around this island is steep enough (see the map in Fig. 3.4). The ground uplift at Lale village was the highest in Ranongga Island (Fig. 3.4). According to inhabitants of Lale (Ranongga Island), the ground was uplifted by about 7 m just after the earthquake. At the same time the sea level started to rise up gradually, and then tsunami came. Two international tsunami survey teams studied results of the earthquake, and the first report [33] was dispatched within a week of the earthquake. This team found that the southern part of Ranongga Island was uplifted by up 3.6 m. Less than 1 m uplift was determined on the northwest tip of Vella Lavella and on Vonavona Islands. The maximum tsunami height of 8.6 m was measured at Tapurai village in Simbo Island. Thus the height of tsunami was greater than the maximum ground uplift.

Cracks in the ground were observed near Lale village (Fig. 4.26). The maximum ground deformation during the earthquake occurred between Ranongga Island and Simbo Island (see Fig. 3.4). The islands were uplifted during the earthquake prior to tsunami arrival, significantly reducing the tsunami impact.

The second team arrived on 16 April [34], and it found that the greatest uplift 2.46 ± 0.14 m was on the southern part of Ranongga. Thus, it appears that during the week the uplift was reduced from 3.6 to 2.6 m.

Of course, the behaviour of the offshore sediments depends on the history. In particular, properties of the materials before and after the earthquake may be quite different. During an earthquake the volume of the bubbles increases. As a result, the soil loosens and its surface is uplifted. After the earthquake the gas volume slowly

Fig. 4.27 Schematic presentation of gassy soil as saturated soil containing gas bubbles (*left*), soil sample (*right*) [35]

reduces and the soil surface moves down. These effects define surface dynamics of the islands, described by Darwin.

The fissure of the island Earthquake-induced oscillations of the small island of Quiriquina attracted Darwin's attention. The width of the island is approximately 500 m, and length 4,000 m. Of course, during the last 2 centuries the form of the island form has changed. Therefore, in Fig. 4.25 the island is presented according to an old map from [36] (left) and a new map (right). The height of the ridge on the island is approximately 140 m.

Darwin described the result of the earthquake-induced vibrations of this island *...The ground was fissured in many parts, in north and south lines; which direction perhaps was caused by the yielding of the parallel and steep sides of the narrow island. Some of the fissures near the cliffs were a yard wide*: ... [1, p. 370]. Joint action of earthquakes, weather, seismic, and ocean waves can change the form of an island. Darwin noted *...many enormous masses had already fallen on the beach; and the inhabitants thought, that when the rains commenced, even much greater slips would happen. ...I believe this convulsion has been more effectual in lessening the size of the island of Quiriquina, than the ordinary wear and tear of the weather and the sea during the course of an entire century...* [1, pp. 370–371]. Coast sediments and soil of islands often contain gas (methane) bubbles formed by biological and thermal processes, and these materials have low strength (Fig. 4.27).

It is known that fractures and cracks may be generated, when a wave reflects from a free surface (see Fig. 4.28). Because of the sharp impedance contrast between the island material and the water, an earthquake-induced wave loses only a small part of energy because of reflections from the cliffs. Thus, the reflection from the cliffs recalls the compression wave reflection from free boundaries. In the last case a rarefaction wave is generated, and fractures occur as a result.

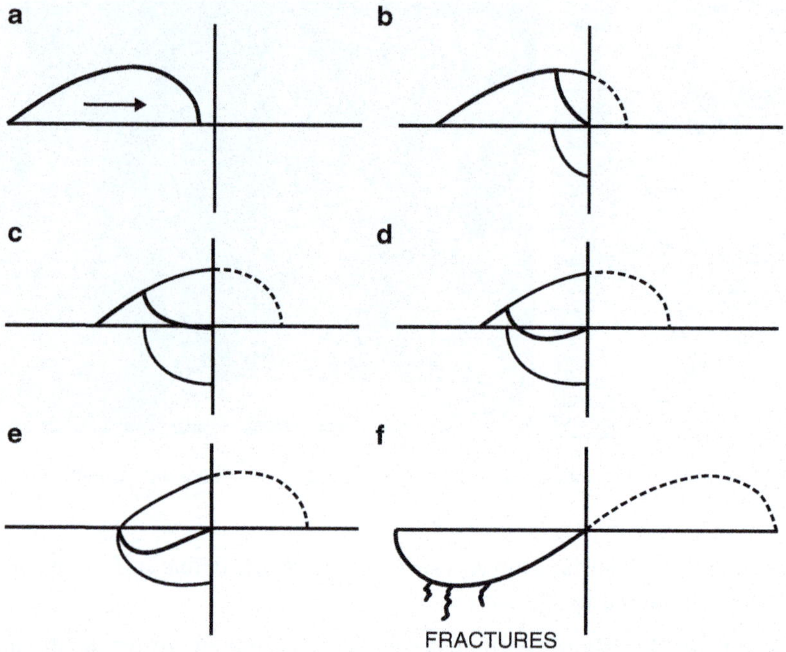

Fig. 4.28 Reflection of a surface-wave from a free boundary and the generation of surface fractures [30]

Figure 4.28 illustrates qualitatively the distribution at various stages of the horizontal surface displacement, when a compression surface wave is reflected from a steep cliff of an island (the free boundary). Following [30], we assume that the displacement at any point near the cliff is the sum of the displacements of the incident and reflected waves, which are shown by the thin lines in Fig. 4.28. The resultant displacements are shown by the thickest lines in each diagram, whilst the broken thick parts of these lines correspond to the portions of the wave which have already been reflected. In Fig. 4.28, (a) shows the wave approaching the free boundary; (b) shows the distribution when a part of the wave has been reflected, but the displacement still corresponds to the compression of the surface material; (c) is a slightly-later stage when some tension has been set up near the boundary; (d) this tension has spread, and (e) half the wave has been reflected and the displacement curve corresponds almost completely to the tension wave. In (f) the reflection has finished, and fractures (cracks) have developed where the tensional displacement exceeded some critical level. Thus, a low level of strength and the presence of gas bubbles can have an essential effect on the behaviour of the seabed, coastal materials, and small ocean islands during earthquakes.

Conclusion We think that the appearance of tension waves, the topographical effect and the small strength of surface materials explain the results (noted above) of the 1835 catastrophic Chilean earthquake.

4.2 Darwin's Ideas About an Intimate Connection Between Volcanic and Elevatory Forces

In 1838 Darwin emphasised ... *the intimate connexion between the volcanic and elevatory forces*; ... [4, p. 54]. In 1998 Linde and Sacks [37] showed clearly that a strong earthquake can be the starting mechanism for big eruptions, and in particular, they have found a sharp peak in the number of eruptions on the same day as the earthquake. The cause of the eruptions induced by earthquakes is not clear, although a few interesting mechanisms are proposed [26, 37–46]. Earthquakes can also generate eruption of geysers, and other displays of seismic activity, as far as 1,000 km from them [47].

At the same time there is also a reverse influence. Darwin wrote that ... *the great columns of smoke shot forth from the tall chimneys of the Andes, relieved the trembling ground, which at that moment was convulsed over the whole surrounding country* ... [4, p. 58]. Thus, he believed that the interaction of a seismic wave of the big amplitude with erupting volcanoes can cause a local change of character of seismic oscillations.

Earthquakes accompanied by eruptions of volcanoes are rare events. Darwin wrote ... *Although it is known that earthquakes have been felt over enormous space,...yet few cases are on record of volcanoes, very far distant from each other, bursting out at the same moment of time...* [1, p. 379–380]. In the connection of earthquakes and eruptions he had seen an important reason for evolution of the face of the Earth. We have already considered some of his judgements on this problem in the Chap. 1 and in subsection 3.3.5. Here we will continue the analysis, illustrating ideas of Darwin by data of modern observations, experiments and theoretical constructions.

4.2.1 Earthquake-Induced Ground Elevation as a Triggering Mechanism for Large-Scale Volcanic Eruptions

We have quoted Darwin's words devoted to destruction of mountain materials, to emphasise his interest to this phenomenon. Here are some further words by him ... *In the morning we climbed up the rough mass of greenstone which crowns the summit. This rock,...was much shattered and broken into huge angular fragments. I observed, however, one remarkable circumstance, namely, that many of the surfaces presented every degree of freshness - some appearing as if broken the day before, ... I so fully believed that this was owing to the frequent earthquakes,* ... [1, p. 313]. For volcanoes similar destructions can lead to large eruptions. Now a similar phenomenon has been well documented for the case of the explosive eruption of the volcano Mount St. Helens

On March 27, 1980 the first emission of gas appeared over the Mount St Helens. Hence, pressure of gas and magma in the volcano had started to come to a critical point, and the ensuing earthquakes indicated that magma had started to move in the volcano. The northern slope of the volcano began to move outward at a rate of more than a metre per day as the magma pressure increased inside. The sequence of earthquakes increased in activity until April 28, and then their activity decreased. On May 18, a magnitude 5.1 earthquake caused a huge landslip of the mountain's slope. That had the same effect as extraction of the cork from a champagne bottle, and all the pent-up gases and magma then exploded (see Fig. 3.7). Jonientz-Trisler (USGS) wrote 'On May 18, the magma and gases were not ready to explode by themselves. But after the landslide, the magma and gases were no longer contained and under pressure. They were then able to move up in the system.'

We emphasise that only after the earthquake and the following landslip did the external pressure fall, so that magma and gases then started to be thrown up from the volcano conduit. Thus, the very fast unloading of the top of the volcano of Mount St. Helens generated the large eruption. Something similar occurs during the reflexion of a vertical seismic wave of compression from a volcano top (see, also, Figs. 4.19, 4.20, 4.21, 4.22 and 4.23). A volcano is often formed by products of previous eruptions, and it grows up as a result of them. If the volcano periodically erupts lava and ashes, then it is formed by layers of these materials. Thus, these volcanoes are formed by porous materials which have low tensile strength.

Our attention will now focus on the documented simultaneous eruptions of 4 volcanoes during the 1835 Chilean earthquake [41]. Darwin [1, p. 380] reported that ... *several of the great chimneys in the Cordillera of central Chile commenced a fresh period of activity* ... Let us consider these eruptions, taking into account a volcano shape and a conduit. Three of the volcanoes (Minchinmavida (2,404 m), Cerro Yanteles (2,050 m), and Peteroa (3,603 m)) are stratovolcanos which are formed as symmetrical cones with steep sides. Robinson Crusoe (922 m) is a shield volcano, which is formed as a cone with gently sloping sides – shield volcanoes are not very active. The latest large eruptions of Minchinmavida, Cerro Yanteles, and Robinson Crusoe were in 1835. Peteroa erupted in 1837 and 1937, and now it has a small crater lake. Darwin reported that the craters of Minchinmavida and Yanteles were covered by snow [1, p. 276].

Conical volcanoes often have a central conduit connecting the magma chamber with the surface. In calm volcanoes the conduit is usually closed by sedimentary beds or lava. If the magma is viscous, then often a dome appears above the conduit. The dome or sedimentary beds form a 'plug' closing the conduit (Fig. 4.29). Sometimes gases and magma can knock the plug out from the conduit vent, and that leads to eruptions of explosive type. Conical volcanoes have usually have craters like a funnel, but during a strong explosive eruption the crater can be destroyed.

We may surmise that the vents of those 4 volcanoes each had a sealing plug (vent filling) in 1835. And each of these volcanoes is conical. These common features are important for our triggering model, which is discussed below. The vent-filling material usually has high porosity and very low tensile strength, and so it can easily be fragmented by tension waves. In particular, the porosity of the lava dome of the

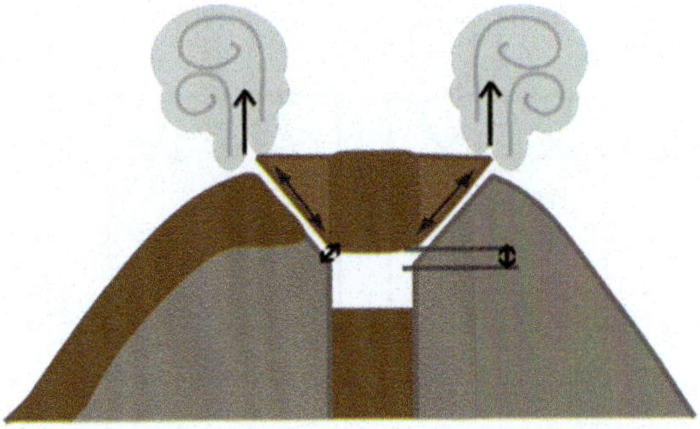

Fig. 4.29 The 'plug' closing the conduit and gas eruptions [48]

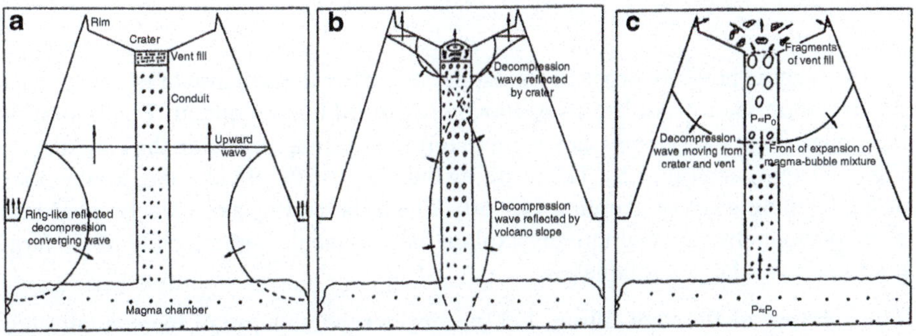

Fig. 4.30 Dynamics of the interaction of a conical volcano truncated by a crater with an upward-going wave. (**a**) A scheme of the conical volcano system (mountain, crater, vent filling, conduit, and magma chamber) and the waves induced by rapid land elevation. (**b**) The tension wave reflected from the crater surface, the beginning of vent filling collapse, the generation of the eruption, the growth of bubbles in magma due to the tension. (**c**) The reflected wave, the generation of the converging radial surface wave moving from the rim, the beginning of large-scale eruption is the magma chamber pressure [26]

Soufriere Hills volcano in Montserrat, West Indies, is 33 %. A vent filling and its crater material have similar levels of porosity.

The action of a severe earthquake on a volcanic base may be similar to the action of a nuclear underground explosion. It was emphasised above, that vertical motion and surface fractures in a cone-type mountain were observed after the 1.7 kiloton underground nuclear explosion [31, 32] (see, also, Fig. 4.23). Analogous phenomena may be generated as a result of a severe earthquake. The volcanic base undergoes great earthquake-induced vertical acceleration, and a compression wave begins to propagate through the volcano body (Fig. 4.30a). Since we are considering a conical volcano, the interaction of this wave with the free surface of the volcano is approximately the same as was shown in Figs. 4.20, 4.21 and 4.22. The reflection of the upward-going wave from the volcano slope produces tensile stresses within the volcano (Fig. 4.30a). The conduit magma is held back at high pressure by the weight and the strength of the vent filling. This filling might collapse (Fig. 4.30b) and get thrown out (Fig. 4.30c), when the upward wave is reflected by the volcano crater.

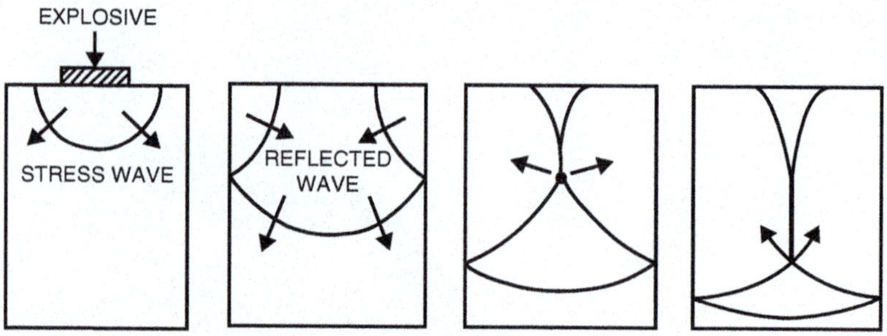

Fig. 4.31 Propagation and refraction of a wave generated by a surface explosion. The central fracture forms due to the interaction of the reflected tension waves [29]

The magma pressure drops to atmospheric P_a after the collapse of the plug. Then the decompression front begins to move downward in the conduit (Fig. 4.30c). In particular, large gas bubbles can begin to form in the magma within the conduit (see Fig. 1.7 (right) and Fig. 2.8). The resulting bubble growth provides the driving force at the beginning of the eruption [49–60]. Thus, the earthquake-induced nonlinear wave phenomena can qualitatively explain the spectacular simultaneity of the large eruptions after large earthquakes.

The collapse of the vent filling Let us consider the collapse of the conduit plug. Possible processes and results of the collapse are illustrated by Figs. 4.31, 4.32 and 4.33. The plug is modelled as a short cylinder or a plate. The surface explosion models an action of the underground shock [60].

The wave dynamics of Fig. 4.30b corresponds qualitatively to the scheme of Fig. 4.31. That last figure gives a simplified scheme of distribution and interaction of the waves produced by a surface explosion. According to Fig. 4.31, on an axis of symmetry of the samples (the vent fillings or plugs) cracks can occur (see, also, Figs. 4.20, 4.21, and 4.22).

The generation, propagation, and reflection of the waves, described above, are illustrated well by the stress waves in the Perspex specimen, shown in Fig. 4.32.

First, the compression wave is reflected from the free surface of the specimen (plug) as the tension waves. The interference of the reflected waves may produce stresses, which are sufficiently large that they can cause fractures, although the amplitude of incident compression wave was too small to do it in this case. Similar fractures (cracks) are shown in Fig. 4.33.

Once a fracture has occurred then the tail part of the incident compression wave will be reflected as a tension wave from this fracture. During this reflection new fractures may occur. The total effect may be a series of more or less parallel cracks. Locations and dimensions of cracks, and also, their directions, often depend on many parameters.

Qualitatively, the failures shown in Fig. 4.33 correspond to Figs. 4.30, 4.31 and 4.32. These Figures give simplified schemes of propagation, distribution, and

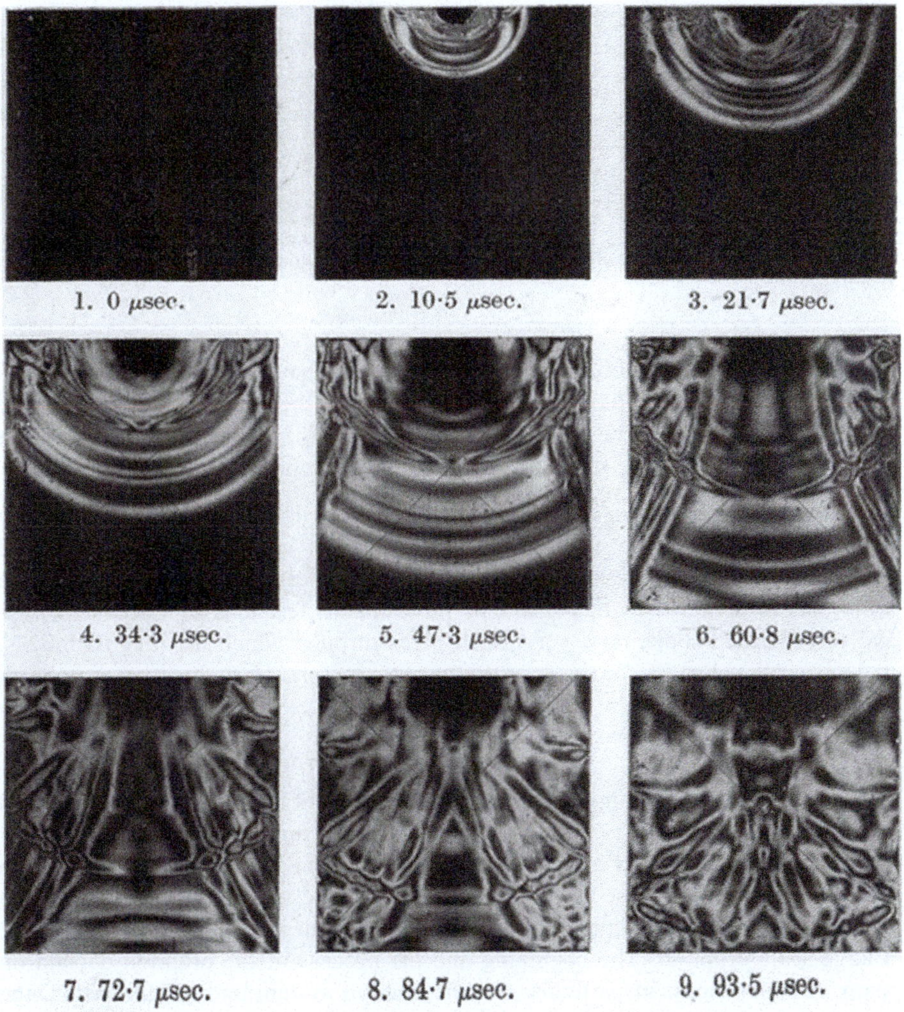

Fig. 4.32 Stress waves produced in inch Perspex plate by a 0.12 gramme charge of lead oxide detonated in the centre of the upper edge. Times given are measured from the instant of detonation of the charge [30]

superposition of the waves generated within the vent filling (plug). The schemes demonstrate reflected tension waves and cracks near the free surface and in the central axis of the material. It is clear that the cracks promote the beginning of a large eruption.

Start of an eruption and nonlinear waves in the magma of the conduit The volcanic shape is a cone truncated by the crater (funnel). The crater surface is formed by weakly-cohesive media (snow, ice, granular materials, sediments, water, and lava). The crater is connected with the base of the volcano by the conduit,

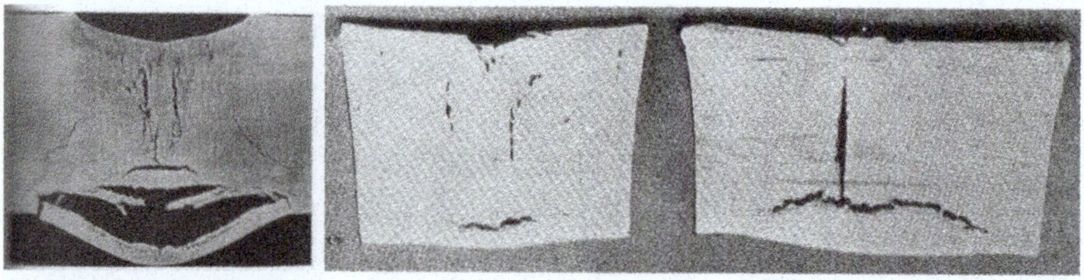

Fig. 4.33 Cross-sections of mild steel plate (*left*) and steel cylinders after the surface explosion (*right*) [29]

containing liquid-like bubbly magma. For example, the Soufriere Hills volcano in Montserrat, West Indies, has a 30 m diameter conduit, the crater has diameter 300 m and depth 100 m, and the volcano vent filling is estimated as 20 m thick. An axisymmetric topography surrounds the vent (~22° slope) [46].

In the *Seattle Post-Intelligencer* of Tuesday, April 21, 1987, Debera Carlton wrote that Jim Zollweg of the U.S. Geological Survey explained: 'If you . . .could look inside the mountain, you'd see a nearly vertical structure – a big pipe,. . .. The width has to be very small. It can't be more than a few tens of yards across. . .. We've described some very, very small pipes through which the magma travels, and the earthquakes are the result of this excess pressure, like when you try to push too much water through a pipe. If extra pressure is exerted, the pipe will really bulge. And since the pipe can't really bulge in this case, the only thing it can do is break the rock around it'.

If the vent filling collapses, the decompression front begins to move downward in the conduit (Fig. 4.30 *c*). The amplitude of the decompression wave is of order P_a. This wave is reflected by the high-pressure magma chamber as a compression wave with amplitude close to P_0 (Fig. 4.30 *c*). As a result, the pressure difference between the region of low pressure (atmosphere) and the magma chamber can cause a large-scale eruption. The beginning and the process of the eruption depend on many circumstances; in particular, on the conduit system and its dimension, the chamber size and the pressure, the magma viscosity and the gas concentration in it.

As a whole, a conical volcano can be schematized as a vertical pipe topped by the funnel with a plug; but the conduit is a complex system of little conduits and pipes, the geometry of which is very important and very difficult to determine. However, very roughly, we can model the conduit as a single pipe connecting the magma chamber and the vent. In this case, we consider the conduit as the resonant pipe in an organ (see, also, Fig. 5.13 and Sect. 7.4).

In particular, due to reverberations and transformations of the wave in the conduit, the eruption can become pulsatory [54–58].

Recently [58], it was experimentally demonstrated that the geometry of the conduit is the key parameter governing the flow of viscous magma, which acts as a trigger/source of volcanic seismic waves.

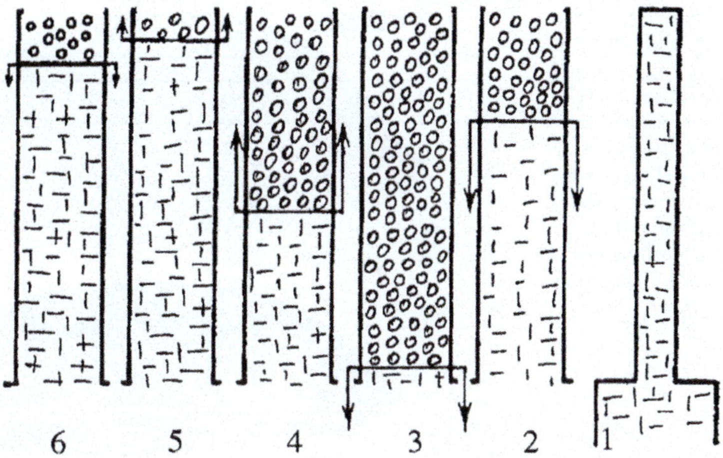

Fig. 4.34 Simple scheme of the wave process, which can be initiated by vent collapse. (*1*) Closed conduit; (*2–6*) Strongly-nonlinear wave, triggered by collapse of the vent filling (See, also, Fig. 4.30)

The strongly-nonlinear waves propagating in the conduit and reflected by the atmosphere and the magma chamber (Fig. 4.34) can explain the volcanic tremor (rhythmic ground shaking, different from the discrete sharp jolts characteristic of earthquakes). The resonant free oscillations in the conduit may continue for a long time, since they are fed by the magma chamber pressure. The behaviour of the system strongly depends on the magma viscosity. Gas can escape from the bubbles more easily in the case of low-viscosity magma. However, if the magma is very viscous and the gas cannot escape easily, then the bubbles grow very quickly only near the vent. The effects of this growth can resemble an explosion [53, 59].

Thus, the collapse of the conduit vent of a volcano by seismic shock can cause a large eruption. On the other hand, the collapse can induce periodic or pulse oscillations of the volcano defined by the properties of the magma, the conduit, and the magmatic chamber, and practically independent of the initiating seismic shock.

Let us estimate the period of the pulsations. The conduit magma contains a lot of gas, therefore the wave speed can vary from about 1,000 m per second to several tens of metres per second [52]. When the conduit height is of the order of 10 km, then the period of the pulsatory explosions can lie in the range from minutes to hours. For example, if $C_m \approx 20$ m/s and the conduit height is 20 km, then the period is about 0.5 h.

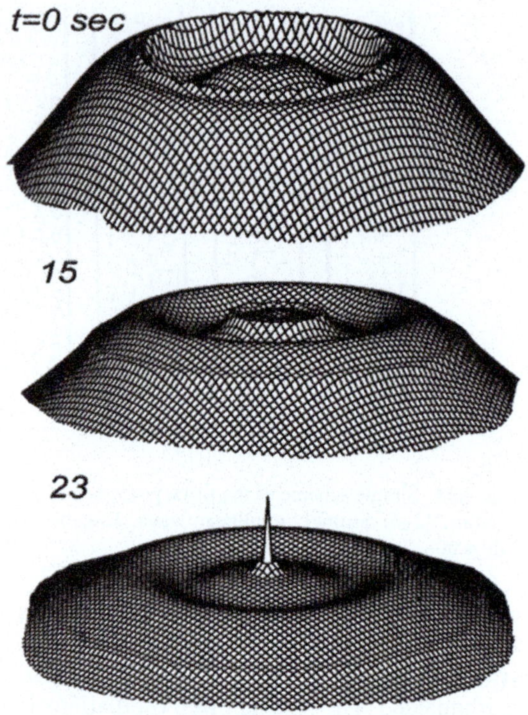

Fig. 4.35 Simulation of results of the interaction of an upward-going wave and the volcano crater. t = 0: Generation of the surface waves at the rim. t = 15: Converging and diverging radial waves. t = 23: The material column (jet-like eruption) induced by the focusing of the converging radial wave [26]

4.2.2 *Surface Waves in the Crater and Short-Time Volcanic Eruptions*

Darwin reported that... *a train of volcanoes situated in the Andes ... instantaneously spouted out a dark column of smoke, ...* [1, p. 380]. Taking into account this note, I suggested [26] that there could be a mechanical mechanism focusing the surface material of the crater along the axis of symmetry. As a result, *...a dark column...* is formed. This mechanism is the interaction of the volcano crater with upward waves, generated by vibrations of the volcano base. Converging waves and diverging waves of the surface material are being formed in the volcano top. The converging waves can form a vertical '*...a dark column of smoke...*' in the centre of the crater. The described processes are qualitatively illustrated by Fig. 4.35.

Other mechanism of formation of '*a dark column*' can be connected with the collapse of the lava crust. Darwin wrote ... *If the earthquake or trembling of the ground (which, however, we have seen was less near these volcanoes than elsewhere) had acted in no other way, than in merely breaking the crust over the lava within the craters, a few jets of smoke might have been emitted, but it could not have given rise to a prolonged and vigorous period of activity* ... [4, p. 61]. For example, it can happen that seismic shock creates a crack in the crust, which then 'is healed' after the emission of gas from the magma conduit. Since the seismic shock does not repeat, the next emission does not repeat either.

Often ...*the crust over the lava within the craters*... is very thin like a plate or a membrane, or the crust is absent. In the last case the volcano magma is directly connected with atmosphere. Certainly, the intensity of the seismic wave necessary to start an eruption will be different in the cases of open and closed conduit. For example, in Ruapehu (National Park Tongariro, New Zealand), eruptions took place at earthquakes of magnitude 2 in case of the open conduit and magnitude 3.4 in case of the closed conduit [61].

To discuss these phenomena we shall consider the dome (or plug) as a cover closing the conduit. When the gas pressure in the conduit exceeds some critical level the cover lifts and a short eruption occurs. Then gas pressure in the conduit reduces, and the cover drops down. As a result, converging radial waves and diverging radial waves of the surface deformation are being formed.

Similar processes were observed in the volcano Santiaguito, Guatemala (Fig. 4.36), where the local peak of the vertical acceleration of the crater surface was up to 3 g [19].

When the gas pressure near the volcano vent increases above some critical level then the cover is lifted up, and a smoke column erupts. This fast lifting is accompanied by an explosive eruption of a mixture of gas and ashes, which forms a ring on the conduit cover. During a few seconds, the 20–80 m thick dome (whose radius is about 200 m) is being lifted by the gas pressure up to several tens of centimetres. These large and rapid uplifts occur 1–2 times per hour, together with explosive ash-rich eruptions. Then the pressure reduces near the vent and the cover drops down. Thus, the phenomena were excited as a result of the periodical degassing of the conduit magma. Waves in the crater were measured – their speed was about 30–50 m/s, which is too small for typical elastic waves within solid or fluid magma. The short-duration events and the dome/crater collision generate the periodical earthquakes or trembling of the ground (see Fig. 4.37).

A ring-like eruption of the volcano Santiaguito is shown in Fig. 4.36 (top). The eruption is a result of the dome uplift. Then the converging wave propagating in atmosphere is formed, and it focuses in the dome centre. Fig. 4.36 (bottom) shows the top of the volcano Santiaguito between the eruptions. Thus, on the vibrating surface of the volcano top there may be a complex interaction of atmosphere, cracks, and boiling magma. As a result of the interaction, a short-time eruption can take place. Because of the sharp vertical motion of the crater surface, the converging radial wave of air can form within the crater. Thus, complex processes of the interaction of atmosphere, surface waves, cracks and volcanic gases can form ...*a dark column*... at the volcano top, but no long eruptions. Let us dare to think that something like these processes are schematically described by the last remark of Darwin quoted in this subsection. The diverging seismic surface waves at the volcano Santiaguito were studied [38].

A seismogram of these waves is presented in Fig. 4.37. The interval between eruptions was 20–40 min. Considering this and the remark presented in the end of the previous subsection, it can be assumed that the average speed of waves in

Fig. 4.36 Focusing of the short-time periodical eruption of the volcano Santiaguito, in Guatemala (After http://www.photovolcanica.com/VolcanoInfo/Santiaguito/Santiaguito.html)

magma of Santiago is of the order of 10 m/s, and the height of the magma conduit is about 10 km.

Strongly-nonlinear vertical waves may be generated in the volcano conduit because of periodical opening of the vent. The simplest scheme of this process is

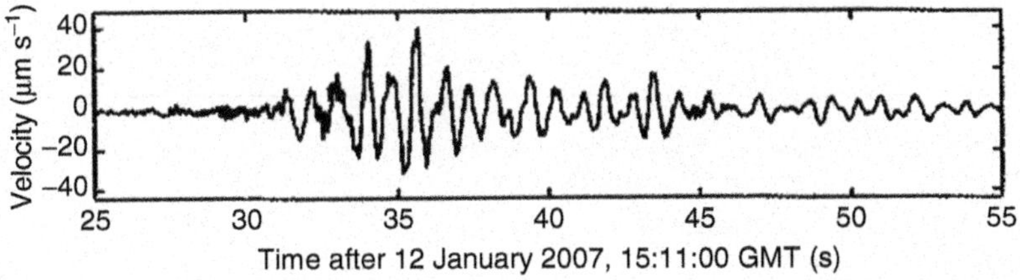

Fig. 4.37 The vertical velocity seismogram (trembling of the ground), which was recorded at a station 3 km from the vent of the volcano Santiaguito [38]

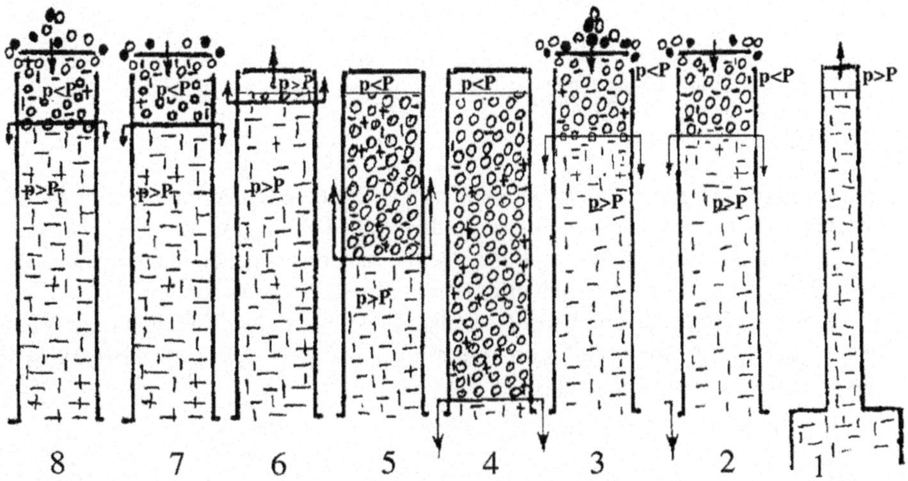

Fig. 4.38 Simple scheme of the strongly-nonlinear wave process, which can be initiated by periodical opening of the volcano conduit: In (*2, 3, 7, 8*), partly-opened conduit when the dome lifts up. In (*1, 4–6*), closed conduit (see, also, Fig. 4.34)

presented in Fig. 4.38, where p is the pressure, and P is the critical value of the pressure.

The cover (vent filling or dome) of the conduit lifts up when a pressure p reaches the critical value P, and then magma begins to boil and erupt. As a result, a wave of expansion (depression) is generated at the vent, the pressure drops down, and then the cover closes the conduit. The process repeats when p under the cover increases above P.

On the whole, the process resembles boiling in a kettle. Its cover rises when the pressure of the steam increases above some critical value, and then the steam pressure decreases. As a result, the water in the kettle starts to boil more strongly, and steam and foam erupt from the kettle. When the pressure drops below some critical value the cover closes the kettle, and then the process repeats.

Thus, the volcano oscillations strongly depend on the behaviour of magma in waves of expansion (or depression). Therefore, this behaviour has been studied in many researches. We shall concentrate on results from [62], where a high-pressure

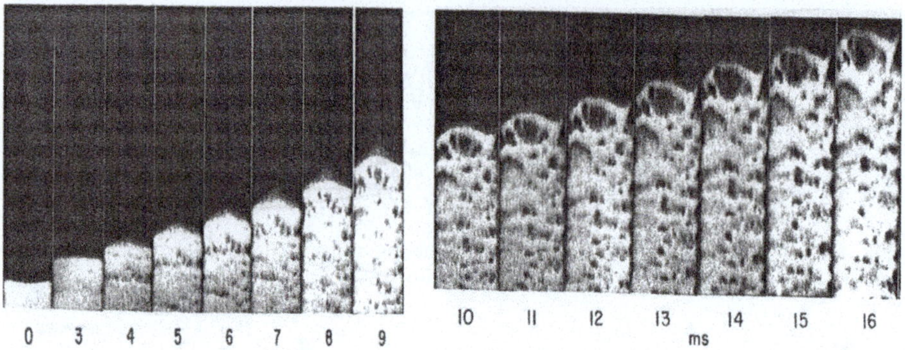

Fig. 4.39 Wave phenomena during fast depressurization of a high-pressure pipe containing gas and granulated material. Expansion and rising of the surface of the layer of 0.25-mm glass balls. The initial layer thickness is 17 cm [62]

pipe was used. The pipe models a conduit of a volcano, and it contains gas and solid particles (Fig. 4.39). When the pipe is rapidly opened an expansion wave is formed, which moves downwards. Within this wave the material expands like magma at the top of the conduit before a strong eruption. Figure 4.39 shows the formation of big bubbles within the expansion wave (see, also, Figs. 1.7 (right), 2.5 (right) and 7.15). Similar fast growth of bubbles in magma qualitatively explains the explosive eruptions of volcanoes [53, 62].

4.2.3 Short-Time Eruptions from Craters

Vibrations in a volcano can generate avalanches inside the crater. It is known that snow avalanches occur in very cold, dry climates; in particular, in the Andes of South America, the Himalayas, and the Rocky Mountains of North America (Fig. 4.40). Light powdery snow grains do not stick together well. An avalanche occurs, when the stress of gravity trying to pull snow downhill exceeds the strength of the bonds among the snow crystals of the snow cover. About 90 per cent of all avalanches start on slopes of 30°–45° (coincidentally, that is the best angle for skiing). Four ingredients are required to produce a snow avalanche: snow, a steep slope, a weak layer in the snow cover, and an external force. An earthquake can be a forcing mechanism for an avalanche.

A strong-enough earthquake can cause this snow to move down to the crater centre. A powerful flow of compressed air can move ahead of the avalanche – a so-called 'air wave' or 'avalanche blizzard' [63]. At the centre of the crater the air wave and then the avalanche are focussed, with the result that a mixture of snow, ice, and solid grains gets formed like a huge cloud (column) rising from the crater. Perhaps, a similar cloud was described by Darwin as *...a dark column....* Darwin

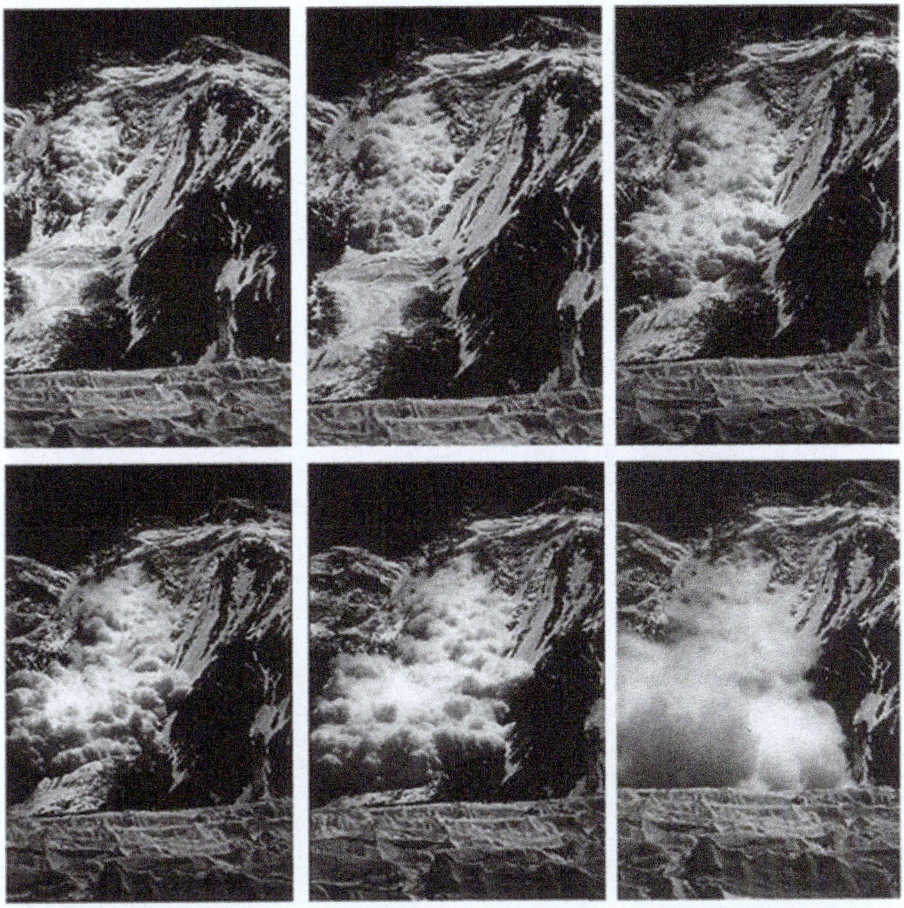

Fig. 4.40 A sequence of snapshots of a powder snow avalanche in the Himalayas [63]

personally did not observe the simultaneous eruptions from the craters of many volcanoes.

It was assumed [26] that the above processes may explain qualitatively the simultaneous short-time eruptions of many volcanoes.

Avalanches and rockslides in craters are also rarely-observed events. In particular, avalanches and eruptions were not observed during the catastrophic Chilean earthquake on February, 27th, 2010. Probably, it is connected with dangers and difficulties of observations of volcano craters. The observations are simpler for the volcano Mount St. Helens. We have already noticed the connection of the beginning of the catastrophic 1980 eruption with the local magnitude 5.1 earthquake. However, the latest observations of the Mount St. Helens crater show that an earthquake can cause avalanches and columns of snow dust in the crater. Similar phenomena could have happened during the Chilean 1835 earthquake.

Fig. 4.41 Example of *...a dark column...* from the crater. That cloud arises because of an avalanche from the lava dome, located in the centre of the crater of Mount St. Helens (see, also, Figs. 3.7 and 4.36) (Internet. https://sites.google.com/site/thehumanitiesempanadas2/volcanoes)

Fig. 4.42 Another (darker) example of the eruption of *...a dark column...* from the crater of Mount St. Helens (7/24/2006) [9]

In the centre of the crater of the Mount St. Helens volcano there is a fairly high lava dome, which is usually covered with snow. During snow avalanches a white cloud rises, and it can be seen above the crater (Figs. 4.41 and 4.42). The reader can see this process in Internet: http://www.youtube.com/watch?v=Zwga-NdU8-8. Similar columns above volcanoes could have happened in 1835.

We found also a description, which may be connected with the avalanche in a volcano crater and the granular jet from its centre, in Kilauea volcano, Hawaii. It was reported, that the Halema'uma'u Crater had a strong shock at 20.03.2008, when stones flew up to the viewing terrace. These were old rocks, and not fresh lava.

Further, we shall consider some data which indirectly confirms the possibility of *...a dark column of smoke...* in a crater.

Interesting examples of an eruption of sand Generally speaking, the eruption of water or granular jet from surface cavities and craters is a well-known phenomenon. We present below some experimental data, which demonstrates the process of focussing converging surface waves in craters. The wave is generated due to different reasons. The shape, dynamics, and components of the corresponding jet are a problem still in debate, and they depend on many parameters [64–67]. But we think that a similar process might take place sometimes during an earthquake-induced sharp acceleration of a volcano crater.

When a dense small body falls quickly onto a layer of fine, loose sand, a surprisingly vigorous jet shoots upward from the surface (Fig. 4.43). It is similar to what happens in a liquid. Formation of craters and a following eruption of granular jets from the craters were observed.

Fig. 4.43 Formation of jets of the granular material when a heavy sphere is dropped onto a bed of loose, fine sand. After the crater formation and a crown-shaped splash of grains from the bed's surface, the jet erupts from the crater [66]

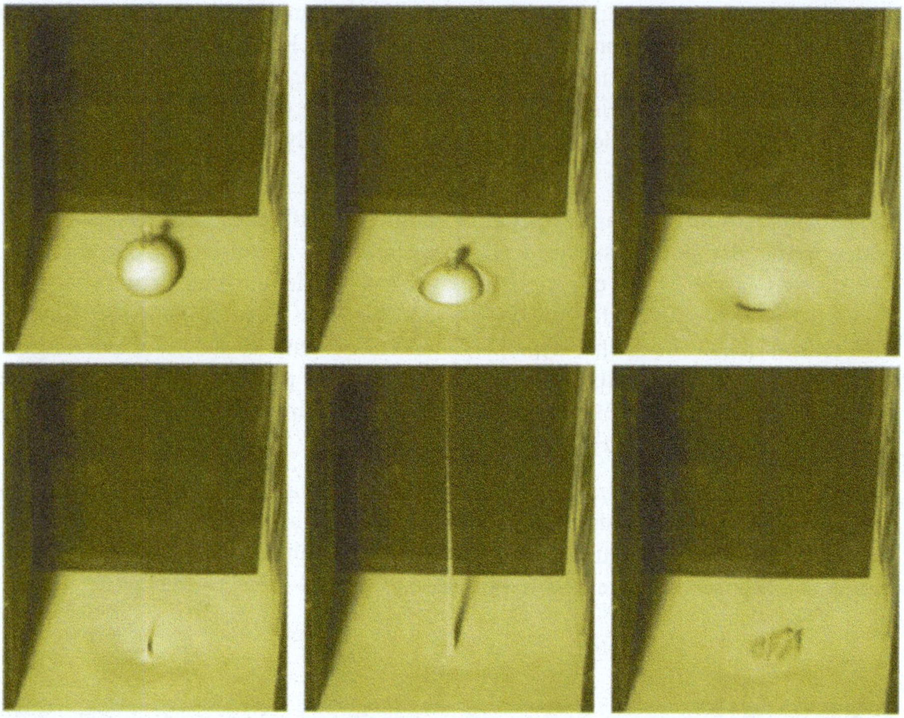

Fig. 4.44 Snapshots of the sinking-ball experiment. At time $t = 0$ ms the ball (mass 133 g) is released, and it immediately starts to sink into the sand ($t = 60$ ms); at $t = 126$ ms, a crater forms; at $t = 147$ ms, a sand jet begin to emerge. It was shown at $t = 251$ ms. After $t \approx 600$ ms, the trapped air bubble reaches the surface. It was shown at $t = 700$ ms [67]

Let us consider the high-velocity impact (Fig. 4.43). In this case an axisymmetric rarefaction wave is being generated in atmosphere, and ejected sand forms a crown-shaped splash. This surface is shown in Fig. 4.43 (middle). We can, approximately, assume that this surface is some analogue of the crater surface material. Subsequent transient radial and vertical motions of air and the ejected sand form a jet along the axis of symmetry (right). We think that this is some analogue of the crater avalanche dynamics with the wave phenomena, illustrated by Fig. 4.43.

The effect of the sand splash was excluded in [67]. A ping-pong ball of radius 2 cm, partly filled with bronze grains, was suspended above 'dry quicksand' (not to be confused with natural quicksand, which is a mixture of sand, clay, and water) so that the ball was just touching the surface. To release the ball without introducing any vibrations, the thin rope supporting the ball, was burned, causing the ball to sink instantaneously into the sand. In contrast with Fig. 4.43, there was in this case no splash, but a straight jet of the sand shot violently into the air after about 100 ms (Fig. 4.44).

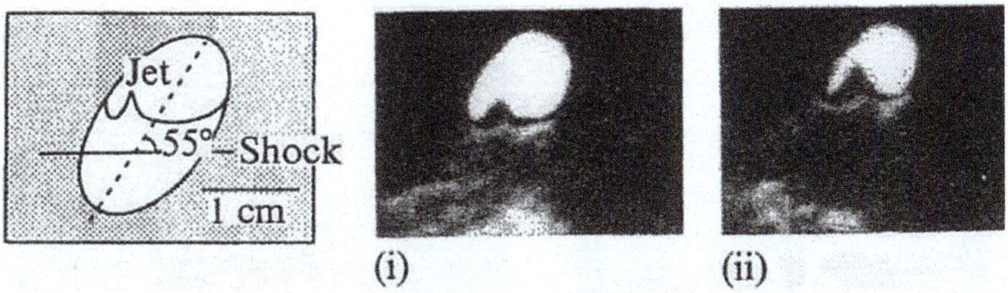

Fig. 4.45 An elliptical cavity collapsing in gelatine. The cavity is inclined at 55° to the incident shock as shown in the scheme. The jet is perpendicular to the shock front [68]

Thus, the formation of the crater column can depend on many circumstances. The curvature of the crater surface, the initial vertical acceleration, and the inertial forces play important roles.

Interesting examples of an eruption of jets of water by curved boundaries Experiments with cavities (craters) on the surface of liquid were described in [68]. The inner shock wave loads craters. As a result jets (plumes) are formed. It was found that the jet velocity increases with the surface curvature and/or the intensity of the shock. The form of the cavity can be very important. Figure 4.45 shows a result of the interaction of a shock-wave with an elliptic bubble. In this case, one jet is formed.

The experiments with semicircular and square cavities demonstrate that jets may be generated near the cavity corners (see Figs. 4.46 and 4.47). These jets move one towards another, and at the same time air shock waves are generated within the cavities.

The jet motion and then the collision of them were observed when a square cavity collapsed. A diagram, showing wave-fronts and moving jets in the cavity, is presented in Fig. 4.47.

The generation of a vertical column may be demonstrated using a tube filled with water. When the tube falls from a height of 10–20 cm and the bottom receives a shock loading, the centre of the water surface ejects a thin vertical column [69]. This is a result of the interaction of the concave meniscus of the water with the upward-wave of the acceleration. The height of the ejected column can be more than 1 m. The deformation of the concave surface and the formation of the column during similar experiments are shown in Fig. 4.48.

The experiments discussed herein, have demonstrated that a crater jet can occur, if there is a strong enough earthquake-induced vertical acceleration of the volcano base. As a result, strongly-nonlinear surface waves, short-duration eruptions, and crater avalanches may occur. The mixture of the surface materials, the ash, and gases focuses in the crater centre and forms the vertical ...*a dark column*..., resembling that described by Darwin.

Thus, according to these much-idealized models presented above, the earthquake-induced volcanoes, separated by very long distances, could instantaneously eject the jets. The fluidization of the crater surface, the cracks, and the

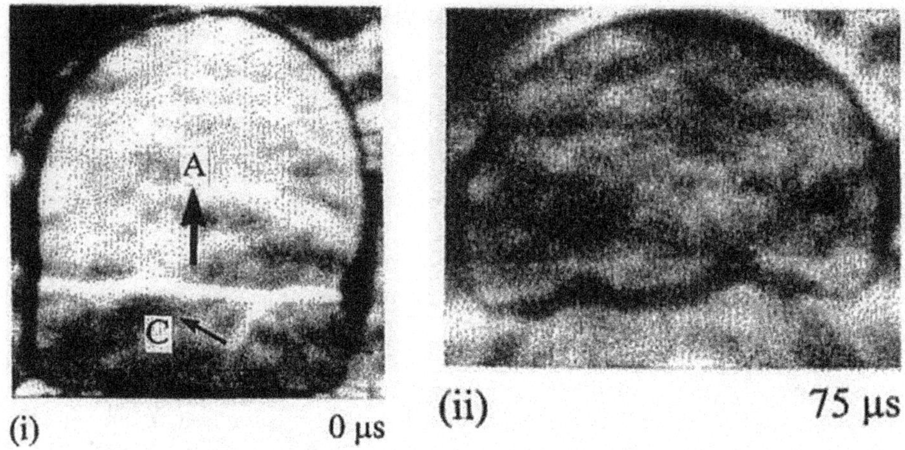

Fig. 4.46 Collapse of a semicircular cavity. (i) In the air the plane shock-wave A and the corner wave C are formed. (ii) Two converging jet-like waves move from the walls [68]

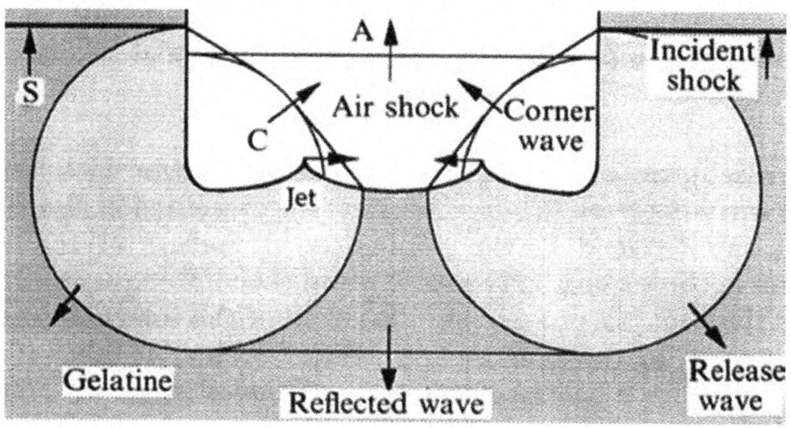

Fig. 4.47 Interaction of a flat-walled cavity with a shock. Waves include the incident shock S, the air shock A, and the corner waves C. The jets, moving from the cones, are shown [68]

fountain-like eruption absorb considerable seismic energy. Therefore, immediately after the eruption, surface oscillations near the volcano may be reduced.

Remark The presented results may be treated as the surface Faraday waves. In particular, the craters and jets shown in Figs. 4.43, 4.44 and 4.48 resemble the craters and jets presented in Figs 5.20, 5.21 and 5.34. On the other hand Figs. 4.43, 4.44 and 4.48 can qualitatively illustrate the birth of the particles during scalar field evolution (see Chap. 7).

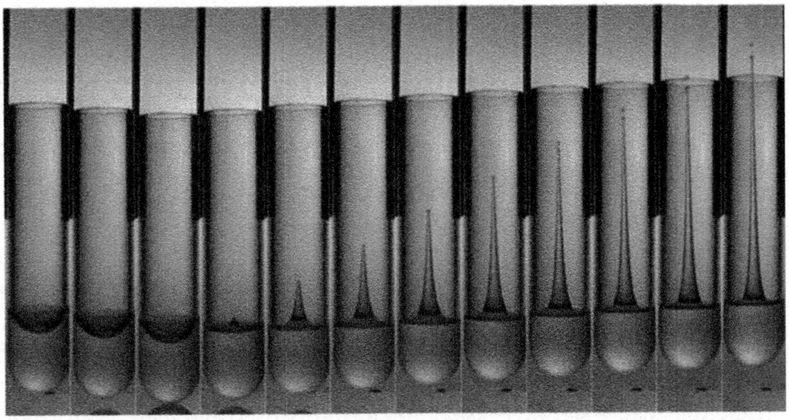

Fig. 4.48 Axial impact of a cylindrical tube falling under gravity and filled with a liquid wetting the tube wall. The curvature of the spherical meniscus (crater) is reversed violently. As a result, a rapidly-ascending jet is formed [70, 71]

4.2.4 Discussion of Earthquake-Induced Volcanic Eruptions

We have reported many observations and experiments. In this subsection we briefly summarize what has been told above.

The model of the Sect. 4.2.1 The connection of seismic and volcanic phenomena is a subject of continued researches. Probably, there are many mechanisms of this connection. However it is already clear that big earthquakes can be the starting mechanism for big eruptions [37–39].

According to Darwin [1, 2] we have considered the large-scale eruptions as a result of brief vertical elevation of the land. The model takes into account the amplification of the upward seismic wave near the top of conical volcanoes, and the interaction of this wave with the crater surface. The elements of the model are: fragmentation of the crater by the tension wave, magma connecting with the atmosphere, and growth of bubbles (cavitation). We think that the shape and the state of the 4 volcanoes during the 1835 Chilean earthquake were such, that the vent fillings were fragmented. As a result, the earthquake opened the conduits and the magma chambers, and magma began to erupt like champagne from a bottle.

Liquefaction of surface material of a crater, generation of cracks, and eruptions each absorb considerable seismic energy. Therefore, immediately after the eruption, oscillations decreased near the volcano.

It is interesting that a majority of volcanoes presented in the list of earthquake-triggered eruption pairs [37] are truncated by large calderas. This observation agrees with Darwin's note [1, p. 380] about *...great chimneys....*

Models of the Sects. 4.2.2 and 4.2.3 The main seismic shock can collapse the crater. However, if the magma system is not ready for a strong eruption, only a brief eruption of gas from the crater can take place. This case resembles the dynamics of the volcano Santiaguito, Guatemala. Fast lifting of its dome was accompanied by the formation of a stream of gas (Figs. 4.35, 4.36 and 4.38). At the same time, *...a dark column of smoke...* can instantly be formed in craters of conical volcanoes after a seismic shock, if it is accompanied by an avalanche in the crater and focusing of the surface wave in the crater centre.

The general discussion In [26] some analogy was noted between parameters typical for an underwater explosion, and parameters typical for a wave/crater interaction. In particular, it was assumed that the explosion power corresponds to the earthquake power. The thickness of the water layer between the point of the explosion and the free water surface corresponds to the thickness of the conduit plug. A shock wave from the explosion corresponds to a seismic wave moving to the volcano top. The water layer is completely collapsed and there is an eruption of products of the explosion into the atmosphere (Fig. 4.49) in the case of the near-surface explosion. This process corresponds to the beginning of a strong eruption.

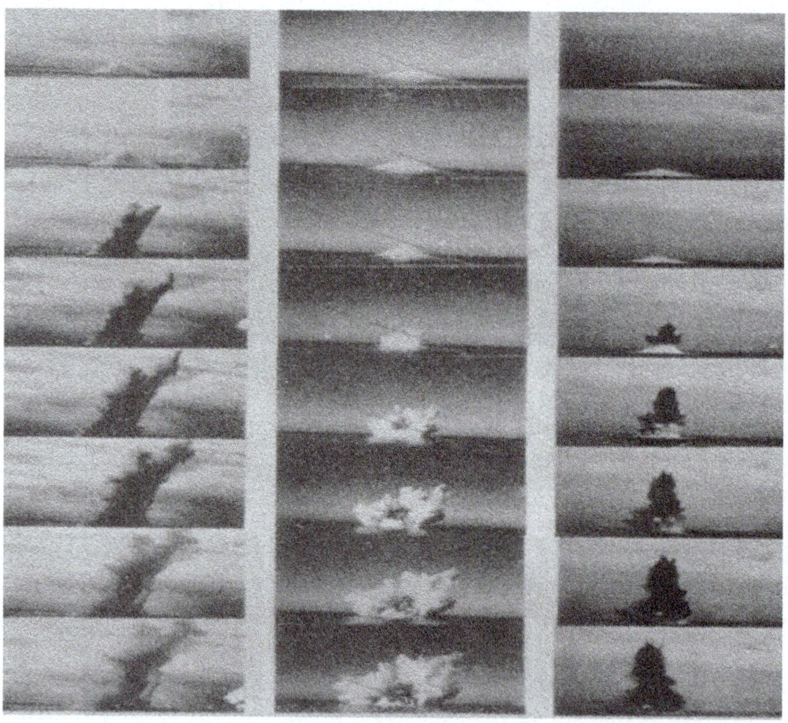

Fig. 4.49 Formation of water domes and eruptions (jets) of a mixture of water and explosion products after underwater explosions. Explosions of the same power took place at depths of 8 m (*left*), 12 m (*centre*) and 19.5 m (*right column*). The rows of 3 photographs correspond to the following periods after the explosion (from top to bottom): 0.5, 0.9, 1.3, 1.7, 2.1, 2.5, 2.9 and 3.3 s [72]

In the case of a deep and powerful explosion the interaction of waves from the explosion with the free water surface leads to a formation of a surface crater. A jet forms along the crater axis, due to the converging rush of the surface water [26, 73]. This jet corresponds to *...a dark column of smoke...* from the craters.

4.3 Amplification of the Earthquake Convulsion. Effects of the Geology and Relief

The configuration and geology of surface materials often determine the characteristics of seismic waves. At the same time, the mechanical properties of surface materials depend on an earthquake. An earthquake usually lasts enough to change significantly the mechanical properties of some soils. For example, the first movements were felt at Concepcion at 11.40 a.m. Slight at first, they increased rapidly in strength until, after half a minute, the convulsive movements were so strong that it was difficult to stand. Buildings swayed and tottered. Then 'suddenly an awful overpowering shock caused universal destruction – and in less than 6 s the city was in ruins' [10, p. 254]. This part of the shock lasted with convulsive movements for nearly 2 min, so that (including the final vibrations) the total duration was about 4 min [36]. That period is enough for liquefaction and failure of soils (see Sect. 4.1.2).

Darwin understood well the impact character of earthquakes. He wrote about *...convulsions...* [4, p. 59, 61], *... shocks...* [4, p. 71, 74] and *...a paroxysmal movements,. ...the most violent explosion... are parts of one common action, only modified by local circumstances...* [1, p. 380–381]. Following Darwin and Captain FitzRoy I underlined many times in this book and in [9, 23–26, 74] the influence of these circumstances. In particular, Robert FitzRoy reported [10, p. 256] that 'The city of Concepcion stands upon a plain, very little higher that the level of the river Bio Bio. The soil is loose and alluvial. To the eastward and northward lie rocky irregular hills: from the foot of which the loose earth was every where parted by the great convulsion, large cracks being left, from an inch to more than a foot in width. It seemed as if the low land had been separated from the hills, having been more disturbed by the shock' and 'At Talcahuano the great earthquake was felt as severely on the 20th February as in the city of Concepcion. It took place at the same time, and in a precisely similar manner: three houses only, upon a rocky foundation, escaped the fate of all those standing upon the loose sandy soil, which lies between the sea-beach and hills' [10, p. 257]. A clear example of the importance of geology for seismic waves in the layer and site effects was demonstrated by Pedersen et al. [75].

Seismometers were placed across a 300-m-high linear ridge in France (Fig. 4.50). The stations noted as S2 and S3 were established on hard limestone, whereas the other 3 stations were placed on unconsolidated sediments. The greatest amplification took place at S5, which was near, but not very close to the edge of the sediment layer. At this point there was an amplification of waves travelling to the

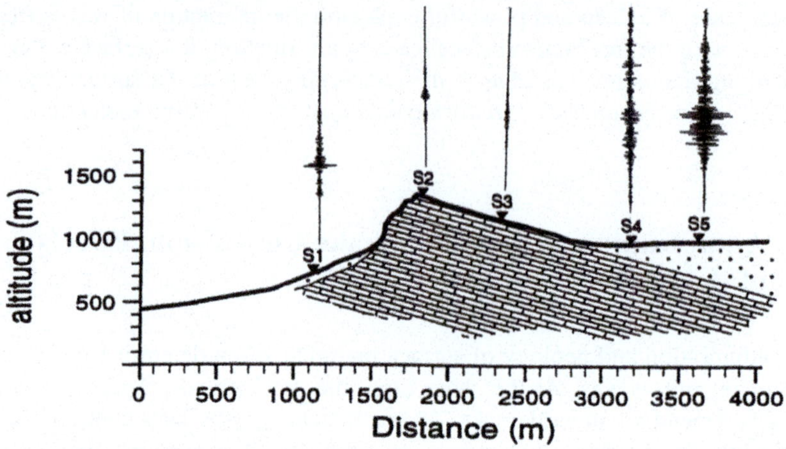

Fig. 4.50 Evolution and amplification of a seismic wave at the sediment layer edge and within the slope of the ridge [26, 75]

layer end. The seismic energy concentrates near the edge of the layer. This amplification depends on the layer thickness variation, on the spatial variation of motion (horizontal motion transforms partly into vertical motion), on the impedance contrast between the down-lying material and the sediment, and on mechanical properties of the sediment.

In particular, because of the earthquake-induced vibrations, the shear stress drops to zero. The layer material is, practically, transformed into the liquefied state. In this case, there is a strong analogy of the evolution of seismic waves near the layer edge and a tsunami near a coastline. For example, the Alaska 1899 earthquake raised the coast on Disenchantment Bay up to 14.5 m. There were over 50 shocks on September 10, with the first earthquake lasting 90 s. The main earthquake, which caused great topographic changes, occurred at 21:41 UTC (see http://neic.usgs.gov/neis/eqlists/USA /1899_09_10.html). The author thinks that they are the result of resonance of seismic waves in the liquefied coastal layer.

It is known that the speed of a shear wave can drop from 160 m/s to less than 10 m/s (about 2–5 m/s) over a 16 s period, as a result of earthquake-induced vibrations [15]. In this case, the layer material transforms into the liquefied state. When the wave moves to the layer edge, the effect of the dispersion reduces and resonance occurs.

The result, presented in Fig. 4.51, qualitatively describes the wide range of phenomena, which can take place, due to focusing of the wave energy at the edge of the layer. As a result, the end behaviour and its acceleration begin to resemble the dynamics of the end of a whip, which produces a sonic bang when the tip moves supersonically. Also strongly-nonlinear surface waves can be excited in layers of weakly-cohesive material [17, 18, 76]. Sometimes shock-like waves with deep depressions ahead of them have been forced in granular layers under strong vertical excitations [77].

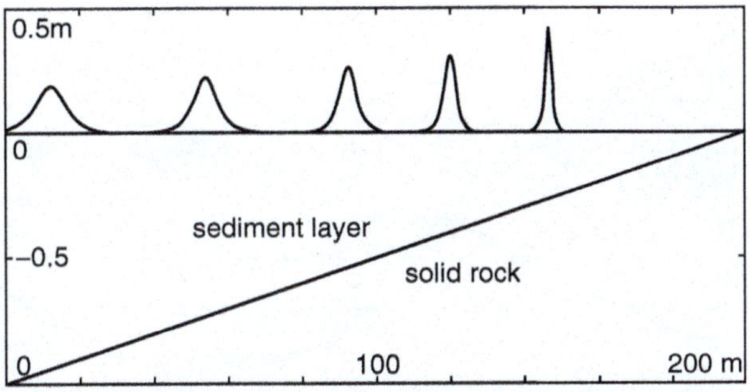

Fig. 4.51 A wave evolution in a shallow sediment layer (the slope is 1:200)

Thus, Darwin described well some effects of the geology and soil properties for seismic phenomena. He wrote ... *That the area thus fissured extends parallel or approximately so, to the neighbouring coast mountains* ... [4, p. 71]. The generation of 'large cracks' may be explained by resonant amplification of seismic waves at 'the low land'. This resonant process resembles the formation of steep fronts of ocean waves and tsunamis travelling to a coastline.

4.4 Darwin on Avalanches as a Cause of Tsunamis

Darwin described the terrific waves which he had observed in Strait of Magellan ...*Great masses of ice frequently fall from these icy cliffs, and the crash reverberates like the broadside of a man-of-war through the lonely channels. These falls, as noticed in the last chapter, produce great waves which break on the adjoining coasts. It is known that earthquakes frequently cause masses of earth to fall from sea-cliffs: how terrific, then, would be the effect of a severe shock (and such occur here...) on a body like a glacier, already in motion and traversed by fissures! I can readily believe that the water would be fairly beaten back out of the deepest channel, and then returning with an overwhelming force, would whirl about huge masses of rock like so much chaff...* [2, p. 298–299] and ... *at intervals violent earthquakes would shoot prodigious masses of ice into the waters below* ... [2, p. 304]

Now the waves, which were described, are well-known, and are often named as a slide-induced local tsunami. On the night of July 9, 1958 an earthquake occurred along the Fairweather Fault in the Alaska Panhandle. This phenomenon was associated with ground motions of high intensity. The earthquake had a magnitude of 7.9, although some sources reported about 8.3. Ground displacements of 1.05 m upward and 6.3 m in the horizontal plane were measured on the surface. Ground motions of such high intensity could have resulted in vertical accelerations up to 0.75 g and horizontal accelerations as much as 1.0 g. A wide variety of geological effects, in addition to faulting, were observed. Among these were avalanches, earth slumps, and a variety of other secondary phenomena. Very high ground accelerations took place near the head of Lituya Bay (Fig. 4.52).

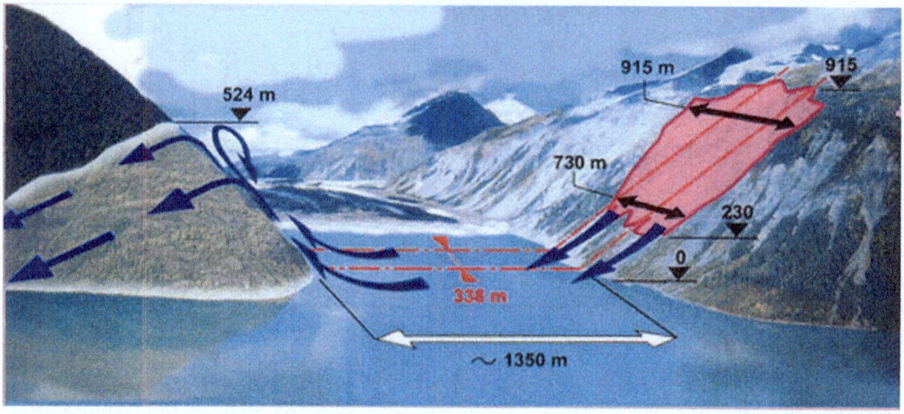

Fig. 4.52 Gilbert Inlet which is located at the head of Lituya Bay. The cliff where the rockslide originated is on the right side of Gilbert Inlet. The opposite valley wall on the left side received the full force of the big wave. The trees and the soil were stripped away, up to 524 m above the surface of Lituya Bay

Fig. 4.53 Lituya Bay is a few weeks after the 1958 tsunami. The areas of destroyed forest along the shorelines are clearly recognizable as the light areas rimming the bay. Gilbert Inlet is seen at the head of the bay (Photo by D.J. Miller, United States Geological Survey)

The earthquake loosened about 30.6 million cubic metres of the rock high above the north-eastern shore of Lituya Bay. This mass of rock plunged from the altitude of approximately 914 m down into the waters of Gilbert Inlet (Fig. 4.52). The impact generated the local tsunami which crashed into the southwest shoreline of Gilbert Inlet. The wave hit with such power that it swept completely over the spur of land that separates Gilbert Inlet from the main body of Lituya Bay (Fig. 4.53).

The wave removed all trees and vegetation on the elevations up to 524 m – that is the highest wave that has ever been known. Millions of trees were uprooted and swept away by the wave. Perhaps, natural disasters similar to that described were imagined by Darwin when the *Beagle* floated along labyrinths of Magellan's Strait.

4.5 Darwin on Transient Cavitation Within Volcanic Bombs

Darwin also reported. . .. *In several places I noticed volcanic bombs; that is, masses of lava which have been shot through the air whilst fluid, and have consequently assumed a spherical or pear shape. Not only their external form, but in several cases their internal structure, shows in a very curious manner that they have revolved in their aërial course. The central part is coarsely cellular, the cells decreasing in size towards the exterior, where there is a shell-like case, about the third of an inch in thickness, of compact stone, which again is overlaid by the outside crust of finely cellular lava. I think there can be little doubt, first, that the external crust cooled rapidly in the state in which we now see it; secondly, that the still fluid lava within was packed by the centrifugal force, generated by the revolving of the bomb, against the external cooled crust, and so produced the solid shell of stone; and lastly, that the centrifugal force, by relieving the pressure in the more central parts of the bomb, allowed the heated vapours to expand their cells, thus forming the coarsely cellular mass of the centre*. . ..[2, p. 589] (Fig. 4.54).

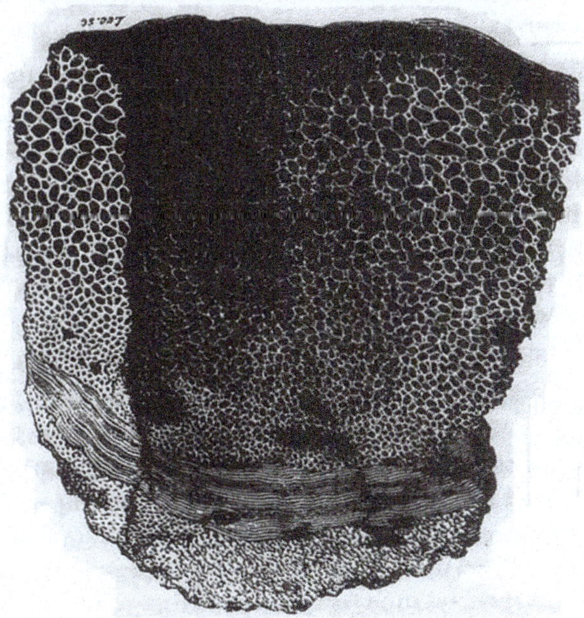

Fig. 4.54 A fragment of the spherical volcanic bomb described by Darwin [2, p. 589]

Thus, Darwin described two processes of the appearance of transient cavitation in condensed media. They were not induced by temperature increasing. This case would be not considered as very curious. The cavitation was induced by sharp decompression in the environment. In fact, it is the same process which is seen when opening a bottle of champagne. In particular, Darwin deservedly felt satisfaction after explaining the surprising structure of volcanic bombs by fast decompression (see Fig. 2.5).

4.6 Darwin on His Theories of Mountain Formation

Approximately 200 years ago Darwin had described the connection of volcanic and seismic forces with a ridge formation ...*the frequent quakings of the earth on this line of coast are caused by the rending of the strata, ...and their injection by fluidified rock. This rending and injection,.. .form a chain of hills*; ... [2, p. 377]. Thus, he thought about the possibility a fast change of the face of the Earth as a result of volcanic and seismic forces. Nowadays, such a point of view is widespread. Huge earthquakes and the most powerful volcanic eruptions could change part of the Earth's surface, its animals and plants, literally within a few years.

Certainly, Darwin was not the first who reflected on the connection of earthquakes with eruptions of volcanoes and evolution of the Earth. Among the first who analyzed this problem were some great philosophers and writers of ancient Greece and Rome. For example, Aristotle thought that earthquakes cause volcano eruptions, and that point of view was widespread for centuries. The well-known poet and scientist of ancient Rome Lucretius explained earthquakes by the subsidence of massive ridges. The geographer Strabo cited examples of both local subsidence of land and raising of a seabed [78, p. 207].

From antique times it was known that the terrestrial surface is not in rest, and is subject to raising and falling. In the Renaissance period Leonardo da Vinci came to the conclusion that the finding of fossilized sea shells at considerable height above the sea level represents a result of rising of land. He believed that catastrophes, like the Biblical flood, cannot explain the appearance of shells at tops of mountains. Discussing another scholastic explanation, da Vinci wrote with some sarcasm 'And if you wish to say that the shells are produced by nature in these mountains by means of the influence of the stars, in what way will you show that this influence produces, in the very same place, shells of varying sizes and varying in age, and of different kinds?' Thus, about 500 years ago da Vinci came to the absolutely correct conclusion: the modern mountains were formerly covered by the sea, after which an elevation of the seabed revealed to us those white chalk-stone layers.

The question about reasons for occurrence of sea shells at mountains far from any water was not left idle during those centuries. It did not correspond to the Bible version about creation of living beings on the Earth during a period of 6 days, 6000 years ago, and that those beings were invariant. During Darwin's journey this question remained topical. Hence, Darwin repeatedly directed the attention of readers to the finds of sea shells at mountains. Generally speaking, the climate of opinion was not favorable to such views. At that time, standard geological history assumed that the land level did not change, whereas the sea level was changing. For example, late in the 19th century Eduard Suess, in his authoritative book **'Das Antlitz der Erde'** denied that the Chilean coasts had been uplifted during the 1835 earthquake or the 1822 earthquake [79].

From the modern point of view this opinion is strange. Indeed, in 1688 the brilliant scientist Robert Hooke described in his book **'Lectures and Discourses of Earthquakes'** both uplift and subsidence of land, seabed and mountains. He explained these phenomena as actions of earthquakes.

Sea shells on mountains confirm a connection of the variation the Earth's surface and the evolution of life. Darwin's thought constantly came back to this connection. His observations convinced him about the connection of earthquakes, eruptions of volcanoes and the evolution of the Earth, and that the age of the Earth was many hundred millions of years. He wrote…*The solitary hovel …has been mentioned by every traveller who has crossed the Andes. …. The geology of the surrounding country is very curious. The Uspallata range is separated from the main Cordillera by a long narrow plain or basin, like those so often mentioned in Chili, but higher, being six thousand feet above the sea. …It consists of various kinds of submarine lava, alternating with volcanic sandstones and other remarkable sedimentary deposits, the whole having a very close resemblance to some of the tertiary beds on the shores of the Pacific. From this resemblance I expected to find silicified wood, which is generally characteristic of those formations. I was gratified in a very extraordinary manner. In the central part of the range, at an elevation of about seven thousand feet, I observed on a bare slope some snow-white projecting columns. … It required little geological practice to interpret the marvellous story which this scene at once unfolded, though I confess I was at first so much astonished that I could scarcely believe the plainest evidence. I saw the spot where a cluster of fine trees once waved their branches on the shores of the Atlantic, when that ocean (now driven back seven hundred miles) came to the foot of the Andes. I saw that they had sprung from a volcanic soil which had been raised above the level of the sea, and that subsequently this dry land, with its upright trees, had been let down into the depths of the ocean. In these depths the formerly dry land was covered by sedimentary beds, and these again by enormous streams of submarine lava – one such mass attaining the thickness of a thousand feet; and these deluges of molten stone and aqueous deposits five times alternately had been spread out. The ocean which received such thick masses must have been profoundly deep; but again the subterranean forces exerted themselves, and I now beheld the bed of that ocean, forming a chain of mountains more than seven thousand feet in height. Nor had those antagonist forces been dormant which are always at work wearing down the*

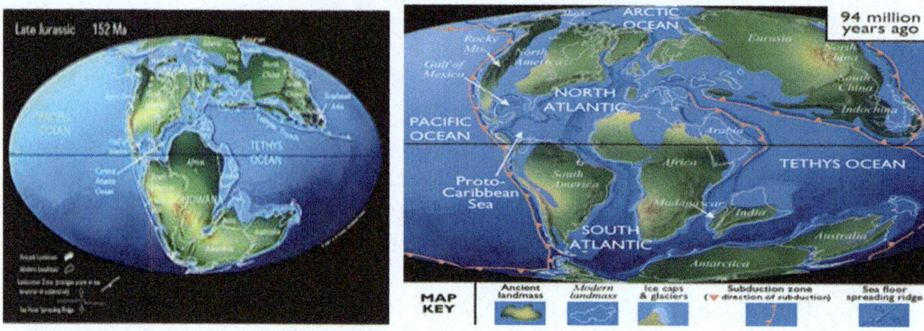

Fig. 4.55 Charts of the terrestrial land existing 152 million years ago (*left*) and 94 million years ago (*right*) years ago. *Thin black lines* define modern contours of the continents (Internet. http://www.scotese.com/cretaceo.htm and http://www.paleoportal.org/index.php?globalnav=time_space&sectionnav=period&period_id=18)

surface of the land. The great piles of strata had been intersected by many wide valleys, and the trees, now changed into silex, were exposed projecting from the volcanic soil, now changed into rock, whence formerly, in a green and budding state, they had raised their lofty heads... [2, p. 400–402].

The stone crust of our planet anywhere and never remains calm. Deeply within the Earth as in a huge boiler, there are continuous changes. The crust thickness is changing. A planetary breathing, muffled and quiet or spasmodic, is felt all the time. As a result, elements of the crust are rising or falling, as Darwin had thought. For example, the speed of rising at some sites of the Russian plain is several millimetres per year. Some mountain areas rise up to 10 cm per year, especially where ridges form a sea coast. As a result, some sites of the Earth can lift up some kilometres during a million years, and the face of the Earth can change markedly over 10 million years (see Fig. 4.55).

Figure 4.55 shows that huge territories which were visited by Darwin (see the Preface, Fig. 1.1) and where he made his paleontological and geological discoveries, had been covered by the ocean in the past. The reader can find other maps on the Internet, which show that the southern extremity of South America had numerous uplifts and subsidences (with reference to ocean level). That information corresponds to the picture described by Darwin of a slow evolution of the Earth's surface.

However the picture needs to be more detailed, since the fossilized trees can, apparently, form only very soon after their immersion in a sea – literally within several days. It could be the result of a strong earthquake. We noticed already that during the Alaska 1899 earthquake, some sites on the coast had risen by 10–16 m. A coastal subsidence of a seabed can also be large. About 250 million years ago 95 % of species living in the ocean and 70 % of species living on land disappeared abruptly. Now there is evidence that it was a period of very powerful eruptions and earthquakes, which probably broke up the supercontinent Pangaea.

Almost incredible data is presented in [9]. During a seaquake in the Adriatic sea in 1873 a cable was torn, 10 km from the island Zante (this island is near the island Ithaca – Odysseus's native land). The cable had been laid at a depth about 400 m, but the ruptured cable was found at a depth over 600 m. The cable had been heaped up by stones. In 1886 there was a new rupture of that cable, after which it was found at depth about 400 m. The 1923 Tokyo earthquake was followed by 27 seabed uplifts and subsidences up to 100 m, and by 6 uplifts and subsidences up to 200 m. This data can be explained by huge landslips and currents occurring at a seabed during earthquakes. Usually, uplifts or subsidences of an ocean bed are not more than a few metres.

Thus, the data of some observations agrees quite well with the evolution which determined the formation of those fossilized trees, according to Darwin. To illustrate some elements of this evolution, we have reproduced on Fig. 4.56 an Internet diagram of the evolution of a crust subsidence into mountains. (That diagram is taken from the geographical dictionary www.ecosystema.ru). Darwin described the formation of ridges and volcanoes by following words. . .*in order to break up and throw over portions of very thick crust, as in Diagram 3, there must have been great horizontal extension, and this, if sudden, would have caused as many continuous outbursts of volcanic matter*. ... and. . . *if the force had acted suddenly, these portions of the earth's crust would have been absolutely blown off*,. . . [4, p. 79]. These thoughts agree with Fig. 4.56.

Thus, Darwin formulated the scheme of the formation of ridges. Basic elements of this scheme were put by James Hall (in 1847) as a basis for his theory on origin of mountains. Elements of Darwin's theory were later developed by James Dwight Dana. He suggested that the thickness of an ocean sediment layer had been increasing, until the layer became unstable (see Sect. 3.3.2 and Darwin's diagram 3). A fast formation of a mountain chain (fast, on a geological time-scale) occurs as a result of sediment layers folding and action of volcanic forces. This theory includes dynamic vertical lifting of a long and narrow strip of the Earth's surface as an essential element.

Let us notice that Dana was in correspondence with Darwin, and he had read **The Voyage of the *Beagle***. In 1859 Darwin sent to Dana a copy of his book **On the Origin of Species by Means of Natural Selection** [80].

During the voyage of the *Beagle*, Darwin understood that the history of a surface of the Earth was very long and very complex. On the whole, he agreed with Charles Lyell that the change of the face of the Earth occurred very slowly. However, he did not exclude the possibility of fast, catastrophic actions of underground forces, which acted. . .*suddenly*. . . During the Chilean 1835 earthquake he had been the action of such forces.

In particular, huge hollows (cavities) can be formed in the Earth crust. For example, crystalline materials can be transformed into a new denser phase by high pressure and temperatures, and so hollows can form within the Earth. On the other hand, some matter of the Earth can cool down and be compressed as a result of thermal convection. Also, volcanoes eject huge volumes of solid materials and gases. Therefore, huge parts of the land surface can suddenly fall into the hollows.

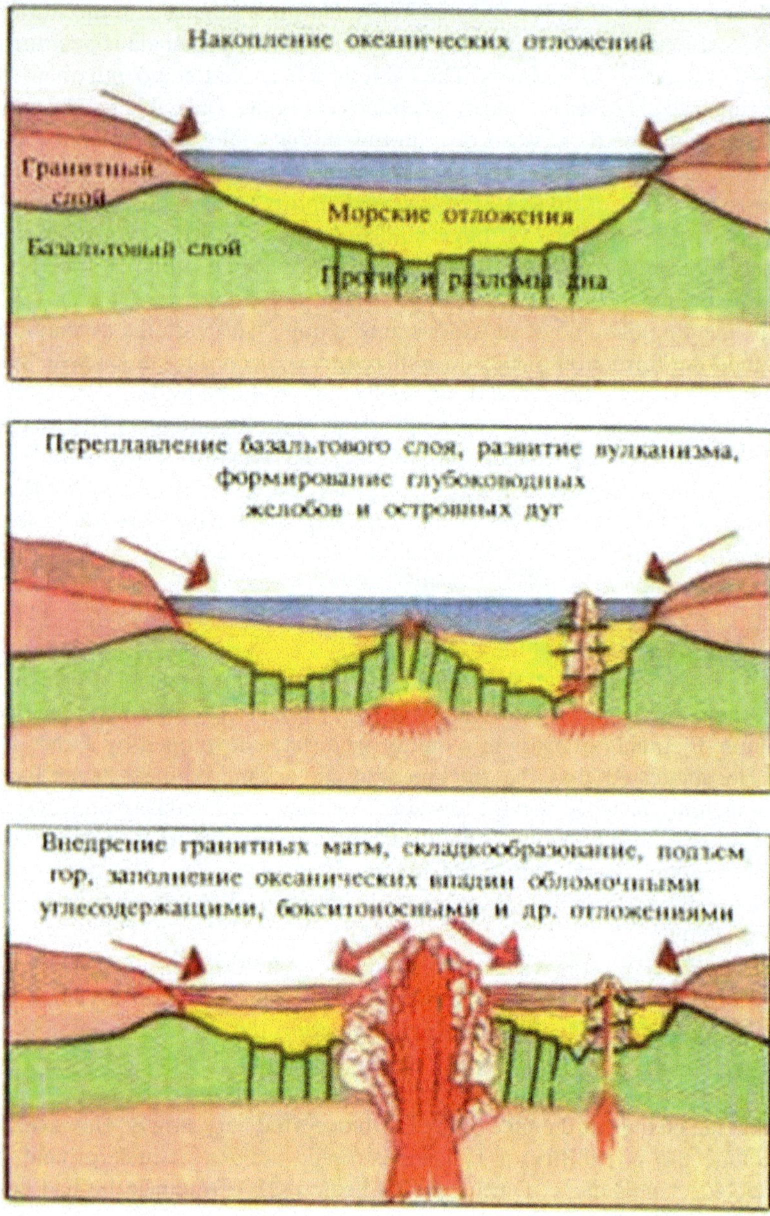

Fig. 4.56 Diagram of a break of the earth crust, the accumulation of sediments and the formation of mountains. The diagram is taken from the Internet (geographical dictionary www.ecosystema.ru)

Darwin wrote... *hollows would be actually left, into which the unsupported masses might be precipitated with the violence of an explosion...* [4, p. 74]. Let us note also that under tectonic plates there are huge thermal sources ('plumes', see Fig. 3.1). They can sometimes reach the earth crust and lift up it on a considerable height (kilometre and more), and subsequently the crust drops down (Fig. 1.4). That process can repeat with the period about a million years. Climate changes greatly influence geography too. Thus, the land surface is dynamic at all times. This corresponds to Darwin's views, which were stated above.

4.7 Dynamic Instability and ... *A Vorticose Movement* ... Within the Surface Layers

Perhaps, Darwin was the first who discussed the instability of upper layers of the Earth. The dynamic instability of an interface of two liquids is a well-known phenomenon, which often leads to a generation of vortices. It is known less in the case of solid media. Therefore it is important for us to note that Darwin, analyzing the motion of the Earth surface during an earthquake, reflected on the possibility the of vortex formation under the upper layers.... *Some square ornaments on the coping of these same walls were moved by the earthquake into a diagonal position. The buttresses of the church of La Merced, at Valparaiso, and some heavy pieces of furniture in the rooms, were similarly affected by the shock of 1822.. .Mr. Lyell... has also given a drawing of an obelisk in Calabria, of which the separate stones were partially turned round. In these instances, the displacement at first appears to be owing to a vorticose movement beneath each point thus affected; but such can hardly be the case...* [1, p. 376]. Darwin's doubt is quite understandable. But the occurrence of vortex motion under soil is apparently quite probable. The initial stages of this motion are illustrated in Fig. 4.57 [81].

The dynamical instabilities in fluid mechanics are responsible for a variety of important common phenomena. A simple example of instabilities in fluid mechanics is Kelvin-Helmholtz instability, where a flat interface between two fluids flowing one past the other at different velocities is unstable [82]. In granular media dynamical instabilities have just begun to be discovered [83]. Some results of this instability are shown in Fig. 4.57. It is visible that the wave arising on an interface of two layers of the materials corresponds to the evolution of harmonic wave into the catastrophic wave demonstrated in Fig. 4.58.

Probably, the generation of folds of layers near a surface and in depths of the Earth, which are well-known to geologists, can be explained by instability of seismic and heat phenomena within the Earth. And now the plumes, huge vertical mushroom-like waves and vortices within of the Earth which are caused by thermal flows, are being actively studied. They explain the motion of tectonic plates. Probably, Darwin's thought did not extend to such depths. However, his idea

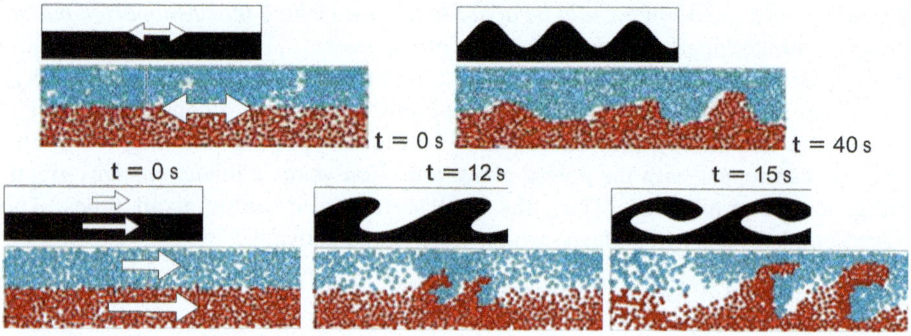

Fig. 4.57 Evolution of harmonic waves forced on the interface of two granular materials into breaking waves [81]

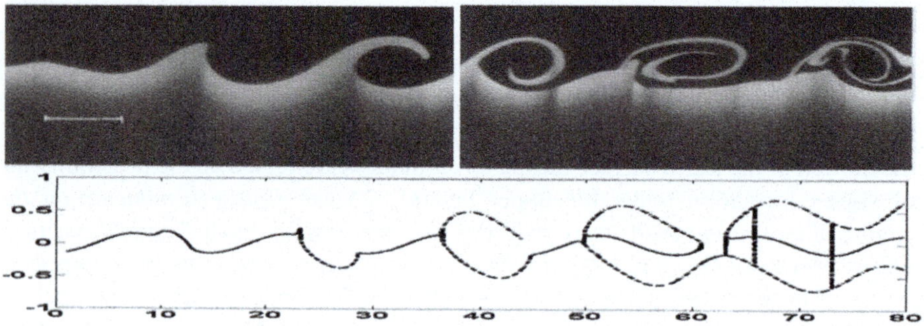

Fig. 4.58 Transresonant evolution of waves into vortices. The experimental data [82, 9] and their modelling [9] (cf. Fig. 4.57)

about the possibility of vortex motions under the Earth surface almost 200 years ago should be considered as amazing.

We underline that ... *a vorticose movement*...can be forced also by vertical earthquakes. In the experiments [84] a tank with glass beads was driven vertically by a shaker. Figure 4.59 shows the velocity field of the granular materials in the bed, under vibrations of 12.5 Hz and 4 g. Two symmetric convection rolls are clearly shown in the figure. The granular materials move upward in the central part and move downward along the side walls of the container. The convection- roll phenomena are localized near the surface. The upper region of a vibrating granular bed is more fluidized and excited than the lower region of a vibrating granular bed.

The growing interest in the study of rotational ground motions induced by earthquakes has led to the development of a new field referred to as rotational seismology. A good review of results in this field was presented in [85].

In the last two sections we have briefly discussed some questions connected with instability of upper layers of the Earth.

Fig. 4.59 The velocity field of glass beads under vertical excitation [84]

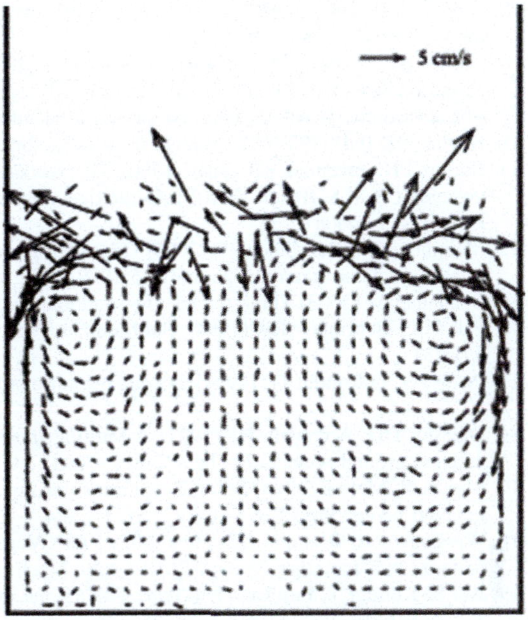

Final remark The aim of this Chapter was to illustrate some of the fundamental problems which excited Darwin, during the days following the 1835 Chile earthquake.

References

1. Darwin C (1839) Journal of researches into the geology and natural history of the various countries visited by H.M.S. Beagle, under the command of Captain FitzRoy, R.N. from 1832 to 1836. Henry Colburn, London
2. Darwin C (1890) Journal of researches into the natural history and geology of the countries visited during the voyage of H.M.S. Beagle around the world, under the command of Captain FitzRoy, R N. T Nelson and Sons, London
3. FitzRoy R (1839) Narrative of the surveying voyages of His Majesty's Ships Adventure and Beagle, between the years 1826 and 1836, describing their examination of the southern shores of South America, and the Beagle's circumnavigation of the globe. In: Proceedings of the second expedition, 1831–1836, under the command of Captain Robert Fitz-Roy R.N., vol 2. Henry Colburn, London
4. Darwin C (1840) On the connexion of certain volcanic phenomena in South America; and on the formation of mountain chains and volcanoes, as the effect of the same power by which continents are elevated. (Read March 7, 1838) In: Barrett PH (ed) (1977) The Collected Papers of Charles Darwin. The University of Chicago Press, Chicago
5. Reiter L (1990) Earthquake hazard analysis. Columbia University Press, New York
6. Robinson A (1993) Earth shock. Thames and Hudson, London
7. Seed HB, Martin PP, Lysmer H (1976) Pore water pressure changes during soil liquefaction. J Geotech Engng ASCE 102:323–346

8. Idriss IM, Boulanger RW (2008) Soil liquefaction during earthquakes. Earthquake Engineering Research Institute, Oakland
9. Galiev ShU (2011) Геофизические Сообщения Чарльза Дарвина как Модели Теории Катастрофических Волн. (Charles Darwin's Geophysical Reports as Models of the Theory of Catastrophic Waves). Centre of Modern Education, Moscow (in Russian)
10. Keynes RD (ed) (1979) The Beagle record. Cambridge University Press, Cambridge
11. Barrett PH, Freeman RB (eds) (1986) The works of Charles Darwin. The geology of the voyage of H.M.S. Beagle. Part III: Geological observations on South America, vol 9. William Pickering, London
12. Majmudar TS, Behringer RP (2005) Contact force measurements and stress-induced anisotropy in granular materials. Nature 435:1079–1082
13. Buchanan M (2003) Think outside the sandbox. Nature 425:556–557
14. Pak HK, Behringer RP (1994) Bubbling in vertically vibrated granular material. Nature 371:231
15. Pease JW, O'Rourke TD (1997) Seismic response of liquefaction sites. J Geotech Engrg ASCE 123:37–45
16. Oglesby DD, Archuleta RJ (1997) A faulting model for the 1992 Petrolia earthquake: can extreme ground acceleration be a source effect? J Geophys Res 102(B6):877–897
17. Mujica N, Melo F (1998) Solid–liquid transition and hydrodynamic surface waves in vibrated granular layers. Phys Rev Let 80(2):5121–5124
18. Wassgren CR, Brennen CE, Hunt ML (1996) Vertical vibration of a deep bed of granular material in a container. Trans ASME J Appl Mech 63:712–719
19. Aoi S, Kunugi T, Fujiwara H (2008) Trampoline effect in extreme ground motion. Science 322:727–730
20. Graizer V (2009) Low-velocity zone and topography as a source of site amplification effect on Tarzana hill, California. Soil Dynam Earthquake Eng 29:324–332
21. Spudich P, Hellweg M, Lee WHK (1996) Directional topographic site response at Tarzana observed in aftershocks of the 1994 Northridge, California, earthquake: implications for mainshock motions. Bull Seis Soc Am 86(1B):S193–S208
22. Rial JA (1996) The anomalous seismic response at the Tarzana hill site during the Northridge 1994 Southern California earthquake: a resonant sliding block. Bull Seism Soc Am 86 (6):1714–1723
23. Galiev SU (1999) Topographic effect in a Faraday experiment. J Phys A: Math Gen 32:6963–7000
24. Galiev ShU (1997) Resonant oscillations governed by the Boussinesq equation with damping. In: Proceedings of 5th international congress on sound and vibration, Adelaide, 1785–1796
25. Galiev SU (1998) Topographic amplification of vertical-induced resonant waves in basins. Advances Hydrosci Eng 3:179–182
26. Galiev SU (2003) The theory of non-linear transresonant wave phenomena and an examination of Charles Darwin's earthquake reports. Geophys J Inter 154:300–354
27. O'Connell DRH (2008) Assessing ground shaking. Science 322:686–687
28. Anderson JG (2007) Physical processes that control strong ground motion. In: Schubert G (ed) Treatise on geophysics. Elsevier, Amsterdam/Boston
29. Johnson W (1972) Impact strength of materials. Edward Arnold, London
30. Kolsky H (1953) Stress waves in solids. Clarendon, Oxford
31. Rinehart JS (1960) The role of stress waves in the comminution of brittle, rocklike materials. In: Davids H (ed) Stress wave propagation in materials. Interscience, New York
32. Whitney JW (2009) Surface-feature indications that Yucca Mountain has not experienced extreme ground motions in the past. U.S. Department of Energy, Las Vegas
33. Fritz H, Kalligeris N (2008) Ancestral heritage saves tribes during 1 April 2007 Solomon Island tsunami. Geophys Res Lett 35, L01607
34. Taylor FW, Briggs RW, Frohlich C, Brown A, Hornbach M, Parabatu AK, Meltzner AJ, Billy D (2008) Rupture across arc segment and plate boundaries in the 1 April 2007 Solomons earthquake. Nat Geosci 1:253–257

35. Sills GC, Wheeler SJ, Thomas SD, Gardner TN (1991) Behaviour of offshore soils containing gas bubbles. Geotechnique 41(2):227–241
36. Davison C (1936) Great earthquakes. Thomas Murby & Co, London
37. Linde AT, Sacks IS (1998) Triggering of volcanic eruptions. Nature 395:888–890
38. Johnson JB, Lees JM, Gerst A, Sahagian D, Varley N (2008) Long-period earthquakes and co-eruptive dome inflation seen with particle image velocimetry. Nature 456:377–381
39. Kawakatsu H, Yamamoto M (2007) Volcano seismology. In: Schubert G (ed) Treatise on geophysics. Elsevier, Amsterdam/Boston
40. Watt SFL, Pyle DM, Mather TA (2009) The influence of great earthquakes on volcanic eruption rate along the Chilean subduction zone. Earth and Plan Sci Lett 277:399–407
41. Sturtevant B, Kanamori H, Brodsky EE (1996) Seismic triggering by rectified diffusion in geothermal systems. J Geophys Res 101(25):269–282
42. Brodsky EE, Sturtevant B, Kanamori H (1998) Earthquakes, volcanoes, and rectified diffusion. J Geophys Res 103(23):827–838
43. Self S, Wilson L, Nairn IA (1997) Vulcanian eruption mechanisms. Nature 277:440–443
44. Gomberg J, Reasenberg PA, Bodin P, Harris RA (2001) Earthquake triggering by seismic waves following the Landers and Hector Mine earthquakes. Nature 411:462–466
45. West M, Sanchez JJ, McNutt SR (2005) Periodical triggered seismicity at Mount Wrangell, Alaska, after the Sumatra earthquake. Science 308:1144–1146
46. Brodsky EE, Prejean SG (2005) New constraints on mechanisms of remotely triggered seismicity at Long Valey Caldera. J Geophys Res 110, B04302
47. Chui G (2009) Shaking up earthquake theory. Nature 461:870–872
48. Lea Scharff (2012) Eruption Dynamics of Vulcanian and Sub-Plinian Volcanoes: From the Generation of Pulses to the Formation of Clouds. Zur Erlangung des Doktorgrades der Naturwissenschaften im Department Geowissenschaften der Universität Hamburg
49. Anilkumar AV, Sparks RSJ, Sturtevant B (1993) Geological implications and applications of high-velocity two-phase flow experiments. J Volcanol Geotherm Res 56:145–160
50. Alidibirov M, Dingwell DB (1996) Magma fragmentation by rapid decompression. Nature 380:146–148
51. Mader HM (1998) Conduit flow and fragmentation. In: Gilbert JS, Sparks RSJ (eds) The physics of explosive volcanic eruptions, vol 145, Geological society, special publications. Geological Society, London, pp 51–71
52. Morrissey MM, Mastin LG (2000) Vulcanian eruption. In: Sigurdsson H (ed) Encyclopedia of volcanoes. Academic, Waltham
53. Clarke AB, Voight B, Neri A, Macedonio G (2002) Transient dynamics of vulcanian explosions and column collapse. Nature 415:897–901
54. Dobran F, Neri A, Macedonio G (1993) Numerical simulations of collapsing volcanic columns. J Geophys Res 98:4231–4259
55. Gonnermann HM, Manga M (2007) The fluid mechanics inside a volcano. Annu Rev Fluid Mech 39:321–356
56. Turcotte DL, Ockendon H, Ockendon JR, Cowley SJ (1990) A mathematical model of vulcanian eruptions. Geophys J Int 103:211–217
57. Voight B, Sparks RSJ, Miller AD, Stewart RC, Hoblitt RP, Clarke A, Ewart J, Aspinall WP, Baptie B, Calder ES, Cole P, Druitt TH, Hartford C, Herd RA, Jackson P, Lejeune AM, Lockhart AB, Loughlin SC, Luckett R, Lynch L (1999) Magma flow instability and cyclic activity at Soufriere Hills Volcano, Montserrat. Science 283:1138–1142
58. James MR, Lane SJ, Chouet BA (2008) Gas slug ascent through changes in conduit diameter: Laboratory insights into a volcano-seismic source process in low-viscosity magma. J Geophys Res 111, B05201
59. Kedrinskii VK (2008) Gas-dynamic signs of explosive eruptions of volcanoes 1. Hydrodynamic analogs of the pre-explosion state of volcanoes, dynamics of three-phase magma state in decompression waves. J Applied Mech Techn Physics 49(6):891–898
60. Prejean SG, Haney MM (2014) Shaking up volcanoes. Science 345:39

61. Latter JH (1981) Volcanic earthquakes, and their relationship to eruptions at Ruapehu and Ngauruhoe volcanoes. J Volcanol Geotherm Res 9:293–309
62. Anilkumar AV (1989) Experimental studies of high-speed dense dusty gases. Dissertation, California Institute of Technology
63. Pudasaini SP, Hutter K (2007) Avalanche dynamics. Springer, Berlin
64. Umbanhowar PB, Melo F, Swinney HL (1996) Localized excitations in a vertically vibrated granular layer. Nature 382:793–796
65. Lioubashevski O, Hamiel Y, Agnon A, Reches Z, Fineberg J (1999) Oscillons and propagating solitary waves in a vertically vibrated colloidal suspension. Phys Rev Lett 83:3190–3193
66. Royer JR, Corwin EI, Flior A, Cordero M-L, Rivers ML, Eng PJ, Jaeger HM (2005) Formation of granular jets observed by high-speed X-ray radiography. Nature Phys 1:164–167
67. Lohse D, Rauhè R, Bergmann R, van der Meer D (2004) Creating a dry variety of quicksand. Nature 432:689–690
68. Bourne NK, Field JE (1992) Shock-induced collapse of single cavities in liquids. J Fluid Mech 244:225–240
69. Lavrentev MA, Shabat BV (1977) Гидродинамические Проблемы и их Математические Модели (Hydrodynamic Problems and Their Mathematical Models). Nauka, Moscow (in Russian)
70. Eggers J, Villermaux E (2008) Physics of liquid jets. Rep Prog Phys 72:036601
71. Antkowiak A, Bremond N, Le Dizes S, Villermaux E (2007) Short-term dynamics of a density interface following an impact. J Fluid Mech 577:241–250
72. Cole RH (1948) Underwater explosions. Princeton University Press, Princeton
73. Weiss R, Wunnemann K (2007) Large waves caused by oceanic impacts of meteorites. In: Kundu A (ed) Tsunami and nonlinear waves. Springer, Berlin
74. Galiev ShU (2009) Modelling of Charles Darwin's earthquake reports as catastrophic wave phenomena. researchspace.auckland.ac.nz/handle/2292/4474
75. Pedersen H, Brun BL, Harzfeld D, Campillo M, Bard PY (1994) Ground motion amplitude across ridges. Bull Seis Soc Am 84(6):1786–1800
76. Jaeger HM, Nagel SR, Behringer RP (1996) Granular solids, liquids, and gases. Rev Mod Phys 68:1259–1273
77. Clèment E, Vanel L, Rajchenbach J, Duran J (1996) Pattern formation in a vibrated granular layer. Phys Rev E 53:2972–2975
78. Hart IB (1961) The world of Leonardo da Vinci. Macdonald, London
79. Vita-Finzi C (1992) Earthquakes. In: Bourriau J (ed) The Darwin college lectures. Understanding catastrophe. Cambridge University Press, Cambridge
80. John Wiester (1998) Paradigm shifts in geology and biology: geosynclinal theory and plate tectonics; Darwinism and Intelligent Design. From Perspectives on Science and Christian Faith. December
81. Ciamara MP, Coniglio A, Nicodemi M (2005) Shear instability in granular mixture. Phys Rev Lett 94(18):188001
82. Schowalter DG, van Atta CW, Lasheras JC (1994) A study of streamwise vortex structure in a stratified shear layer. J Fluid Mech 281:247–291
83. Karsten RH, Swaters GE (2000) Nonlinear effects in two-layer large-amplitude geostrophic dynamics. part 2. The weak-beta case. J Fluid Mech 412:161–196
84. Tai CH, Hsiau SS (2004) Dynamic behaviors of powders in a vibrating bed. Powder Technol 139:221–232
85. Castellani A (2012) Study of rotational ground motion in the near-field region. Dissertation, Politecnico di Milano

Chapter 5
Darwin's Reports on Catastrophic Natural Phenomena and Modern Science: Seaquake-Induced Waves, Atomization and Cavitation

*...It is mere rubbish thinking, at present, of origin of life; one might as well think of origin of matter... (**Darwin.** A letter to Joseph Dalton Hooker on March 29, 1863.)*

Darwin reported that ... *From the great wave not immediately following the earthquake, but sometimes after the interval of even half-an-hour, ...it appears that the wave first rises in the offing; and as this is of general occurrence, the cause must be general. I suspect we must look to the line where the less disturbed waters of the deep ocean join the water nearer the coast, which has partaken of the movements of the land, as the place where the great wave is first generated. It would also appear that the wave is larger or smaller, according to the extent of shoal water which has been agitated together with the bottom on which it rested...* [1, p. 374]. These words, and also the material given in the Chap. 3 show that Darwin had described the formation and the coastal dynamics of ocean waves caused by the earthquake. Obviously, he was the first who had given a qualitative, but scientific, analysis of this phenomenon, and had emphasized the complex character of the coastal development of tsunamis. Describing results of the earthquake Darwin wrote also ... *two explosions, one like a column of smoke, and another like the blowing of a great whale, were seen in the bay of Concepcion. The water also appeared every where to be boiling; and it 'became black, and exhaled a most disagreeable sulphureous smell'.... The two great explosions in the first case must no doubt be connected with deep-seated changes; but the bubbling water, its black colour and fetid smell, the usual concomitants of a severe earthquake, may, I think, be attributed to the disturbance of mud containing organic matter in decay* ... [2, p. 374]. Thus, Darwin had described the behaviour of the shallow sea and the bottom mud during a seaquake. The explosions are results of the passing up of large cavitation zones (bubbles) through the sea/atmosphere surface. These zones were formed because of vertical oscillations of the sea bed. Surface waves and the boiling may be also explained by the seaquake. Apparently, bubbles and surface waves may also appear in the mud because of the underwater earthquake.

Darwin drew very complex pictures of the behaviour of the sea during severe seaquakes, when different kinds of extreme waves may form in the water, with some of them evolving into vortices. Generally speaking, extreme waves similar to

Sh.U. Galiev, *Darwin, Geodynamics and Extreme Waves*,
DOI 10.1007/978-3-319-16994-1_5

those in the ocean might exist in different media and physical fields. In the last sections of the Chap. 7 we consider waves and vortices in scalar fields which resemble excited by seaquakes.

5.1 Darwin's Description of Tsunamis Generated by Coastal Earthquakes

It is known that very long waves can exist on the surface of the ocean. In particular, these are tsunamis formed by underwater earthquakes or by seabed mudslides, and much less often, by volcanic eruptions or by meteorites which strike the ocean surface. The amplitude of a tsunami in open ocean is of an order of 1 m. The wave profile has a very weak slope relative to the ocean surface, and so a tsunami cannot be noticed from ships. However, when tsunamis enter into shallow coastal waters they can be transformed into catastrophic waves, which can reach heights of 40 m and have an overturning front.

In this section we study effects of a coastal depth variation on the evolution of a tsunami.

According to Darwin a tsunami is generated by motions of the seabed caused by earthquake. He wrote ... *The primary shock is caused by a violent rending of the strata, which seems generally to occur at the bottom of the neighbouring sea* ... [3, p. 71]. Let us consider a tsunami generated in the open ocean. How does the tsunami evolve near the shore? In particular, we are interested in the influence of the coastal seabed on the formation of a wall of water at the tsunami front, and the occurrence of an ebb ahead of the tsunami. A description of these phenomena was given by Charles Darwin [2, p. 377]. Darwin emphasized the dependence of earthquake-induced waves upon the form of the coast and the coastal depth:... *and lastly, of its size being modified (as appears to be the case) by the form of the neighbouring coast. For instance, Talcuhano and Callao are situated at the head of great shoaling bays, and they have always suffered from this phenomenon; whereas, the town of Valparaiso, which is seated close on the border of a profound ocean, though shaken by the severest earthquake, has never been overwhelmed by one of these terrific deluges*...[2, p. 378]. A recent instance of these phenomena happened during the catastrophic earthquake on March 11th 2011 in Japan.

The question about the appearance of the ebb ahead of the tsunami was not clear at the time of Darwin. Darwin said about this question ... *During most earthquakes, and especially during those on the west coast of America, it is certain that the first movement of the waters has been a retirement. Some authors have attempted to explain this by supposing that the water retains its level, whilst the land oscillates upwards; but surely the water close to the land, even on a rather steep coast, would partake of the motion of the bottom*... [1, pp. 371–372]. Darwin wrote about an analogy between the ebb because of tsunami and of a water ebb of 2 or 3 ft because of an overturning wave from a steamboat [2, p. 378]. In the last case we cannot

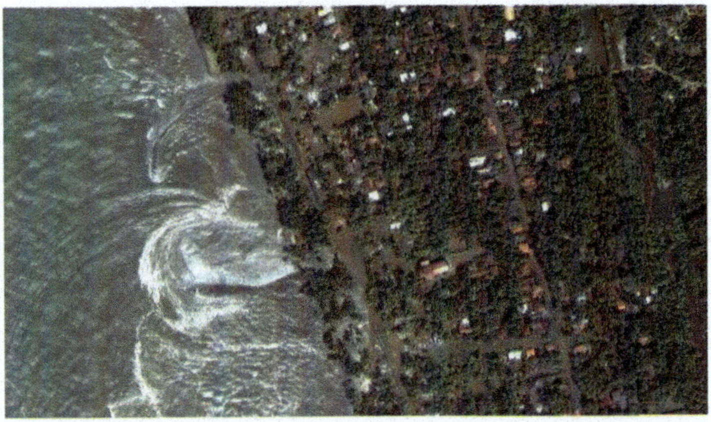

Fig. 5.1 Evolution of the 2004 Indian Ocean tsunami near the coast of Sri Lanka: the form changes from the wall of water (*low part*), through the breaker to the localized jet (*upper part*) [6]

speak about any lifting of the shore land. In our opinion the phenomena described by Darwin are connected with nonlinear coastal dynamics of a tsunami. Indeed, he reported that . . . *the whole body of the sea retires from the coast, and then returns in great waves of overwhelming force* . . . [2, p. 378].

Now these phenomena are well-documented, and in particular the results of the 2004 Sumatra earthquake are well-known. To document the 2004 tsunami many scientists from all over the world visited the affected coasts [4, 5]. The measured tsunami heights in Sumatra Island were sometimes larger than 20 m with the maximum above 30 m (Fig. 5.1) [6].

Sometimes a tsunami with a front depression may form. They were observed and documented during the Great Alaskan Earthquake of 1964 at Kodiak Island, the 1992 Nicaraguan earthquake [7, 8], the 1992 earthquake near the northeastern region of Flores Island, in Indonesia, the 2004 Sumatra earthquake [9], and the 2007 Solomon Islands earthquake [10, 11]. In these cases the waves were caused by earthquakes near the coastline. As a result, the generated wave does not have sufficient distance to evolve into a solitary elevation wave, or a series of solitary waves.

In this section, we, first, study the effect of the coastal depth on the generation of the water wall. Then, it is shown that the front depression of the tsunami can be strongly increased near the coastline.

5.1.1 Effect of the Coast Bottom on the Generation of a Catastrophic Tsunami

The tsunami height is measured from the trough to the peak, whereas the amplitude is measured from the smooth sea level to the peak, either positive or negative. For example, the front amplitude of the 2004 tsunami in Indian Ocean was positive near Maldives as shown in Fig. 5.2.

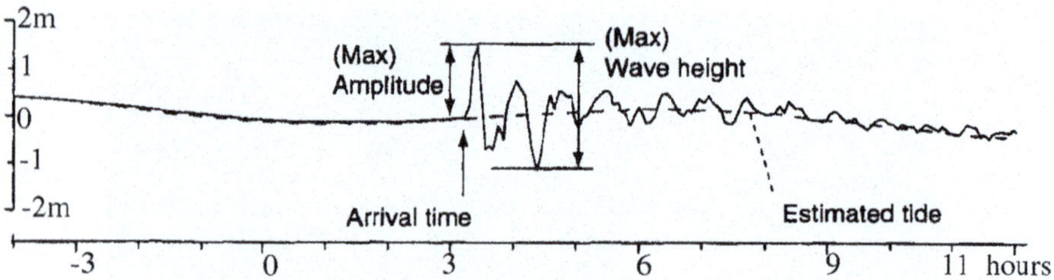

Fig. 5.2 Tsunami waveforms recorded at Male, Maldives (the 2004 Sumatra earthquake) [9, 12]

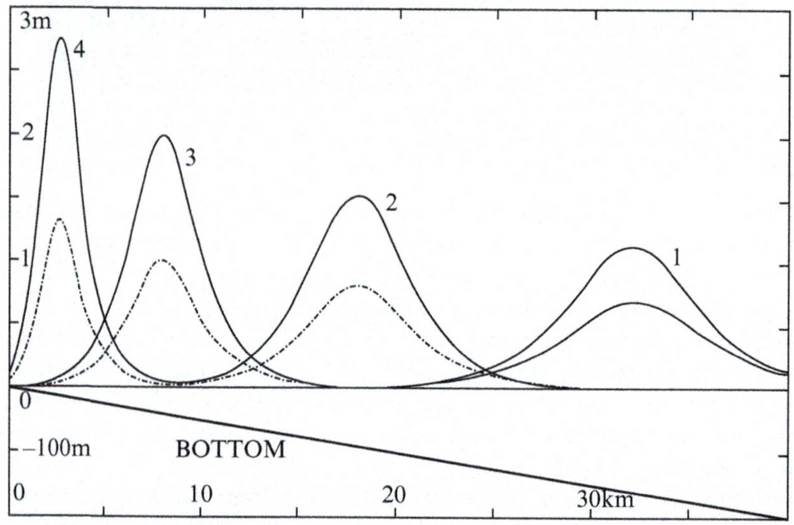

Fig. 5.3 Evolution of a tsunami far from the coast. The profiles 1, 2, 3, 4, and 5 are calculated for t=−800 (curves *1*), −600 (*2*), −400 (*3*), −230 s (*4*), respectively. There is a strong amplification of the wave near the coast [17, 18]

The dynamics of nonlinear ocean-surface waves is complex and some aspects are still not well understood, although the numerical solutions for many three-dimensional problems are now available [5, 12–14]. Therefore up to this moment the shallow-water wave equations [12–15] and their analytic solutions are used for a tsunami evolution description. Despite their severe limitations, these equations repeatedly produced realistic predictions for some nonlinear ocean waves and laboratory data [9]. We will use the shallow-water wave equations (see Sects. 7.1 and 7.5) – the solutions of these equations should describe the coastal tsunami evolution since the tsunami length is much larger than the coastal depth.

The coastal evolution of a tsunami, as described by Darwin, was analytically studied in [16–18]. Here we follow [18]. The results are shown in Figs. 5.3, 5.4 and 5.5. The solid lines determine the results of the nonlinear theory, the broken lines

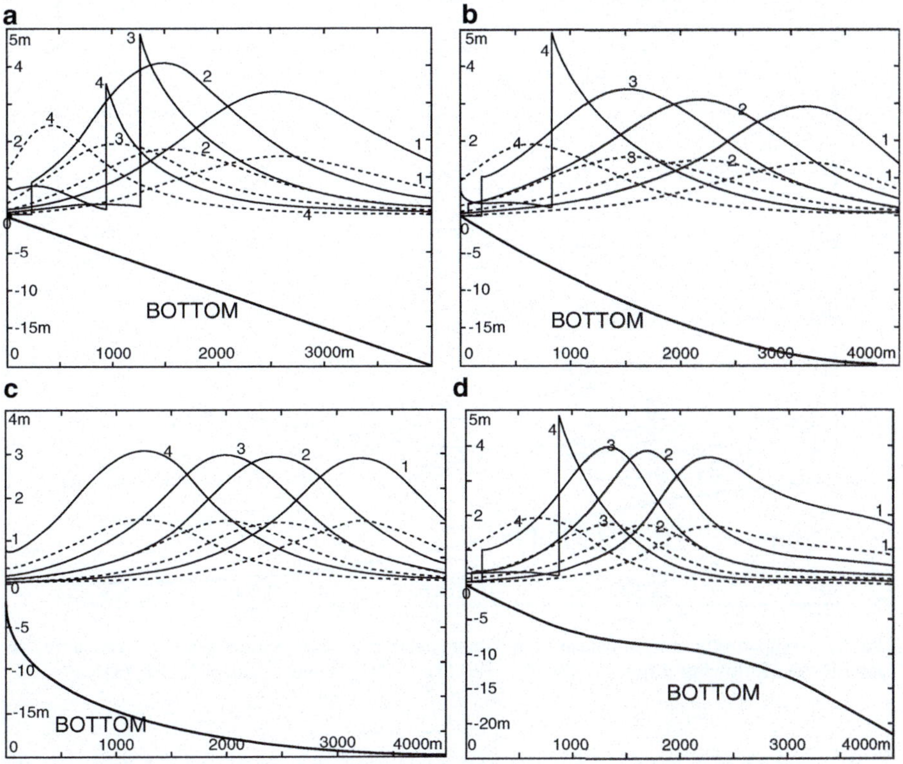

Fig. 5.4 Effect of the seabed and the depth variation on the wave evolution: (**a**) (linear slope of the *bottom*), (**b–d**) (nonlinear slope of the *bottom*). The curves are found for the following times: −230 (*1*), −180 (*2*), −150 (*3*) and −100 s (*4*) [17, 18]

determine the results of the linear theory. Far from the coast the interrupted lines describe approximately the initial wave coming from the deep ocean (Fig. 5.3). The initial wave (tsunami) is simulated as a solitary wave. The reader can see that the length of the tsunami is being reduced. The maximum height of the wave is about 3 m, or about three times of the initial wave height. On the whole, this result agrees with some numerical calculations, which gave the amplification as nearly 3 [19]. Of course, this result depends on the shape and the slope of the seabed and on the initial amplitude and the wavelength of a tsunami.

It is shown that the coastal evolution strongly depends on the profile of the seabed and the distance from the coastline (Fig. 5.4). Far from the shore the wave surface is smooth and the wave is rather long; but the wave profile begins to change quickly when the coastal water begins to be shallow. In this case, a steep front of the tsunami may be formed (Fig. 5.4a, b, d). However, the tsunami, practically, does not change if the coastal water is deep (Fig. 5.4c).

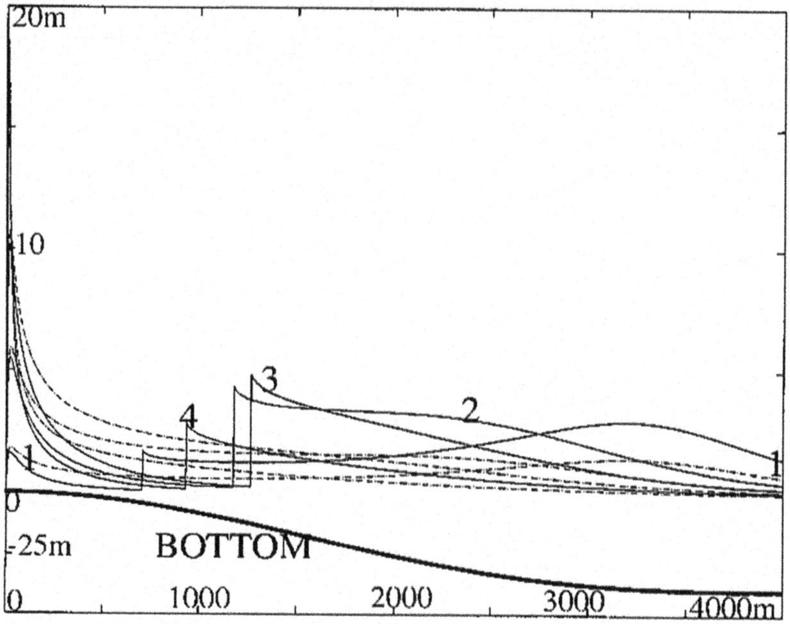

Fig. 5.5 Catastrophic amplification of a tsunami above a very shallow seabed. The curves are found for the following times: −230 (*1*), −180 (*2*), −150 (*3*) and −100 s (*4*) [17, 18]

We studied also effects of very shallow seabed and small coastal depth on the tsunami evolution. The results of the calculations are presented in Fig. 5.5. In this case, the amplification of tsunami may be very large – up to 20 m.

These calculations supported Darwin's tsunami reports. Curves a, b and d (Fig. 5.4) and Fig. 5.5 illustrate the catastrophic effect when a water-wall is generated. The tsunami with three peaks may be formed near the coast (see Figs. 5.5 and 3.9).

As a whole, the change of a coastal depth defines the coastal evolution of a tsunami. In particular, if the depth is small and changes smoothly, the tsunami evolves into the catastrophic wave. In the case of the large coastal depth a threat of such wave is absent. Data of the calculations agree with Darwin's reports about tsunami. Curves a, b and d (Fig. 5.4) and Fig. 5.5 illustrate the catastrophic effects of a formation of the water wall. In particular, near the coast the wave having two or three peaks can be formed (see Captain FitzRoy's description in Sect. 3.3.6).

Results of the linear and nonlinear calculations differ essentially near the coast. There, according to the nonlinear theory, the water level does not increase up to the arrival of the steep front of the wave.

5.1.2 *Effect of the Coast Bottom on the Ocean Ebb and the Steep Front of a Tsunami*

Sometimes tsunamis with a leading depression were observed and documented (Fig. 5.6).

Perhaps, Darwin was the first who published the report about the appearance of an ebb ahead of a tsunami. To simulate this process, we change the form of the initial solitary wave. We assume that a tsunami wave has a central elevation and a depression at the front and at the back of the elevation (see broken lines 1 in Fig. 5.7, which qualitatively describe the tsunami form). According to [8], we shall call this wave an N-wave. Our purpose is to simulate a development of the depression ahead of the tsunami elevation. We studied this process according to [18]. Some results of the calculations are presented in Fig. 5.7. The continuous curves were calculated according to a nonlinear theory, and the broken curves were calculated according to the linear relations.

According to Fig. 5.7, the ebb forms a deep depression with steep walls, which bares the seabed on a distance up to 500 m from the shore. It is possible to speak about the depression wave, which is formed at the edge of the water. Then the steep ebb front starts to move into the deep ocean. The depression amplitude is comparable with the elevation amplitude.

Thus, the coastal evolution of a tsunami depends on the coastal depth. The amplitude of the elevation may be strongly increased. At the same time a water wall may be formed. The steep (discontinuous) front of a tsunami can be generated in shallow coastal water. The water level reduces ahead of the front (Figs. 5.4 and 5.5), or the ebb can appear there (Fig. 5.7). Then this front begins to move from the coast – into the ocean (Figs. 5.4, 5.5 and 5.7). This direction is opposite to the motion of the whole wave. As a result, a tsunami begins to move more slowly. This process explains the catastrophic effect of a tsunami, when a water wall appears instantly instead of the slow elevation of the ocean level, as in the case of deep

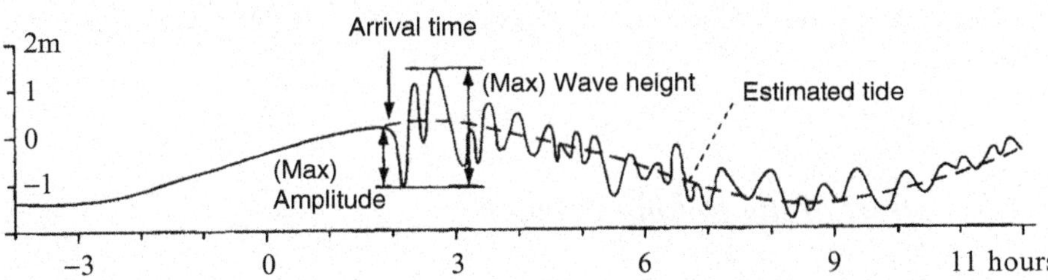

Fig. 5.6 Tsunami waveforms from the 2004 Sumatra earthquake, recorded at Phuket, Thailand [9, 12]

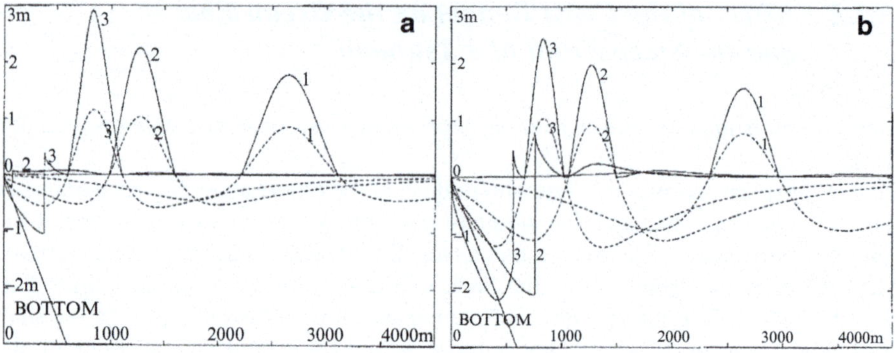

Fig. 5.7 Two examples (**a**, **b**) of the coastal evolution of the N-wave and the formation of the ebb ahead of a tsunami. The curves are found for the following different moments of time: −230 (*1*), −160 (*2*) and −130 s (*3*) [17, 18]

coastal water [16–18]. On the whole, the conclusions agree with Darwin's reports [2, pp. 377–378] and his following comments ... *Shortly after the shock, a great wave was seen from the distance of three or four miles approaching in the middle of the bay with a smooth outline;... At the head of the bay it broke in a fearful line of white breakers, which rushed up to a height of twenty-three vertical feet above the highest spring-tides. Their force must have been prodigious; ... two large vessels anchored near together were whirled about, and their cables were thrice wound round each other: though anchored at a depth of thirty-six feet they were for some minutes aground. The great wave must have travelled slowly, for the inhabitants of Talcahuano had time to run up the hills behind the town; and some sailors pulled out seaward, trusting successfully to their boat riding securely over the swell if they could reach it before it broke* ... [1, pp. 369–370].

Of course, this complex evolution of the tsunami front near the coastline is very sensitive to any variation of parameters. In particular, variations of the amplitude of the initial wave, the seabed slope, or the shape of the coast can change the described process. The tsunami wave can lose its stability at some points of the coastal zone. As a result, a tsunami can have different forms in different points of the coastal zone (Fig. 3.9).

5.1.3 *Effect of the Bottom Friction*

The bottom friction may have a large effect on the coastal wave evolution (see, also, Sect. 7.5). Perhaps, this friction participates in the formation of the 'water wall' of tsunamis (Fig. 5.8). The splash moves ahead of this wall.

Fig. 5.8 Killing tsunami [18]

Here we investigate an influence of the bottom friction, as illustrated by Fig. 5.9. The curves a and b are calculated for the linear changing depth and two values of the bottom friction. These curves correspond to curves Fig. 5.4a, but calculated without the friction.

The curves c, d, e (Fig. 5.10) are drawn for a curvilinear bottom and three values of the bottom friction. These curves correspond to Fig. 5.5, constructed without the friction. In Figs. 5.9 and 5.10 the continuous curves are calculated according to a nonlinear theory, and the broken curves are calculated according to the linear relations.

According to our calculations, the coastal evolution of a tsunami may be accompanied by the occurrence of a steep front and a strong change of the amplitude and the profile of the wave. A tsunami with one top can evolve into a wave with 2 or 3 tops near a coast (see curves c and d in Fig. 5.10). This evolution agrees with Captain FitzRoy's description 'About half an hour after the shock, . . .– an enormous wave was seen forcing its way through the western passage which separates Quiriquina Island from the mainland. . . .It broke over dashed along, . . . and then rushed back. . .. In a few minutes, ...a second great wave was seen approaching, with more noise and impetuosity than the first; but though this was more powerful, its effects were not so considerable – simply because there was less to destroy. . . .After some minutes of awful suspense, a third enormous swell was seen between Quiriquina and the mainland, apparently larger than either of the two former.' (cf. Sect. 3.3.6). According to our calculations the bottom friction changes the amplitude of the wave, but does not change the character of the formation of the steep front and the law of its movement. Thus, tsunami forms are not connected directly with bottom friction, but are related to the seabed relief and nonlinear effects.

Fig. 5.9 The effect of *bottom* friction on the coastal evolution of a tsunami (linear slope of the *bottom*) [18]

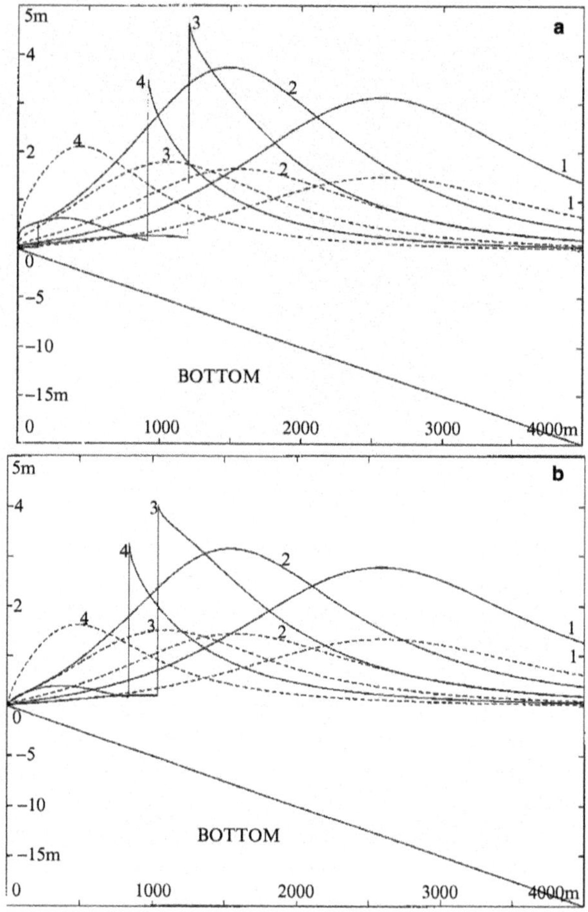

A water splash may appear ahead the tsunami during its evolution, and so splash is a harbinger of the tsunami (Figs. 5.7 and 5.9). Perhaps, this splash corresponds to the splash ahead the tsunami in Fig. 5.8.

Another important result definitely follows from the calculations presented in [16–18]. This result is connected with the well-known opinion that the use of numerical methods eliminates all problems of modelling of a tsunami. It is possible to agree with this opinion in the case of open ocean, where the amplitude of the wave is small in comparison with the depth and nonlinear effects are almost absent. However, near to a coast, if the depth is small and changes very smoothly, there is a resonance zone, where the nonlinearity increases very strongly. There the numerical analysis encounters the complex problem of solution of a large system of highly-nonlinear algebraic equations.

Each of them can have a few very different solutions (see, for example, solutions of the equation 7.70). The number of absolutely different solutions strongly

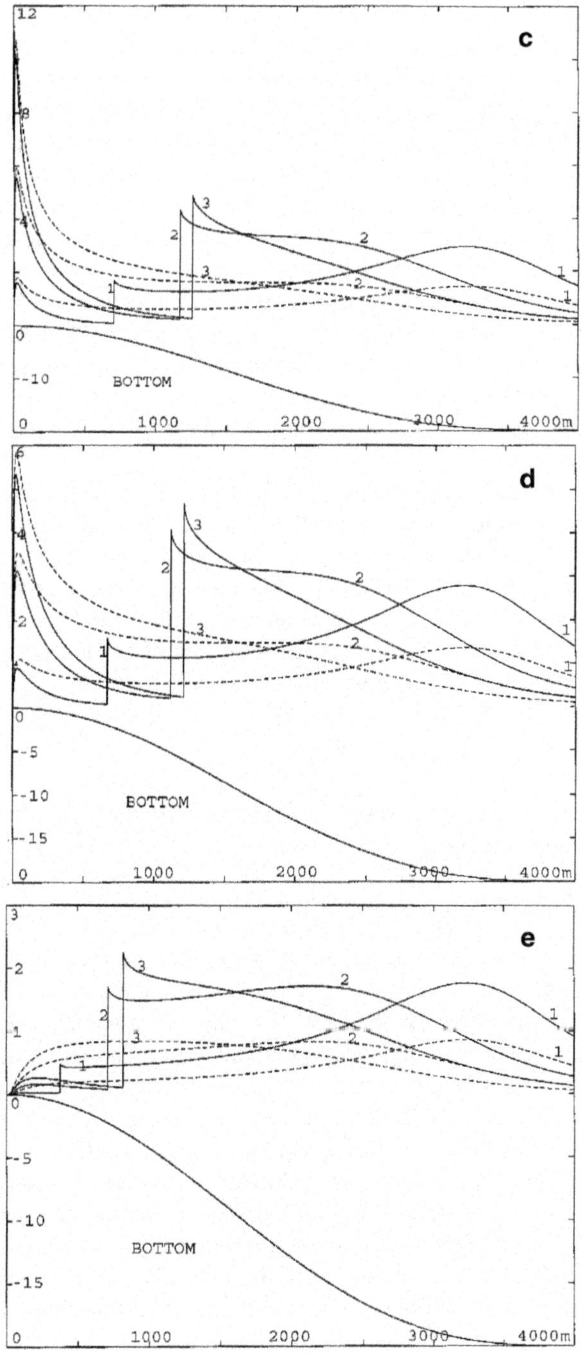

Fig. 5.10 The effect of *bottom* friction on the coastal evolution of a tsunami [18]

increases, together with the number of equations which are considered. Therefore the calculation can yield results which are very far from reality. We emphasize there are practically no numerical calculations in the literature, for the coastal evolution of a tsunami, in the case of a very smoothly-changing seabed, where a tsunami amplitude can increase ten and more times (Figs. 5.5 and 5.10). This case is most interesting for inhabitants of coastal plains. Indeed, according to our theory, the catastrophic effect of a tsunami is determined, mainly not by the energy of an earthquake or the site of its source. Rather, the tsunami threat is determined by the seabed relief near the coastal line, as Darwin had explained.

It is sometimes possible to estimate correctly the tsunami height, in a qualitative sense. Indeed, errors in measuring the intensity of an underwater earthquake can be insignificant for predicting the amplitude of a tsunami near a coast, if the nonlinear effects are large there. To illustrate this thesis we will assume that near the coast the tsunami size is defined by the cube root of the amplitude of the tsunami near its origin.

As an example, we assume amplitudes with values 0.06, 0.2 or 0.6 (dimensionless). Therefore, in the seismic source the maximal difference of the amplitudes is a ratio 10. Each of these waves will give near the coast a tsunami with the amplitudes from 0.4 to 1. Thus, near the coast the amplitude ratio will be approximately two times. Thus, there is some analogy with the topographical effect, which was studied in the Chap. 4. The influence of this effect is frequently more important than the intensity of an earthquake.

5.2 Seaquakes, Transient Cavitation, Internal and Surface Waves

Every day there are earthquakes which cause local oscillations of the ocean surface. If the depth is large, these oscillations can evolve near a coast into a tsunami. However, situations are possible where the amplitude of seabed oscillations may be comparable with the depth, or where the seabed accelerations considerably exceed g, or there is resonance of vertical waves. In these cases there are strongly-nonlinear wave phenomena on the surface and at depth, which may be illustrated qualitatively by Figs. 1.7, 2.2, 2.3, 2.5, 2.6 and 2.7. Such phenomena were observed by Darwin (cf. beginning of this Chapter) [2, p. 374]. The bubbles can appear as a result of local collapses of water within the depression waves (non-stationary cavitation) that take place near the oscillating seabed and the water surface. The explosions accompanied by jets of liquid and gas are connected with bursting on the ocean surface of large volumes of gas. The water becomes stratified and ruptured (Fig. 2.5). Bubbles, craters and jets appear on the water surface.

Apparently, Darwin was the first researcher to describe these phenomena, although during the time of great geographical discoveries some Spanish and

Portuguese sailors had defined such behaviour as 'sea shakes', and some Dutch captains reported 'awful seas'.

Levin [20] described this phenomenon, as 'a quite smooth surface of the sea suddenly was covered by waves. They promptly grew up to a height approximately 8 m and then promptly fell down, forming deep craters... People began to run captured by panic fear along a vessel; and some, having lost their minds, began to jump from the vessel. Suddenly the vessel was shaken by a very strong blow, and some persons were thrown overboard. The blows from the bottom followed one after another, and it seemed that the vessel was banged about a rocky bottom... The bangs then stopped instantly'.

Similar waves are the results of seabed vertical oscillations [21]. The sea starts to be shaken like a layer in Faraday's experiments [22], because of strong vertical oscillations of the bottom (Figs. 2.2 and 2.6).

5.2.1 *Seaquakes and Cavitation*

Transient cavitation processes were briefly considered in the Sect. 4.5. These phenomena explain, possibly, the boiling of seawater during seaquakes and the uplift of a surface of loose sediments during of earthquakes. Here we shall continue the analysis of the corresponding reports by Darwin, about the boiling and the surface explosions. These phenomena are very unexpected and very complex.

Let us discuss the situation using results of Natanzon's experiments (Fig. 5.11) [23–27]. Natanzon studied unexpected cavitation behaviour of a water column excited by a piston. We assume that the seabed oscillations during underwater earthquakes may be modelled as the oscillating piston in Fig. 5.11a, where the upper surface of the column was free. Thus, we consider the analogy of Natanzon's experiments and the simplest scheme of a seaquake. We will use Natanzon's data to understand qualitatively a large-scale phenomenon, which can occur during severe seaquakes.

Figure 5.11a sketches the experimental apparatus which Natanzon used. Experiments were conducted using a vertical metal tube having an inner diameter of 240 mm, and the tube-wall thickness was 2.5 mm. The piston produced vertical oscillations of the water. The frequency of harmonic vibrations of the piston could be smoothly changed from 0.5 to 50 Hz, and the amplitude A from 0 to 15 mm. The pressure on the free surface of the water column was atmospheric pressure, and the speed of sound in the tube was 750 m/s. The length of the water column was varied from 4 to 7.5 m. Therefore, the fundamental frequency of the water column was varied from 300 to 160 Hz approximately. During the experiments the piston displacement and the pressure near the piston were measured. Experiments showed that large-amplitude steep-front waves (shock waves) were generated in the tube as a result of the piston/water collision, when the piston acceleration exceeded g.

According to simple calculations, if $A = 0.002$ m then a cavitation zone forms when the angular frequency of the piston is 11 Hz. If $A = 0.0011$ m, then it forms

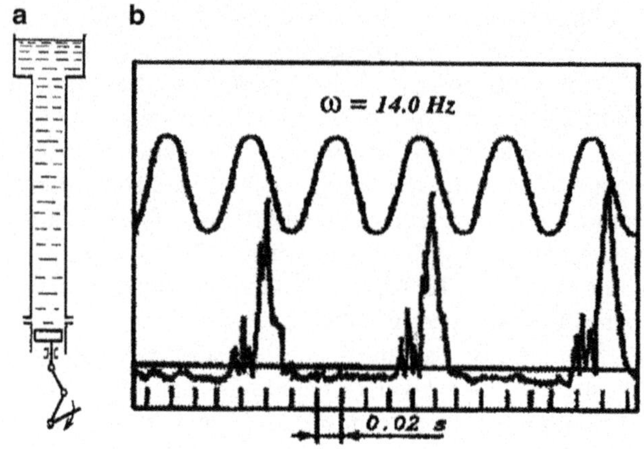

Fig. 5.11 Sketch of Natanzon's experimental apparatus (**a**). Strongly-nonlinear oscillations of the water column (the exciting amplitude is $l = 0.004$ m, the column length is 6 m, the forcing frequency is 14 Hz). The flight time is so large that sub-harmonic oscillations of the water column are excited (**b**) [23, 26]

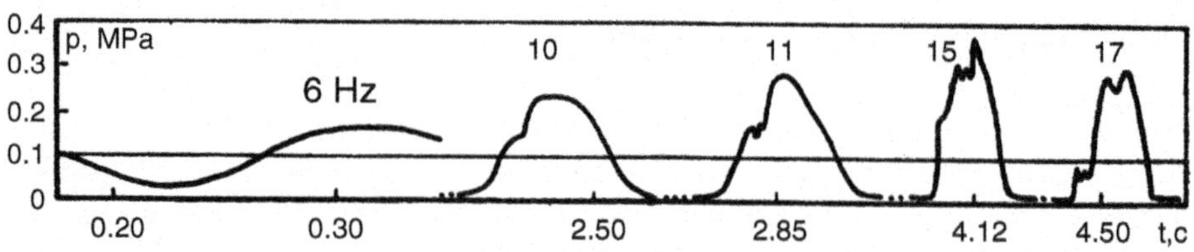

Fig. 5.12 Change of the wave happens near the piston when the forcing frequency crosses the resonance [24]

when the angular frequency of the piston is 15 Hz. In both cases, the piston acceleration slightly exceeds *g*. As a result of periodic collisions of the water column by the piston, shock-like waves were excited near the piston rather sinusoidal acoustic waves (Figs. 2.3 and 5.11b). In Figs. 2.3 and 5.11b the oscillograms of the water pressure are presented, and the sinusoidal curves show the piston position. The smooth parts of the oscillograms may be explained by the periodical appearances of the gap (cavitation zone) in the water near the piston. The distance between the peaks determines the free flight time of the column. The flight time of the water column is determined by the amplitude and the forcing frequency. The pressure peaks are the result of collapse of the gap and the column/piston collision.

A 1-dimensional model was used [24], and the pressure curves calculated for different forcing frequencies near the piston are shown in Fig. 5.12. Far from the resonance (6 Hz) the pressure changes smoothly. When the frequency increases, the pressure curves begin to resemble periodical solitary waves observed in Natanzon's experiments (Figs. 2.3 and 5.11b). At the resonance (15 Hz) the wave-forms have very steep slopes. The formation of these slopes is explained by a sharp rupture of the water within the cavitation zones.

The formation of bubbles (cavitation zones) in the vertical tubes was studied in [28]. The mechanics of forming bubbles was investigated by using glass beads lying on a vertical vibrating base of the tube. The aim of the experiments was to

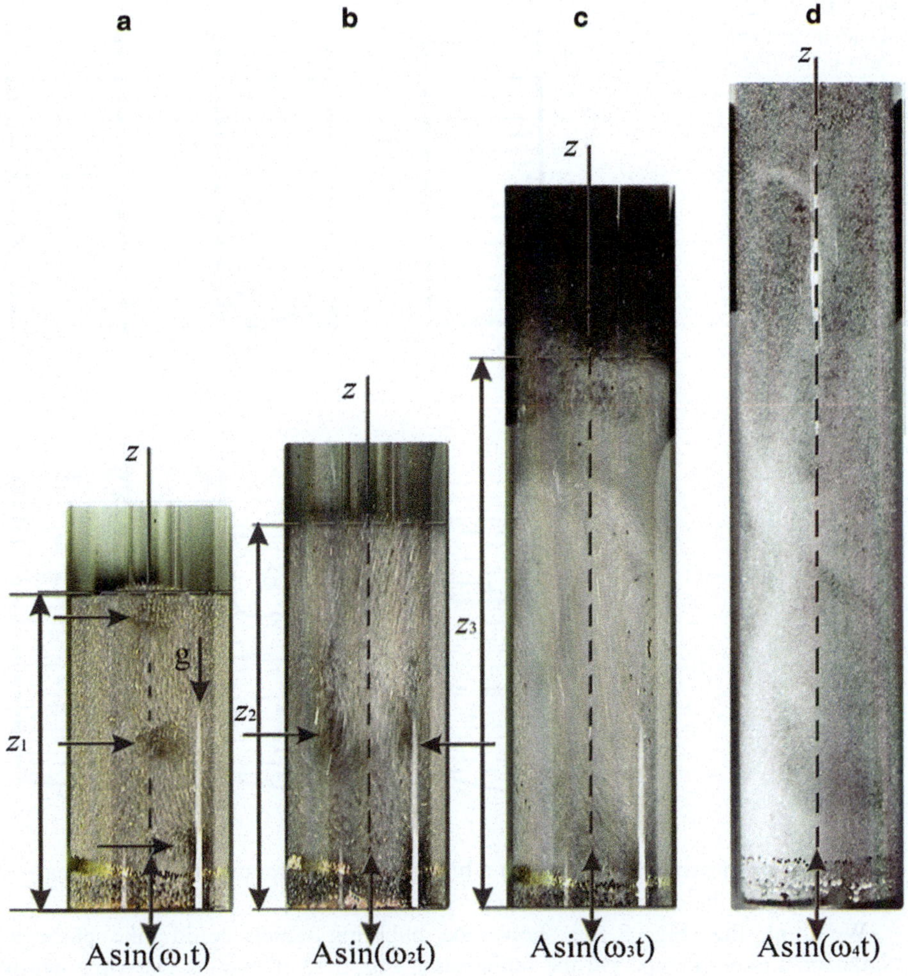

Fig. 5.13 Results of experiments in which bubbles are produced in vertically-shaken granular materials [28]

determine whether bubbling is possible in the granular beds which are subjected to vertical vibrations.

Figure 5.13a shows an image of the bed in which a bubble is created near the base, a second bubble is located in the middle of the bed, and a third bubble is approaching the free surface of the system. As the vertical acceleration increases the bubbles become larger as shown in Fig. 5.13b. Bubbles close to each other may coalesce and form a larger bubble, and as a result the bed grows in height. A further increase of the acceleration results in elongated bubbles, and the bed grows further in height (Fig. 5.13c). When the acceleration exceeds $10\,g$, the elongated bubbles

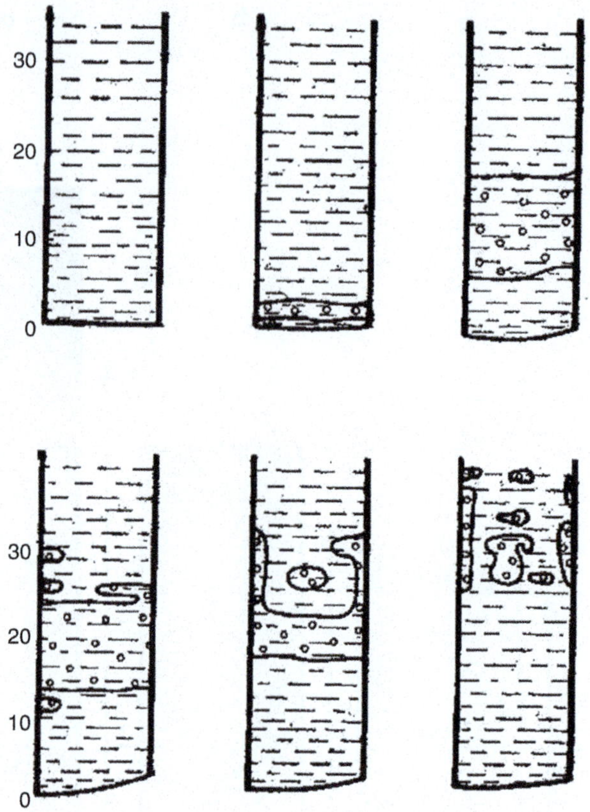

Fig. 5.14 Growth and evolution of cavitation zones in a cylindrical tank

begin to break down, which leads to a highly-agitated granular bed. In this case a spiral shape can be observed in 5.13d.

We think that Fig. 5.13a shows the bubbling which could take place in Natanzon's experiments. On the other hand, Fig. 5.13 illustrates Darwin's words about the processes in the sea and the bottom mud which accompanied the seaquake (see, also, Fig. 7.15).

The waves, as shown in Figs. 2.3 and 5.11b, 5.12 and 5.13, can be formed in different surface layers during great earthquakes and seaquakes. The calculation of the appearance and the evolution of the depression and cavitation zones during these great events is not a simple problem.

Simple results of 2- dimensional calculations for a finite water tank are shown in Fig. 5.14.

The reader can see the generation and evolution of cavitation zones in the cylindrical tank, because of transient downward motion of the tank bottom. The initial state of the tank is shown in the left picture of the upper row, where the full section of the tank is presented. Then 5 moments of the generation and the evolution of the cavitation zones are shown. For these moments only half of the tank section is presented. We think that Fig. 5.14. gives qualitative understanding about the

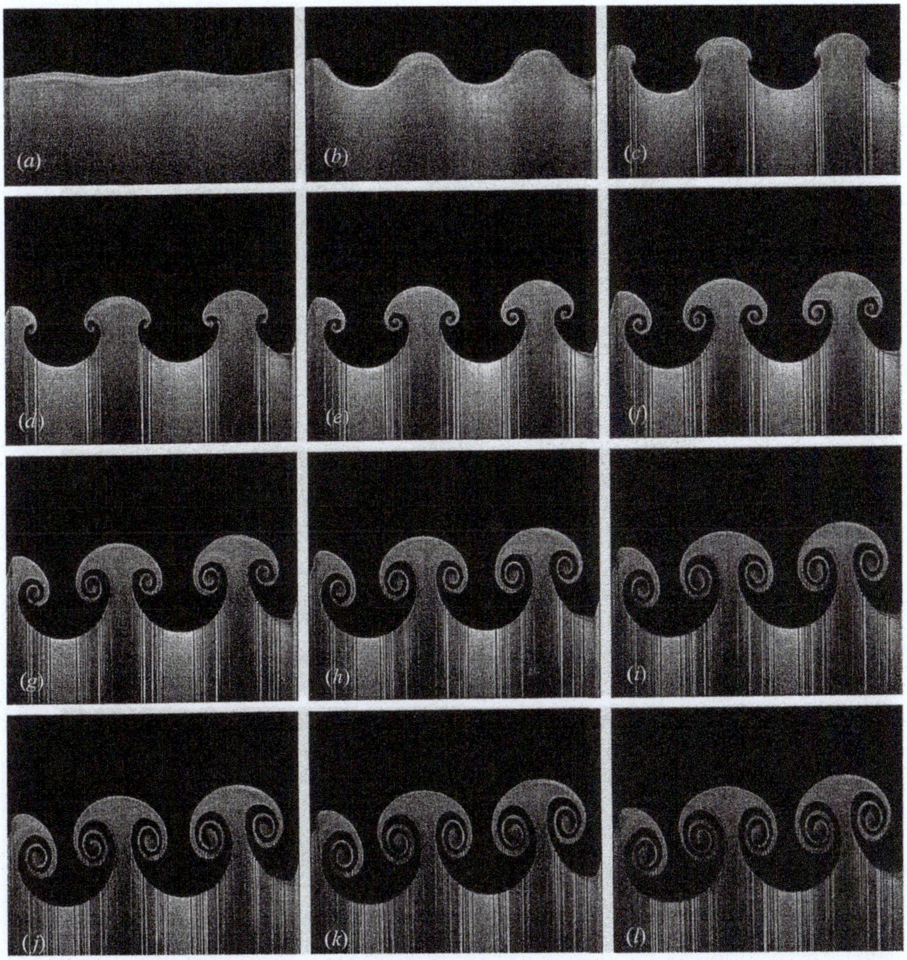

Fig. 5.15 An example of highly-nonlinear standing waves appearing on the interface of two liquids with different densities. The interface has the low-frequency sinusoidal perturbation [30]

formation of giant bubbles in water during seaquakes, coastal earthquakes and oscillations of coastal volcanoes.

5.2.2 Seaquakes and Internal Waves

Vertical seaquakes can also form giant internal ocean waves. In particular, these waves may appear on the interface of cold and warm ocean water. An example of the appearance and dynamics of similar waves is shown in Fig. 2.7 [29]. The process of the formation of these waves resembles that shown in Figs. 5.15 and 5.16, where data

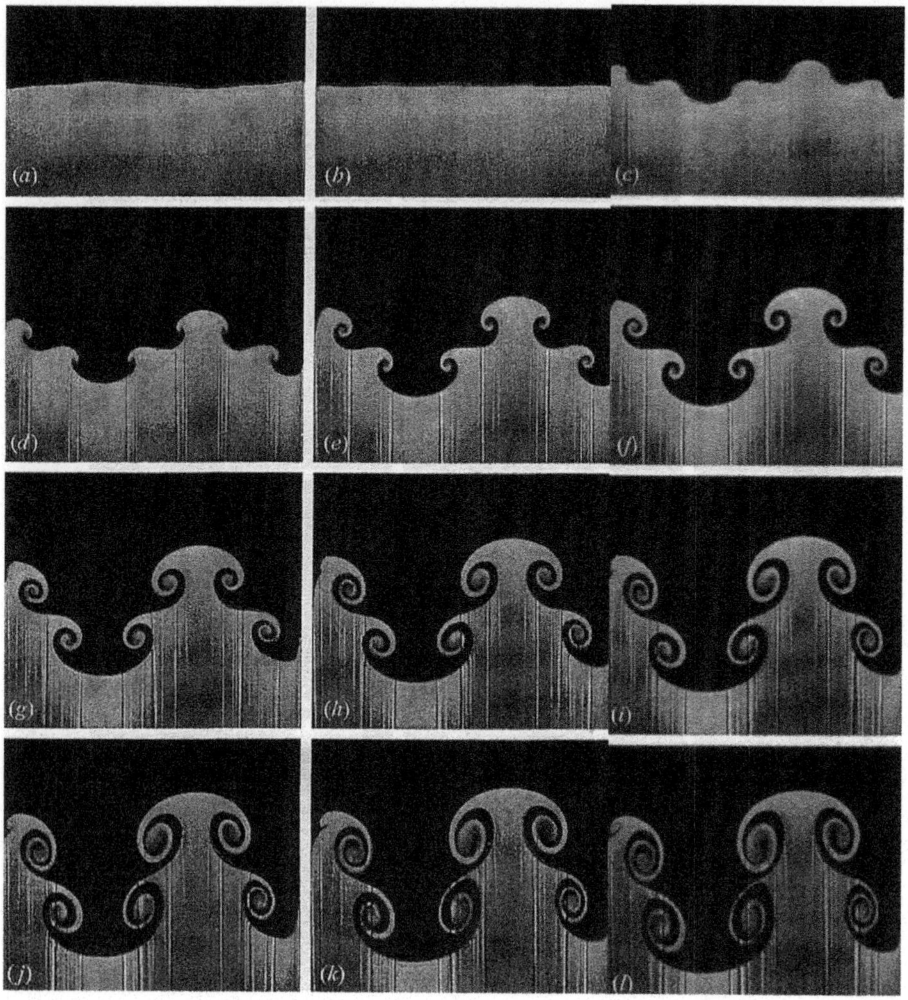

Fig. 5.16 An example of highly-nonlinear standing waves appearing on the interface of two liquids with different densities. The interface has the high-frequency sinusoidal perturbation [30]

of experiments is presented from [30]. The experiments were conducted using a 3 m drop tower. A Plexiglas tank containing two liquids with different densities was mounted on a linear rail system, constraining its main motion to the vertical direction. The instability of the interface is generated by dropping the sled onto a coil spring, producing a nearly impulsive upward acceleration. The subsequent free-fall that occurs as the container travels upward and then downward on the rail allows the instability to evolve in the absence of gravity. The interface initially has a well-defined sinusoidal perturbation that quickly inverts and then grows in amplitude, after undergoing the impulsive acceleration.

Internal ocean waves may be similar to those in Figs. 5.15 and 5.16. They may be forced by seaquakes on ocean thermoclines, or on the interface of water layers having different salinity.

According to [31], the surface separating the upper light water from denser water is located near the free surface, and in some ocean areas it is close to 50 м depth. In principle, the thermocline layer can rise in places with horizontal and vertical currents. As a result, some areas appear with the thermocline layer rising and the thickness of the warm light water reducing. Probably, such areas cannot exist for long periods, Figs. 2.5, 2.6, 2.7 and 5.13, 5.14, 5.15 and 5.16 give some information about bubbles and internal waves rising during earthquakes and seaquakes. In particular, they give some basic understanding of the bubbly boiling described by Darwin. However, the complex phenomena on the free surface of water are not quite clear; but it appears that these phenomena are determined by the Faraday effect.

Vertically-induced nonlinear surface waves are considered in the next subsection. According to experiments, their appearance can be accompanied by jets, drops and collapsing craters. The specified phenomena were noted by many observers of the sea boiling during vibrations caused by seaquakes, volcanoes and earthquakes. Perhaps, these complex wave phenomena may occur on the surface of magma during severe earthquakes, and such boiling of the magma in volcano conduit may form … *a dark column* … [1, p. 380].

5.2.3 Strongly-Nonlinear Resonant Waves and Surface Atomization: Experiments

During the last two decades, some surprising nonlinear wave phenomena have been observed on free surfaces of some vertically excited media. We have earlier considered some of those phenomena. Now we continue our consideration, by studying surface forced waves in resonators.

Atomization of water surface because of resonances Here we consider data from experiments [32–36]. In particular, those researches showed the possibility of pyramidal waves forming on a liquid surface. As an example we give results of an experiment, in which a cylindrical column of liquid 18.5 cm high and with a diameter 16.45 cm was oscillated vertically [32].

Crests of the pyramidal waves can sometimes erupt drops and jets (or streams), as in Fig. 5.17a. Sometimes jets (streams) on the surface can appear without visible waves, as in Fig. 5.17b. Perhaps these jets form during collapse of bubbles or of surface craters which are not visible in Fig. 5.17b.

Important characteristics of sloshing motion of a vertically-forced tank with internal dimensions 1,480 × 400 × 750 mm (length × width × height) are shown in Fig. 5.18 [33]. Those may be induced by the instabilities of a parametrically-driven

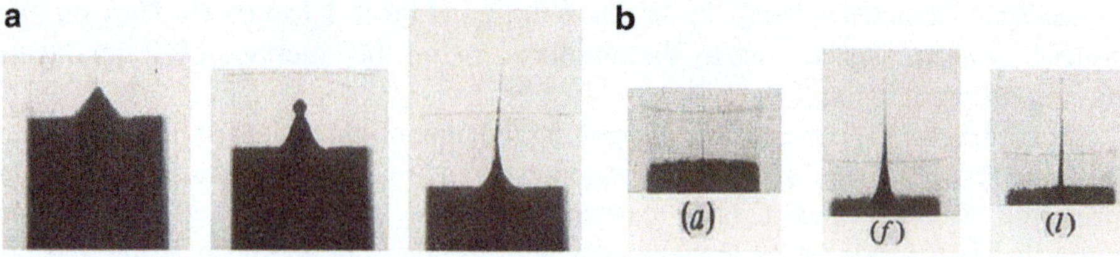

Fig. 5.17 Formation of pyramidal waves, drops and streams on a surface of a cylindrical column of vertically-excited liquid [32]

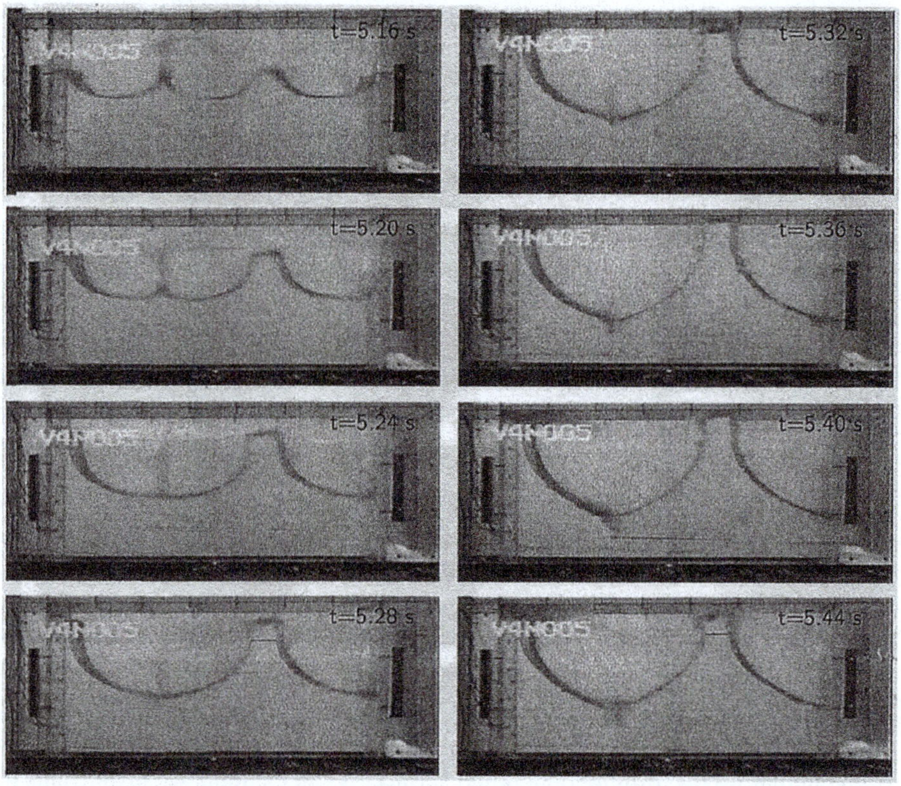

Fig. 5.18 Evolution of flat-topped waves in the resonator, accompanied by the appearance of jet, crater and deep troughs [33]

liquid surface when subjected to sinusoidal oscillations. Video snapshots show the evolution of flat-topped waves and surface jets. The time for each snapshot is given in the upper right corner of each frame. The water depth in the tank was 400 mm.

The flat-topped shape develops during the upward motion (frames 5.16–5.28 s). The final four frames show the transition from upward to downward motion. Also, frames 5.16–5.44 s demonstrate the evolution of surface jets into surface craters.

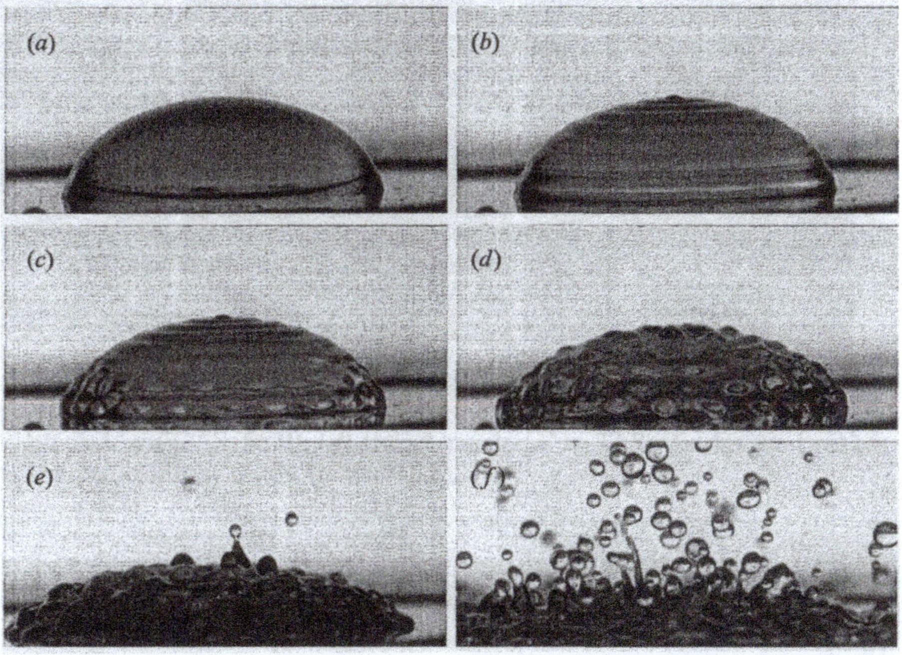

Fig. 5.19 Examples of atomization and surface highly-nonlinear waves. Vertically-excited waves and drops on the surface of a 100 μl liquid drop (diameter about 8 mm). (**a**) The undisturbed drop, (**b**) axisymmetric standing waves on the free surface at small excitation amplitude, (**c**) the appearance of stationary and then slowly-rotating azimuthal waves on the free surface, (**d**) intense time-dependent wave motion over the entire drop free surface, (**e**) the appearance of time-dependent craters, liquid spikes, and pyramid-like waves, (**f**) the rapid–ejection process of small secondary droplets (bursting) and the formation of mushroom-like waves [34]

Figure 5.19 shows a sequence of video frames of a drop which was vibrated in the vertical direction in its fundamental axisymmetric mode [34]. The excited amplitude is slowly increased from image (a) to image (f). Then the amplitude held fixed for a period long enough so that transient motions, associated with the increasing amplitude, can disappear. The frequency is fixed at 987 Hz. We think that the forms of the surface waves and the small drops, shown on Fig. 5.19, are determined by surface tension and some resonant conditions.

The process of the atomization, shown in Fig. 5.19, corresponds to a frequency which is close to the resonance frequency of waves on a drop surface. It can serve as some model for many natural wave phenomena. In particular, we can image the appearance of similar waves, craters and spikes on the sea surface during seaquakes.

Resonant particles, drops, jets, surface craters and bubbles We have considered the appearance of counterintuitive wave phenomena on liquid surfaces observed in the experiments. The results obtained, on the whole, agree with extraordinary waves excited in other conditions. In particular, parametrically

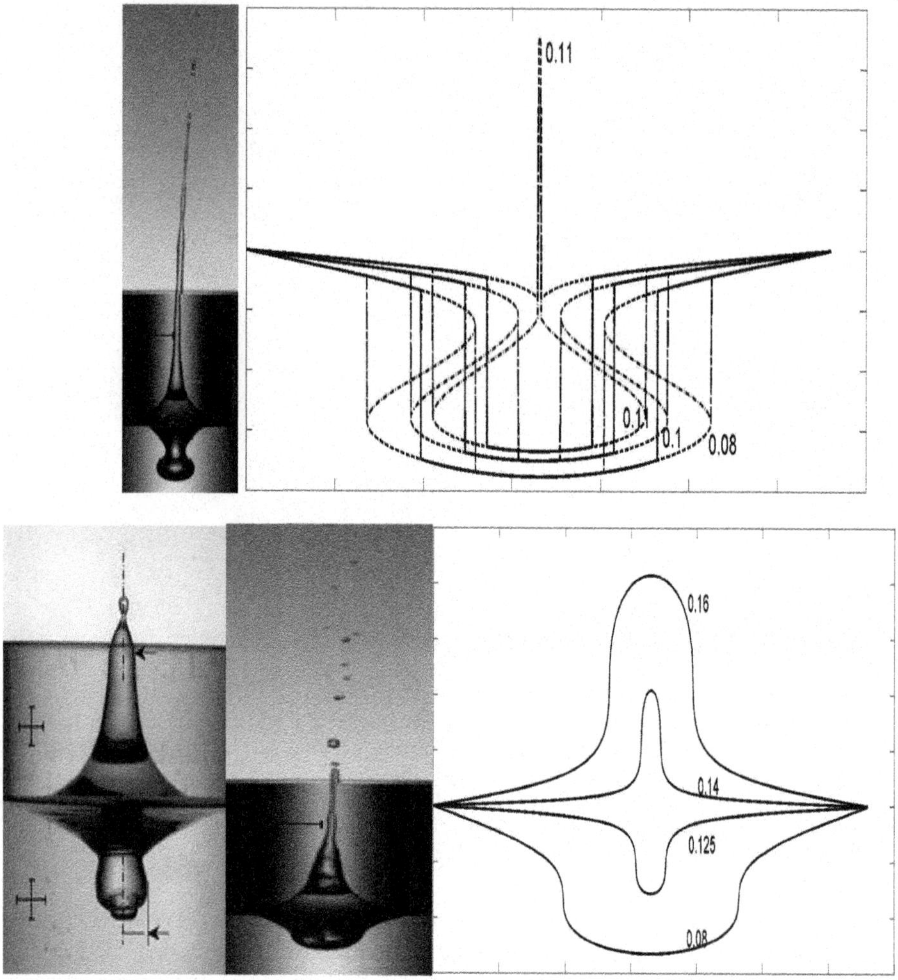

Fig. 5.20 Photographs showing the axisymmetric wave depressions (surface cavities) and jets [35]. Results of calculations [18] (right; see, also, Figs. 5.34, 4.35 and 6.48). The figures near the calculated curves determine dimensionless time

(vertically) forced surface waves, known as Faraday waves, were studied in circular containers [35, 36], and examples of these waves are presented in Fig. 5.20. The figure shows the formation of jets of a liquid during the collapse of the surface depression or cavity caused by the large-amplitude axisymmetric standing wave [35].

The experiments were conducted in circular cylindrical Plexiglas containers having radius $R = 5$ cm filled to a depth 6 cm. Images of the wave depression (lower part of the images) and of the jet formation (upper part of the images) are shown for two different forcing conditions. The experiments (left) are illustrated by

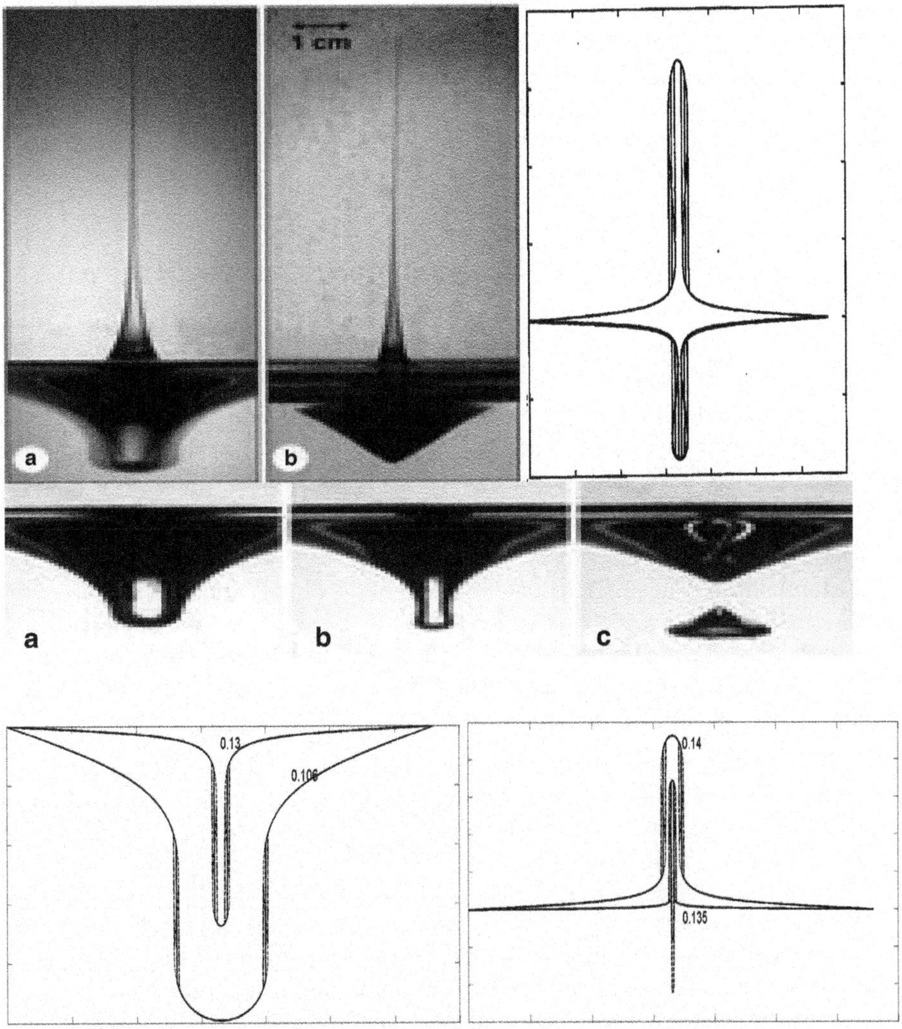

Fig. 5.21 Photomontages of the surface collapsing craters, a bubble near the liquid surface and surface jets [36]. The calculated curves illustrate the photomontages [18] (see, also, Figs. 5.34 and 5.35). The figures near the calculated curves determine dimensionless time

calculations (right). The appearance of craters and jets are illustrated additionally in Figs. 5.21, 5.23, 5.34, 5.35 and 5.38 (see, also, Chap. 7).

Another examples of these waves are presented in Fig. 5.21, in which the photomontages show the formation of jets of a liquid during collapse of the surface craters.

The processes shown in Fig. 5.21 were observed during the vertical resonant excitation of a circular basin with a liquid [36]. The photos demonstrate various forms of the craters arising on the liquid surface (a–c) and a bubble (c). The calculated curves are found for different dimensionless time. The general agreement of experiments and calculations is visible.

Those experiments and calculations help us to understand the seaquake-induced processes as resonant phenomena, connected with transient cavitation in the sea water and the interaction of nonlinear waves, craters and jets on the sea surface.

5.2.4 *Strongly-Nonlinear Resonant Waves and the Elastica Forms*

Resonance is a classical problem of great practical import. This phenomenon is usually associated with oscillations of a point-like mass. For example, the vertical motion of an upper layer may be analysed using the model of a ball. However, some experiments and observations showed that such a model cannot describe the complex phenomena within the layer and on its boundaries [37–40]. Sometimes wave properties must be taken into account.

The atomization The process of atomization of the liquid surface as a result of the transresonant process was modelled in [17, 18, 26, 27, 37, 40]. A parameter R which was introduced there determined the width of the resonant band as R varied from 1 to -1, with $R=0$ at resonance. The results of the calculations for 4 values R located very near $R=0$ are presented in Fig. 5.22.

It is emphasized [17, 18, 40] that Fig. 5.22 had described qualitatively some peculiarities of wave processes shown in Fig. 5.19. In particular, the eruption of water jets from the wave tops is determined by some resonant condition. However, the reader can see from Figs. 5.19, 5.20, 5.21 and 5.23 that the jets are formed by resonances and collapses of water craters (see, also, Figs. 5.34, 5.35 and Chap. 7).

When droplet ejection begins from the free drop surface (Fig. 5.19), the wave motion on it appears to be mostly chaotic. During this motion on the free surface a depression, or crater, precedes the rising liquid spike that ultimately ejects the secondary droplet (Fig. 5.23). This is similar to the spike observed after the collapse of a cavity caused by an air bubble that has just broken through the free surface of a liquid layer from below. Another phenomenon which can be compared with the bursting is the impact of a drop on a solid surface or a liquid layer (see, also, Sect. 4.2.3). In either case a crater is formed, followed by a jet that may eject several secondary droplets.

We think that the appearance of surface bubbles and craters, the collapse of them and the formation of surface jets may happen on the ocean surface during seaquakes.

Of course, multivalued waves, such as those in Figs. 5.15 and 5.23, are possible in the most various environments and fields, wherever there is enough strong interaction of media together with the manifestation of highly-nonlinear effects [18]. In certain cases the surface tension is responsible for appearance of such

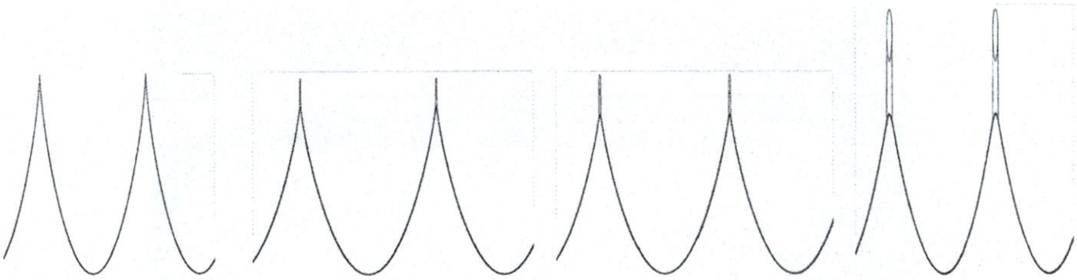

Fig. 5.22 The eruption of particles and jets (streams) from tops of waves when the transresonant parameter R varies from positive to negative values very near $R = 0$ [17, 18, 40]

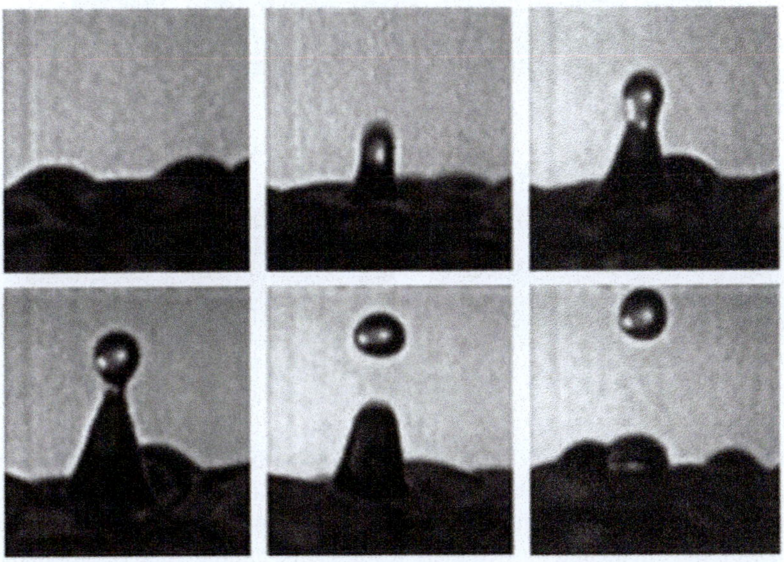

Fig. 5.23 A time sequence of images of a small part of the drop surface (2×2 mm) in Fig. 5.19 [34]

waves. Namely, the behaviour of systems consisting of 2 or more adjoining environments (phases) can be determined by a thin layer of the contact between these environments (phases). For example, in the case of very short surface waves the mechanical properties of such surfaces are characterized, first of all, by the surface tension. This tension defines properties of a surface similar to properties of a thread (string).

Experiments of Taylor and the elastica [41–44] Let us consider results of the interesting experiments performed by Sir G. I. Taylor [41]. He wrote in 1969: 'The analogy between the equations of motion of a viscous fluid and that of an incompressible elastica solid is well-known. It was pointed out to me by Brooke Benjamin that when a thin rod is subjected to end compression it forms an elastica and that Love's *Theory of Elasticity* contains a picture of the set of elasticas. I, therefore,

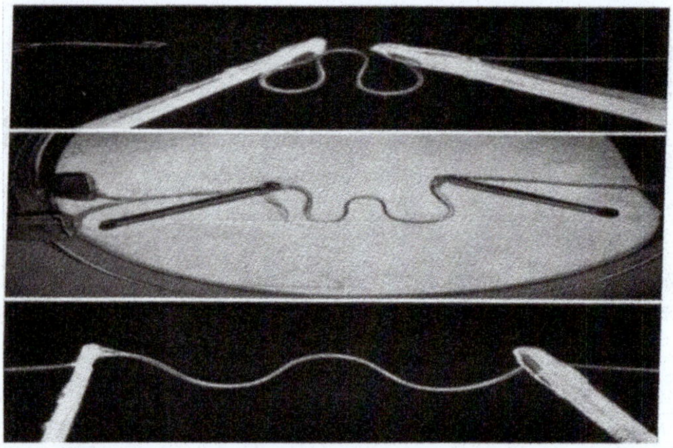

Fig. 5.24 Threads of SAIB, floating on mercury [41]

Fig. 5.25 Flat –topped wave form (*left*) and mushroom-like wave form (*right*) of a thin rod [44]

found a very viscous fluid, known as SAIB (Sucrose Acetate Isobutyrate) and stretched a thread of it between two sticks. I then laid this thread on the surface of a dish of mercury and either pushed the sticks towards one another or laid two matches on the thread and pushed them towards one another, and photographed the thread' (see Fig. 5.24). Taylor observed also a very viscous stream of the fluid falling onto a plate. A short distance above the plate the stream begins to wave or rotate into a spiral form [41]. More complex folding was observed when the stream falls into a fluid of less density than itself. In the last case the stream can fold and form knots.

The folding, knotting and relating spiral forms are well-known processes of the instability of many long flexible systems [44]. In particular, folding and knotting of thin structural elements can take place under an axial impact. These phenomena are explained by dynamic progressive buckling of the structures [45].

Let us note that some wave forms in liquids (Figs. 2.7, 5.15, 5.16 and 5.23) are similar to the forms shown in Figs. 5.24, 5.25 and 5.26. On the whole, they resemble forms which can theoretically arise during longitudinal compression of thin rods. These are so-called the elastica forms, which Taylor wrote about, and some of them were described more than 300 years ago [42, 43]. According to Figs. 2.7, 5.15 and 5.16, these forms can arise also on surfaces of liquids. That does not surprise us,

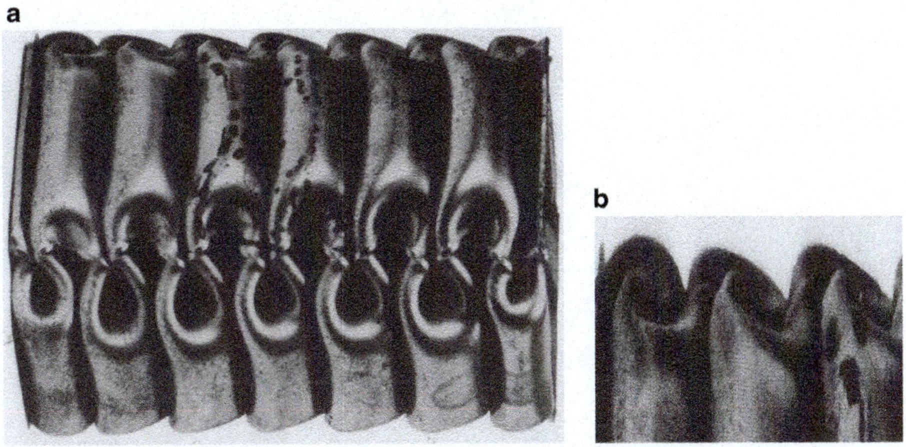

Fig. 5.26 Folding and knotting can take place simultaneously during the crushing of square tubes. The mushroom-like folding (**a**, **b**) and vortex-like folding (**b**) [45]

owing to the analogy existing between wave processes. Apparently, the elastica forms can exist in many highly-nonlinear wave systems. Hereafter we will analyse this assumption.

The elastica forms calculated for different values of the parameter λ: $\lambda = 0.1$, $\lambda = 0.2$, $\lambda = 0.249$, $\lambda = 0.25$, $\lambda = 0.251$, $\lambda = 0.28$, $\lambda = 0.3027$, $\lambda = 0.35$, $\lambda = 0.4$, $\lambda = 0.5$, $\lambda = 1$ and $\lambda = 2$ are shown in Fig. 5.27. This evolution can qualitatively describe the appearance of the surface craters (Fig. 5.27: $\lambda = 0.249$, $\lambda = 0.25$) and the atomization of a free liquid surface due to highly-nonlinear wave processes (Fig. 5.27: $\lambda = 0.251$, $\lambda = 0.28$).

On the other hand, the curves $\lambda = 2$; 1; 0.5; 0.4 and 0.35 qualitatively describe the instability of the interface of two liquids with different densities, which was shown in Fig. 5.28. The reader can see transforms of the harmonic wave ($\lambda = 2$) into the shock-like wave ($\lambda = 0.5$) and the mushroom-like configurations ($\lambda = 0.4$ and 0.35) [46].

The form $\lambda = 0.35$ of Fig. 5.27 corresponds to the tucks arising at axial crushing of the square tube (Fig. 5.26). The forms presented in Figs. 5.25 and 5.27 correspond to the static compression of very flexible rods. On the contrary, we suppose that the elastica forms can describe qualitatively the periodic appearance of splashes, drops and bubbles on a liquid surface, as a result of highly-nonlinear effects. In particular, the forms which are intended to describe different forms of rods may be used for the analysis of highly-nonlinear capillary waves. Indeed, some elastica forms can describe some water surface forms which can result from capillarity. Pierre Simon Laplace discovered this fact in 1807 [43].

Elastica-like forms and the evolution of some vortex structures We think that the elastica-like waves and their dynamics describe processes which have been

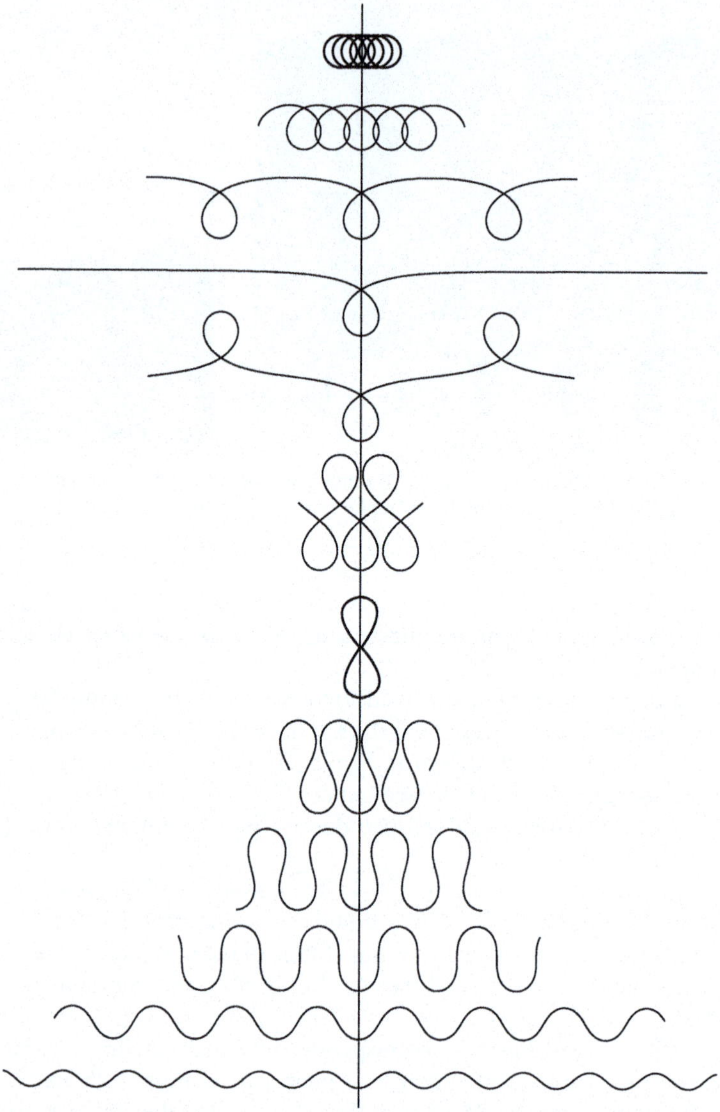

Fig. 5.27 The evolution of the elastica forms as a function of some parameter λ: $\lambda = 0.1$; 0.2; 0.249; 0.25; 0.251; 0.28; 0.3027; 0.35; 0.4; 0.5; 1 and 2 (from *top to bottom* of the picture) [43]

observed in many experiments. In particular, these dynamics can qualitatively describe a transformation of waves into vortices. This transformation resulting from instability in a sinusoidally-oscillating flow was studied in [47]. Some results of the experiments are presented in Fig. 5.29. One can see that some elastica forms describe the transformation. Namely, curves $\lambda = 0.5$; 0.4 and 0.35 in Fig. 5.27 qualitatively describe the evolution shown in Fig. 5.29.

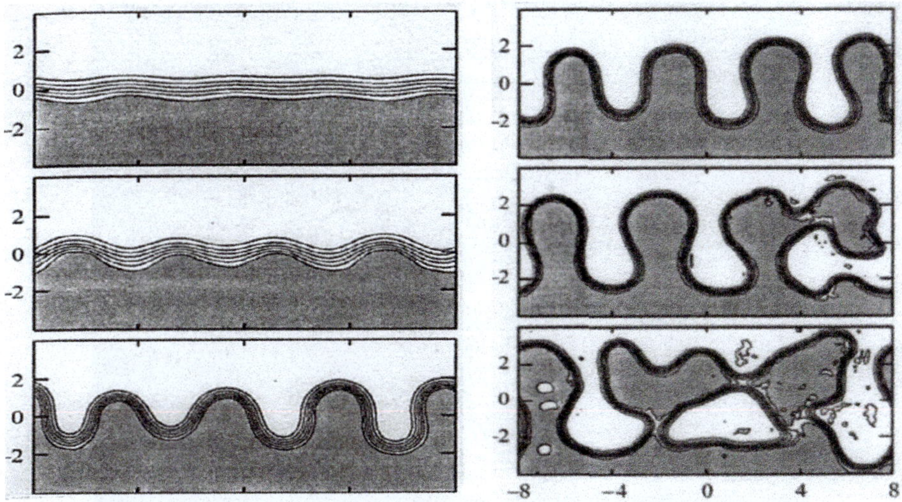

Fig. 5.28 Results of numerical simulation of forms of the thin surface layer [46]

The elastica forms have traditionally been described by complicated mathematical expressions. However, recently, it was found that the elastica forms may be described by solutions of highly-nonlinear algebraic equations [37–39]. On the other hand, these solutions can describe wave forms which have more complex shapes than the elastica forms. Examples of these counterintuitive shapes are presented in Fig. 5.30. It is seen that the mushroom-like wave can evolve into drops and bubbles, which can appear above or below the surface of the wave.

It is possible to give a different interpretation of the presented results. In particular, the dynamics and the disintegration of the mushroom-like waves can, perhaps, describe the evolution of complex vortex structures (Fig. 5.31).

We have given above the simplest interpretation of the elastica forms. More complex interpretation of them includes the possibility of jumps in zones of multivalues of the wave profiles. Thus, we think that the wave forms shown in Figs. 5.27 and 5.30 have much wider applicability than just describing large bending of rods and surface waves.

5.2.5 *Strongly-Nonlinear Resonant Waves, Drops, Bubbles and Craters: Calculations*

According to Fig. 5.27, most of the elastica forms have the fold on its sides. Figure 5.27 show also the transformation from the smooth harmonic wave into the mushroom-like waves, when λ reduces from 2 to 0.35. This evolution is illustrated by Fig. 5.32. It was emphasised in [17, 18] that similar evolution is described by calculations which simulate the crossing the resonant band by the harmonic waves.

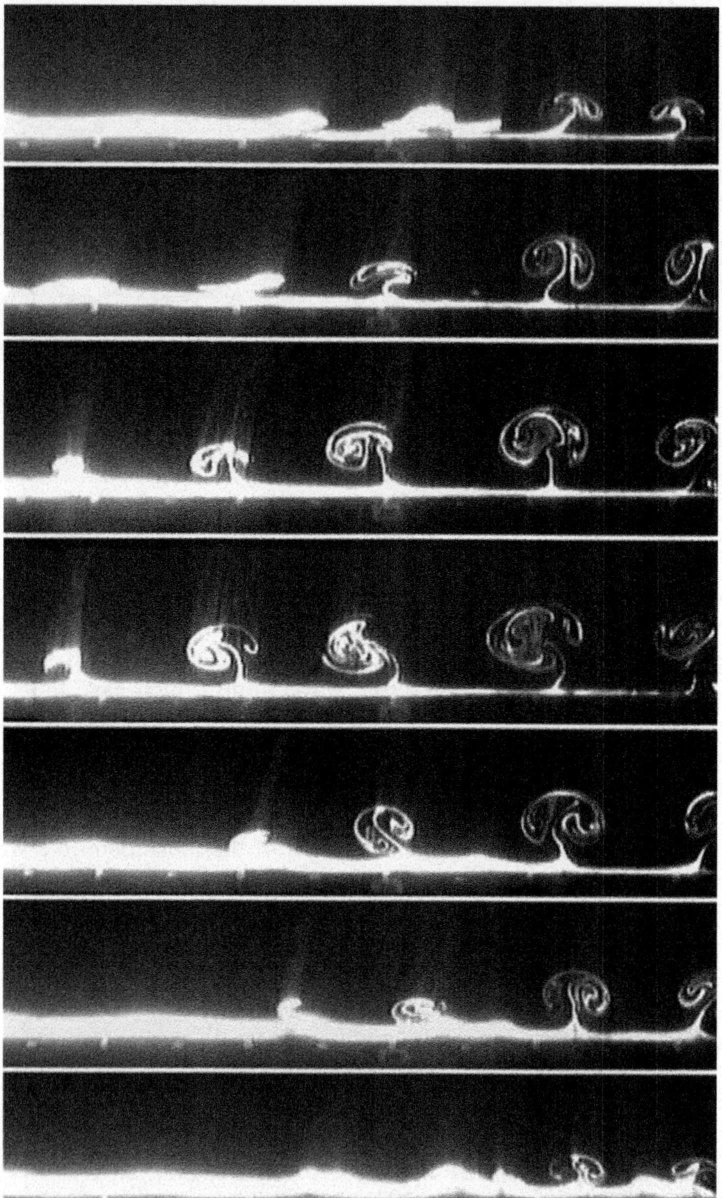

Fig. 5.29 Photographs of some sequences of developments of small disturbances (not for equal time intervals) into mushroom-like structures. They grow in size and number when the flow speed increases [47]

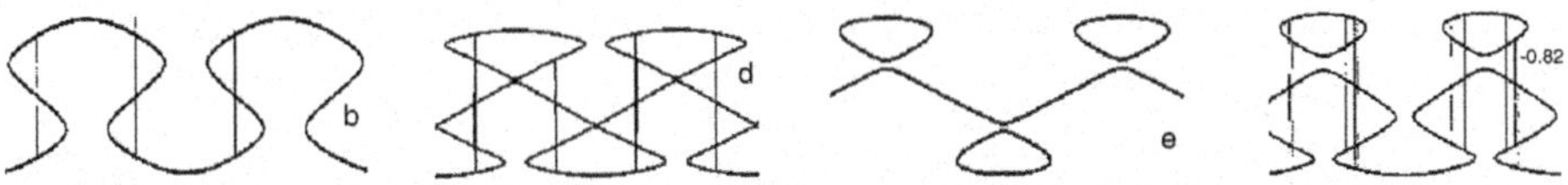

Fig. 5.30 The profiles and the structures are determined by solutions of high-nonlinear algebraic equations [17]

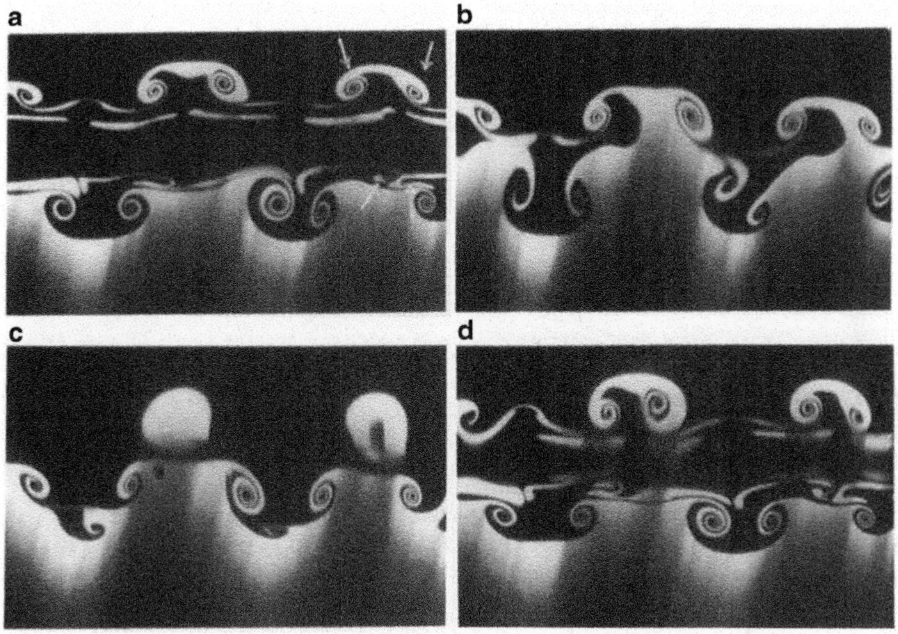

Fig. 5.31 Visualization of the evolution of mushroom–like structures and vortices [48]

The elastica forms can exist in many highly-nonlinear systems. In particular, they can be formed during longitudinal compression of thin rods. About 250 years ago Leonhard Euler has constructed some forms important for us (1744) (Fig. 5.32).

It was found that the front and back sides of the mushroom-like waves are formed by segments of various solutions of a cubic algebraic equation which is valid inside the resonant band [17, 18] (see Sect. 7.10). These segments (solutions) contact continuously and smoothly with each other. The mushroom-like waves may be called also multivalued waves. On the other hand, according to the catastrophe theory [49], these waves may be called catastrophic waves [18].

The transresonance evolution of the harmonic waves is shown in Fig. 5.33. The mushroom-like waves appear inside the interval $0 > R > -1$. The evolution of these forms reminds us of those presented in Fig. 5.32. Cnoidal and pyramid-like waves arise near the resonant band ($R = -1.01$) (Fig. 5.33, top). The calculations have also shown that near $R \approx -1$, at the moment of formation of the pyramidal

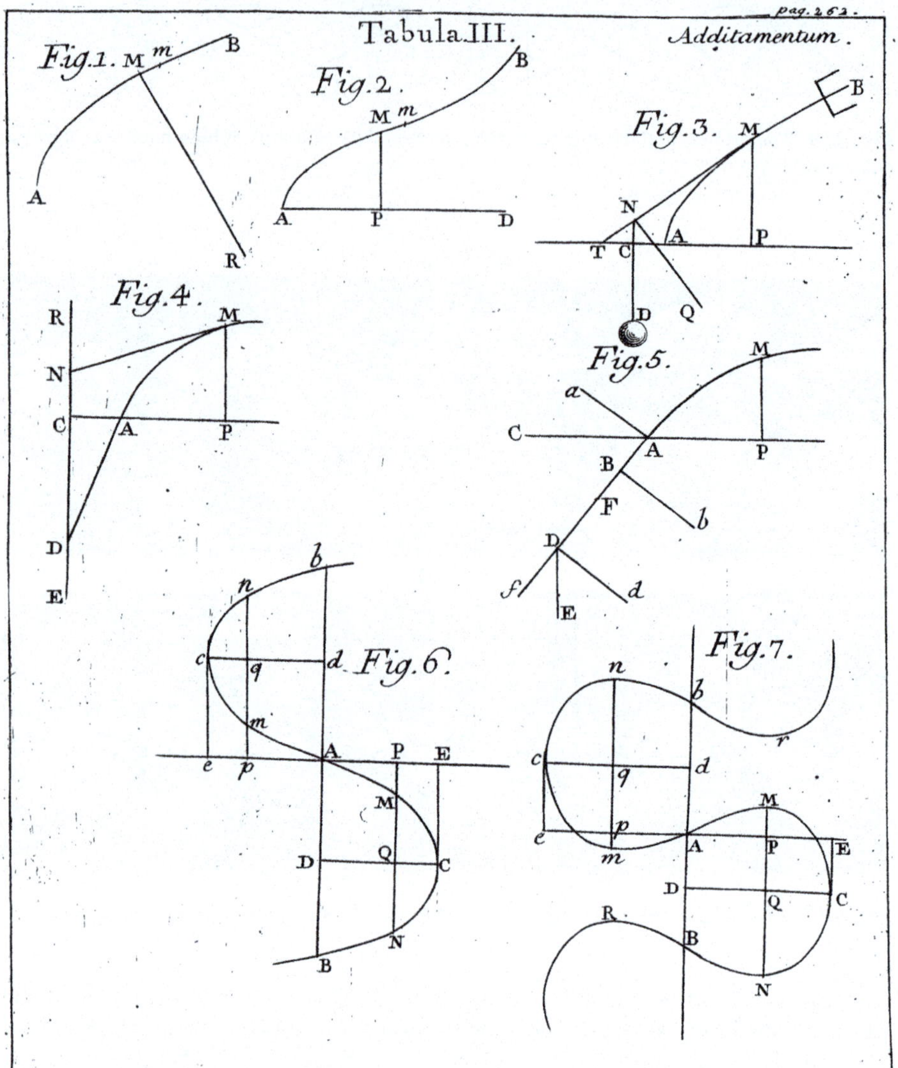

Fig. 5.32 Euler's elastica figures [43]

waves [17, 18, 39], there is a possibility of the eruption of a jet from the wave top. These jets and pyramidal waves were shown in Figs. 5.17, 5.20 and 5.21. Thus, Fig. 5.33 (top) demonstrates the transresonance evolution of harmonic waves into the elastica forms, and the appearance of pyramidal and cnoidal waves near the boundary of the resonant band, where $R \approx -1$.

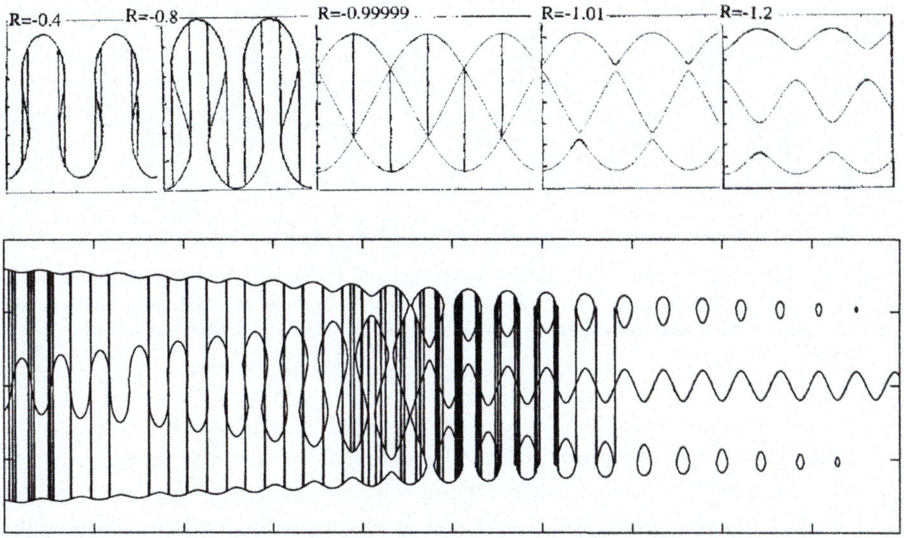

Fig. 5.33 The transresonant evolution of surface waves shown in [37]. The generation (R = −0.4, the transresonant evolution (R = −0.8; −0.999) and the bifurcation (R = −1.01) of the catastrophic waves (*top*)). Growth ripples and generation of mushroom-like waves, and clusters of bubbles of energy (clots of energy) (*bottom*). The *bottom picture* was used so that to illustrate the evolution of ripples, the generation of particles and seeds of galaxies in the early Universe [18, 37] (see Chap. 7).

Another evolution is shown in Fig. 5.33 (bottom). In this case there is the appearance of pyramidal waves and clusters of bubbles near the boundary of the resonant band, where $R \approx -1$. Taking into account certain results of the Chap. 7 we can name the bubbles as bubbles of energy or clots of energy.

Let us consider now the elastica forms corresponding to $\lambda = 0.251$ and 0.28 (Fig. 5.27). We can imagine that these forms describe jets and craters on a certain free surface. These surface elements are formed during the transresonant evolution of initial harmonic waves. It was shown in [18] that the formation of craters, jets and drops on a liquid surface may be calculated as results of vertical oscillations of tanks with liquids. Let us consider the results of those calculations, which are presented in Figs. 5.34 and 5.35. We expect them to help us understand and describe qualitatively the waves observed in experiments, and shown in Figs. 5.15, 5.16, 5.19, 5.20, 5.21 and 5.23.

According to the calculations, generation of the drops can be explained by the resonant process. During the further development of the resonant process (as R reduces) the crater edges converge and form into jets and drops, which separate from the water surface. That process can be seen in Fig. 5.34.

Details of the resonant nonlinear oscillations of surface waves are presented in Fig. 5.35. The full period of oscillations is shown for $R = -0.008$.

Thus, we have considered some very complicated wave processes. For example, the appearance of bubbles, waves, craters and jets as presented in Fig. 5.21, have seemed very counterintuitive [36].

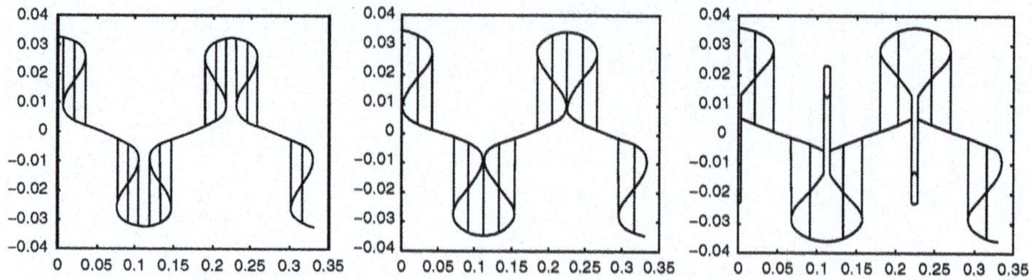

Fig. 5.34 Resonant oscillations of surface waves in finite resonator. The influence of closeness to resonance on the convergence of the surface crater, and on the formation of jets and drops. Pictures calculated for different values of R: $R = -0.0075$, -0.008 and -0.011 (*left to right*) [18]

According to the experiments and the calculations, on free surfaces of the liquid or liquefied environments undergoing vertical excitation, cavities (bubbles), craters and jets of the substance can appear successively. Probably, something similar can occur with very strong seaquakes and earthquakes. In particular, the eruption of liquid from wave crests (Fig. 5.23) strikingly resembles volcano eruptions. On the other hand, Figs. 5.34 and 5.35 can illustrate the appearance of particles of mass and energy at the origin of the Universe (see Chap. 7 and Figs. 7.48, 7.49 and 7.50).

5.3 Evolution of Vertically-Induced Waves on Liquid Surface: From Jets to Breakers and Vibrating Solitons

Here we will continue to study extreme surface waves accompanied by jets, craters and bubbles. The speed of long surface waves can depend on the vertical acceleration according to the expression $\sqrt{\left(g + g_f\right)h}$, where h is the layer thickness and g_f is the forcing acceleration. According to that expression, if $g_f = -g$ at some instant then the wave stops. At other times the speed of the wave depends on the current value of g_f. If the speed becomes imaginary, then it may be considered as equal to zero [18, 26].

Using the above circumstance, I constructed the theory [18, 26] which qualitatively described results of a few experiments. The theory modelled the observed waves as the sum of the standing linear wave and nonlinear travelling ripples. It was supposed that the speed of the ripples depended on the container acceleration. Results of the theory qualitatively correspond to some data of experiments, despite the specified idealizations. Here we present the experimental data and calculations.

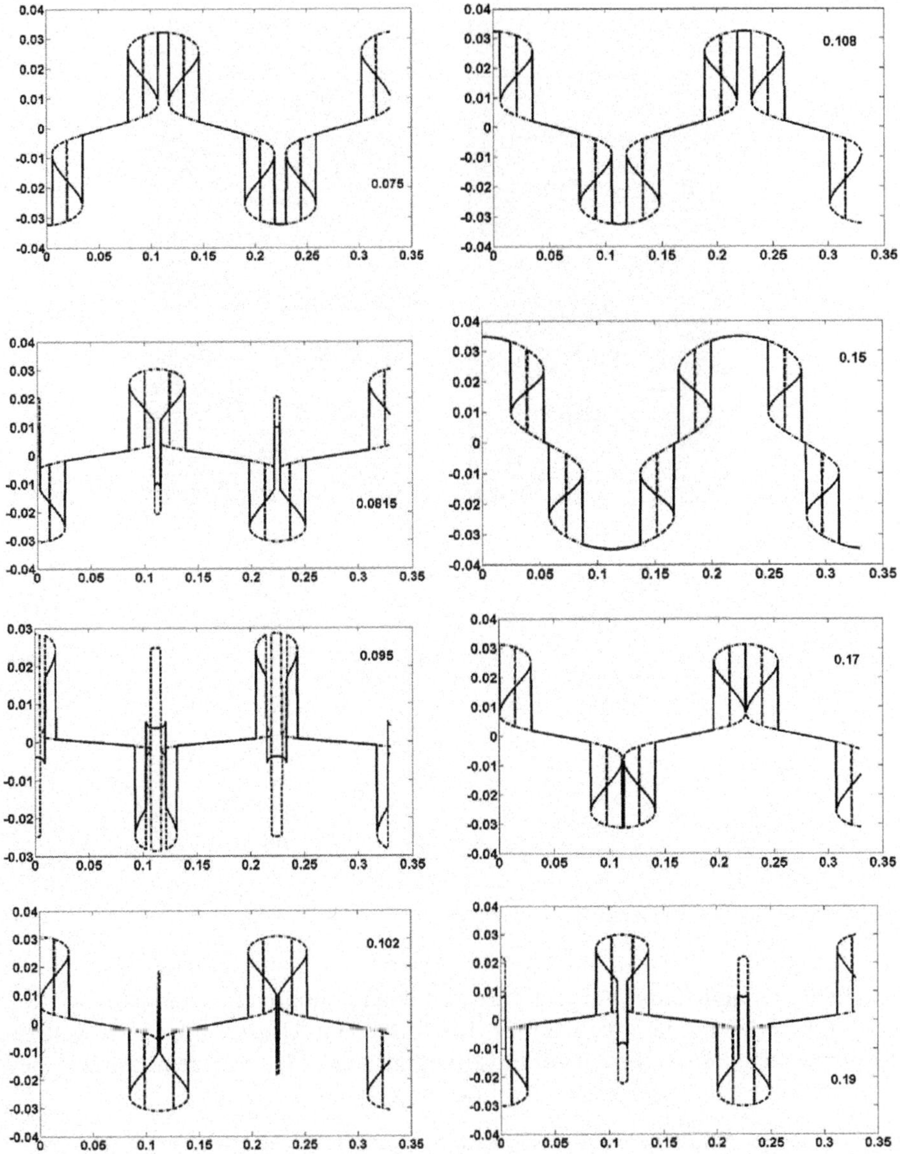

Fig. 5.35 Resonant oscillations in the finite resonator. The period of some oscillations are shown. The wave dynamics is accompanied by the appearance of surface craters and jets from them. The curves are formed by multivalued solutions of some cubic algebraic equations. They are shown by continuous, *dotted* and *dash-dot lines* [18]

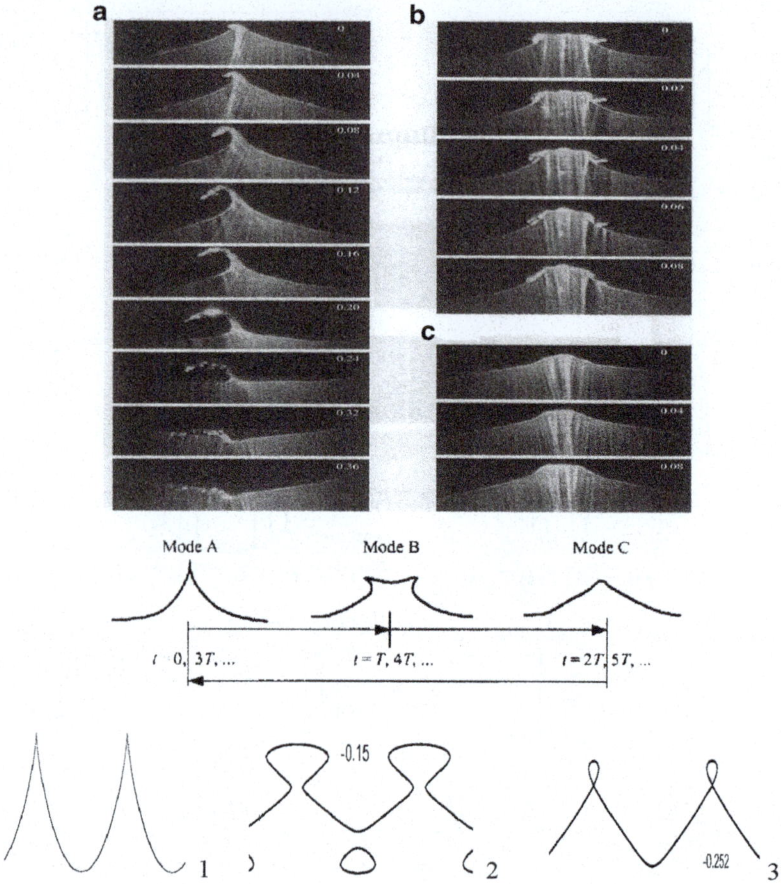

Fig. 5.36 Profiles of counterintuitive waves [18, 50]

Surface jets and folds Vertically-excited counterintuitive waves were observed during experiments described in [50]. The wave crests erupted streams of water, or also they were crowned by small travelling breakers. Fig. 5.36 demonstrates these waves, which can be schematized as modes A, B and C.

The mode A in Fig. 5.36 is defined by a sharp angle at the crest and by the appearance there of a jet. The mode B is characterized by almost-plane or slightly-concave peak, and by the presence of folds on the crest edges. The mode C has a dome-like crest. Counterintuitive profiles of these waves appeared at the container centre.

We have more than once emphasized that inside the resonant bands, counterintuitive waves can occur. Some of them are shown in [16–18, 37, 39]. We have viewed results of the calculations presented there and have found a few curves resembling waves A, B and C – those curves are denoted in Fig. 5.36 by numbers 1, 2 and 3. The curve 1 reproduces one of the curves of Fig. 5.22. The wave

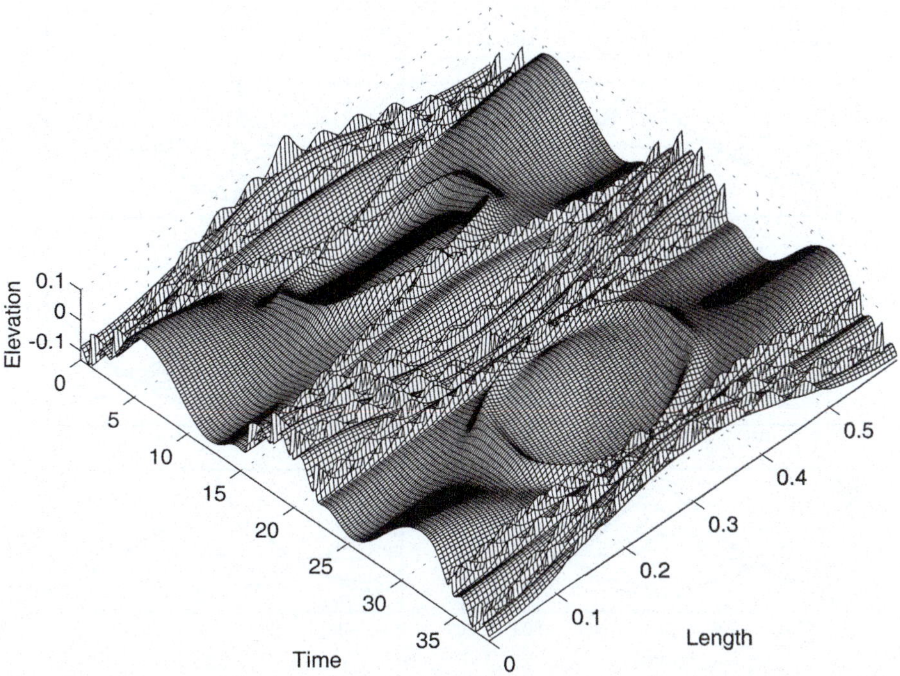

Fig. 5.37 The wave pattern describing qualitatively a period tripling and jets observed in the experiments [27]

2 corresponds to the mode B, but this wave is accompanied by the bubble under the trough. The wave 3 corresponds to the mode C.

The waves presented in Figs. 5.36 were excited in a vertically-oscillating container, with length of 0.6 m, width of 0.06 m, and depth of water was 0.3 m.

We have found that as a result of parametric excitation a system of a standing wave and traveling waves (ripples) appears in the container. The standing wave period is 4 π and a period of the ripples is 12 π. A peak is being formed when the oppositely traveling waves meet each other in the centre. The water elevation is determined by a sum of the standing wave and the ripples if to use the superposition law. The sum is maximal when the amplitudes of the waves are positive and maximal. In this case a largest peak (a sharp crest) occurs in the centre. In Fig. 5.37 these peaks are shown for the dimensionless time approximately equal 6 and 42. On the other hand, the traveling waves meet each other on the trenchs and the slopes of the surface of the standing wave. In the latter cases the flat crests or the non-breaking crests are formed. A formation of the different crests explains the complex wave forms shown in Fig. 5.37.

In general, the results obtained are in agreement with the experimental data shown in Figs. 5.36 and 5.38. Of course, the superposition law was not valid exactly in the experiment. Therefore there are certain discrepancies between the experiments and the calculations.

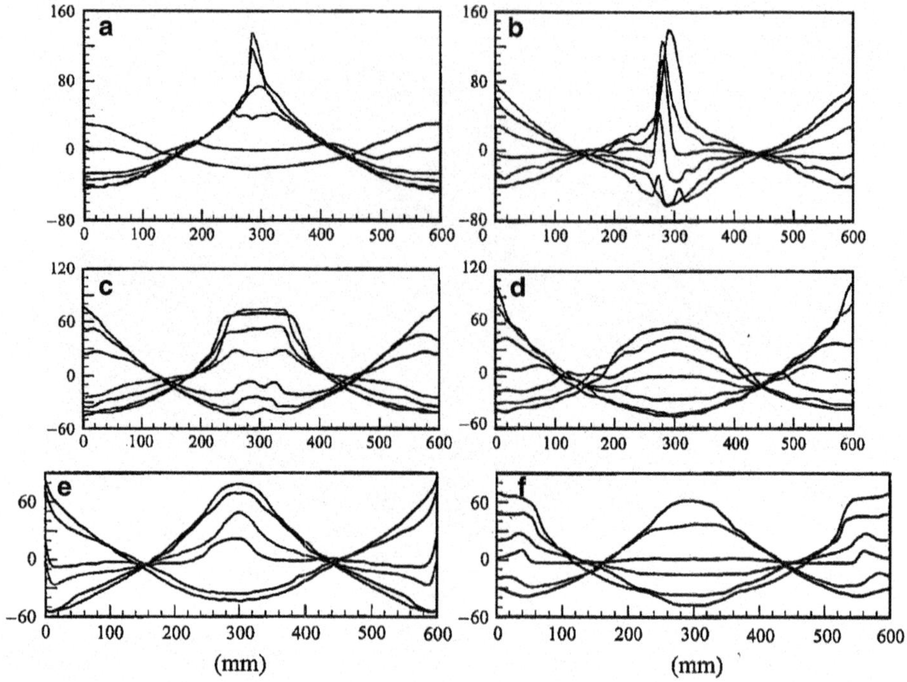

Fig. 5.38 The detected profiles of waves excited during the experiments [50]

5.4 Solitons and Oscillons

Vibrating solitons A soliton is a solitary wave, i.e. a spatially-localized wave with spectacular stability properties. It was first observed in 1834 on the surface of a canal by John Scott Russell [52]. Since that time, solitons have simulated much interest and debates. In particular, it was found that a few very popular mathematical equations do have soliton solutions. This is why solitons have interested mathematicians very much.

However solitons are also of major interest to physicists. In particular, such waves have been observed on the sea surface. We can imagine that soliton-like waves may be excited during seaquakes.

In particular, solitons can be excited on water layers by vertical excitation. These solitons are not travelling waves – they are horizontally or vertically-oscillated objects. Horizontally-oscillating solitons were observed in [53, 54]. Some cases were found when a few solitons interact and form certain systems (Fig. 5.39).

Reasons for the appearance of weakly-oscillating solitary waves on a water surface were studied analytically [26, 27], and results of the calculations are presented in Fig. 5.40. Waves a and b are modelling the behaviour of 2 weakly-interacting solitons (Fig. 5.39b–d), and waves (c) are modelling the behaviour of

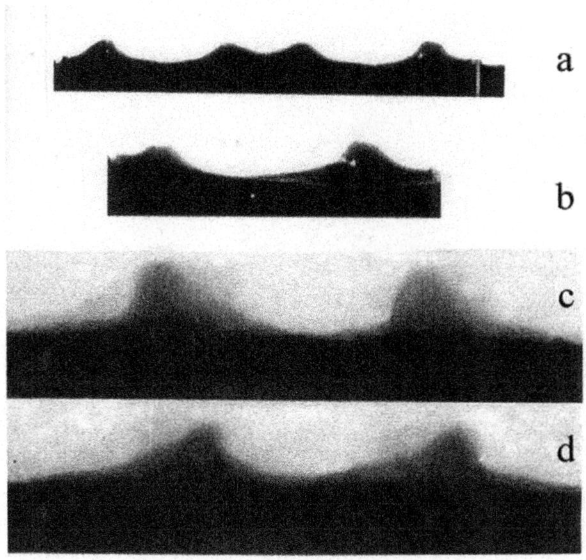

Fig. 5.39 Solitons horizontally vibrating with small amplitude on water surface of a vertically - excited layer (resonator) (**a–d**) [53, 54]

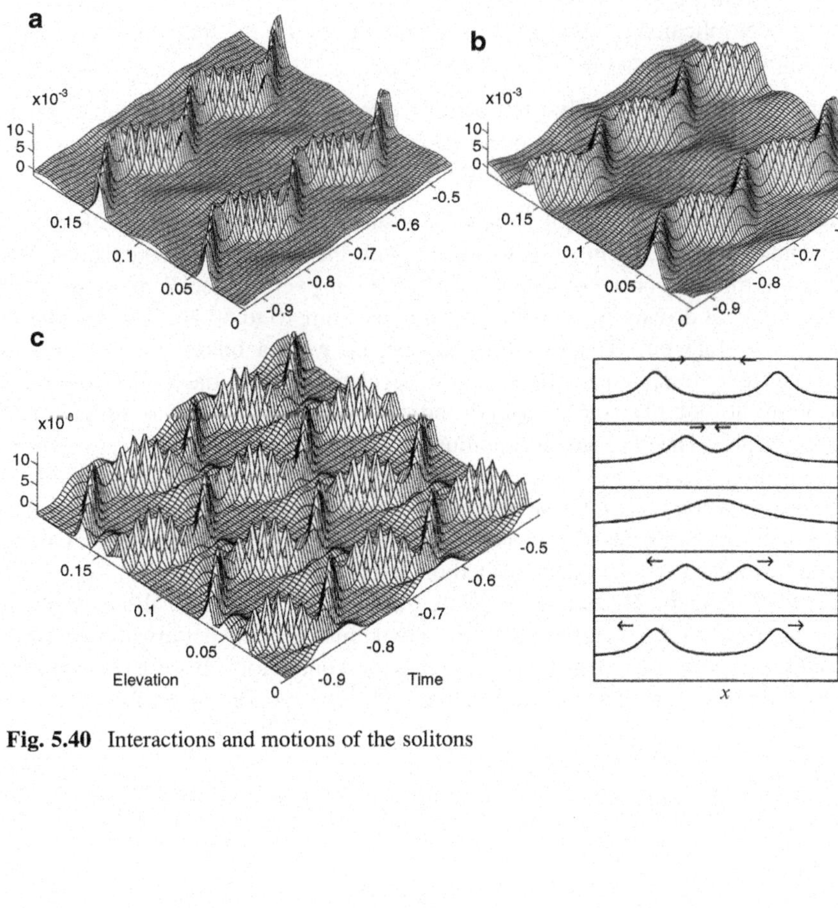

Fig. 5.40 Interactions and motions of the solitons

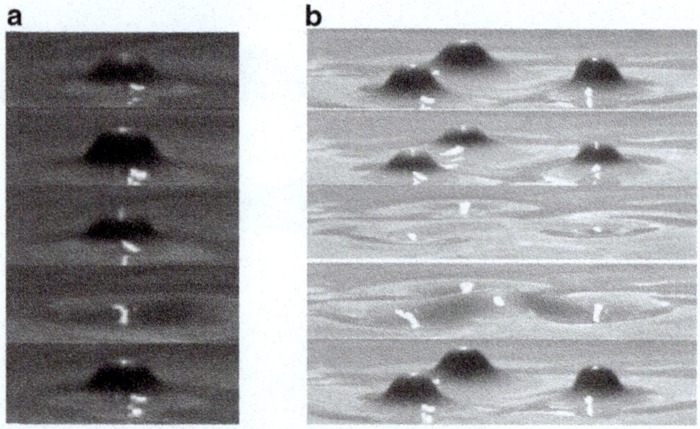

Fig. 5.41 Typical oscillons excited on a suspension surface. One oscillon (**a**) and three oscillons (**b**) [56]

4 weakly-interacting solitons (Fig. 5.39a). The wave oscillations shown in Fig. 5.40 (low right) correspond qualitatively to the waves b. When the frequency decreases, the effects of layer/base collision may increase, and the possibility of wave coupling may also increase. At a certain frequency they start to interact strongly. As a result, very complicated surface patterns may be formed if the frequency decreases but the amplitude is large enough.

Oscillons (localized vertically-oscillating waves) Oscillons on a granular layer were observed in 1996 [55]. Then oscillons were excited on suspension layers. The photos illustrating the last case are presented in Fig. 5.41.

In the experimental system [56] a layer of a suspension was vibrated vertically. A cubical container with 20 cm sides was used, with Plexiglas lateral boundaries. The container was mounted on a mechanical shaker providing vertical acceleration from 0 to 30 g. The range of driving frequencies was limited from 10 to 60 Hz. The maximum amplitude was 1.25 cm. The working suspensions were a mixture of water with commercial clay powder. At a critical value of acceleration the initial spatially-uniform surface loses its stability, and localized vertically-oscillating waves (oscillons) can appear (Fig. 5.41). The experiments are illustrated by Fig. 5.42.

Thus, from 1997 to 1999 some extraordinary surface waves were observed on water and suspension layers. Those observations have evoked much interest. Perhaps, similar waves can be excited by seaquakes, earthquakes and volcano tremors. On the other hand, similar waves may be excited on layers of granular materials.

We underline again 'Different dynamic patterns are yielded by these resonant travelling waves in the x-t and x-y planes. They simulate many patterns observed in liquid and granular layers, optical systems, superconductors, Bose-Einstein condensates, micro- and electron resonators. The harmonic excitation may be

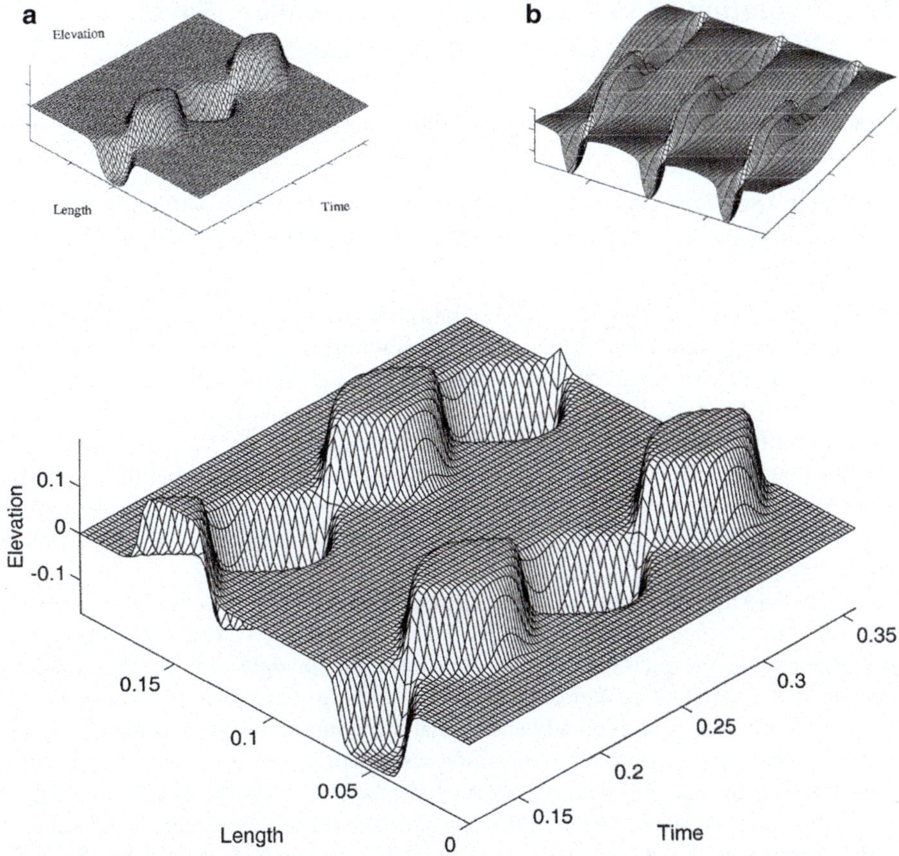

Fig. 5.42 Simulation of step-like standing surface oscillons excited in suspension: single oscillon **a** and oscillon triad **b** and two oscillons [18, 37]

compressed and transformed inside the resonant band into travelling or standing particle-like waves. The area of application of these solutions and results may possibly vary from the generation of nuclear particles, acoustical turbulence and catastrophic seismic waves to the formation of galaxies and the Universe. In particular, the formation of galaxies and galaxy clusters may be connected with nonlinear and resonant phenomena in the early Universe'.

In general, the results constructed some forms important the experimental data shown in Figs. 5.36 and 5.38. Of course, the superposition law was not valid exactly in the experiment. Therefore there are certain discrepancies between the experiments and the calculations.

5.5 Evolution of Vertically-Induced Granular Waves: From Jets to Breakers and Vibrating Solitons

Granular waves may play an important role in natural events such as earthquakes and avalanches.

Granular oscillons On a granular layer, strongly-localized vertically-vibrating waves can occur. Examples of such waves arising on a layer of small balls are shown in Fig. 5.43.

The oscillon is an axisymmetric excitation. It oscillates with half the frequency of the exciting frequency. Thus, it is a parametrically-excited localized wave. During one cycle of the excitation it is a peak; on the next cycle it is a crater. The oscillon may be started by touching the surface of brass balls with a pencil. After formation of the surface crater, the oscillon begins to bounce up and down, while the surrounding material stays in the same place. The oscillon height is usually larger than the layer thickness (Fig. 5.44). Thus, these excitations may be considered as strongly-nonlinear waves, where the vertical motion of the particles is connected with their horizontal motion (Fig. 5.44).

In Fig. 5.44 the horizontal white line defines the position of the base. Waves (a) and (b) (left) are formed immediately after layer/base collision. The travelling waves (c) demonstrate the picture after the collision. On the right side of Fig. 5.44 are presented the result of numerical modelling of the granular waves excited by the vibrating base. Fig. 5.44 explicitly shows that the standing waves existing during the flight are instantly transformed to travelling waves during the collision. Then during a new collision the travelling waves meet each other. As a result, they form granular columns on a new place. Thus, on the layer surface during the different moments of time there are two types of waves: standing waves (mostly with the period of oscillations) and travelling waves (at the moment of the collision and immediately after it).

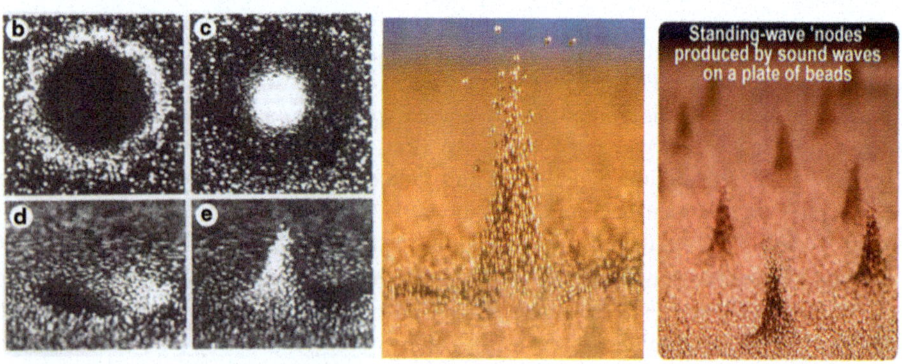

Fig. 5.43 Periodical granular column and crater on the surface of a vertically-excited layer: (**b**, **c**) are a bird view, (**d**, **e**), are a side view [55]. A granular column and two craters near it [57], and granular standing-wave 'nodes' [60]

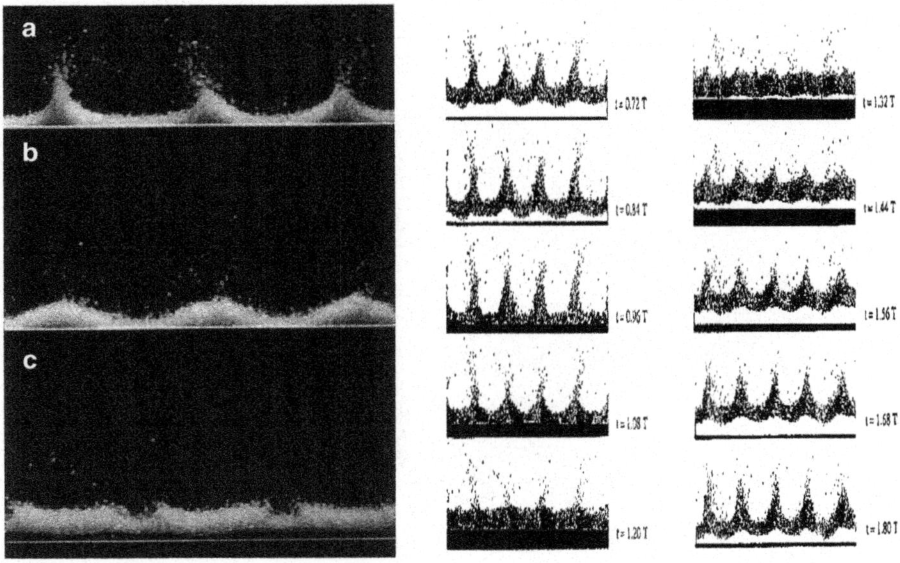

Fig. 5.44 Typical wave formation on granular layers under strong vertical vibrations: Experiment (*left*) [58], and numerical modelling (*right*) [59]

Figure 5.44 can be interpreted also in some other ways. We can assume that on a layer surface there is a wave oscillating with variable speed. Namely, during the moments of collisions the speed of the wave sharply increases. However, during the free flight of the layer this speed drops to zero. Such schematization of the wave pattern enables us to model the results of experiments (Fig. 5.45) [26].

According to the calculations the oscillons appear, when the vertical acceleration much exceeds g, the material separates from the base and free flight begins. All of the dynamics happens during the collision of the layer and the container, when the craters collapse and the material erupts. Surprisingly, but the appearance of oscillons can be explained qualitatively by the interaction of the resonant ripples (Fig. 5.45), which periodically generate peaks and craters on a perturbed surface. Taking into account that told above we can suggest that Fig. 5.35 determines the oscillons excited in the finite resonator.

Surface patterns The localised waves can form surface bands, squares and hexagons (Fig. 5.46) [60].

The area of existence of the specified patterns depend on the vertical acceleration and the forcing frequency [61]. Examples are given in Fig. 5.47 which correspond to certain patterns of Fig. 5.46.

Thus, we have described a class of strongly-localized non-linear waves, whose interaction defines the formation of various dynamic surface patterns. These waves

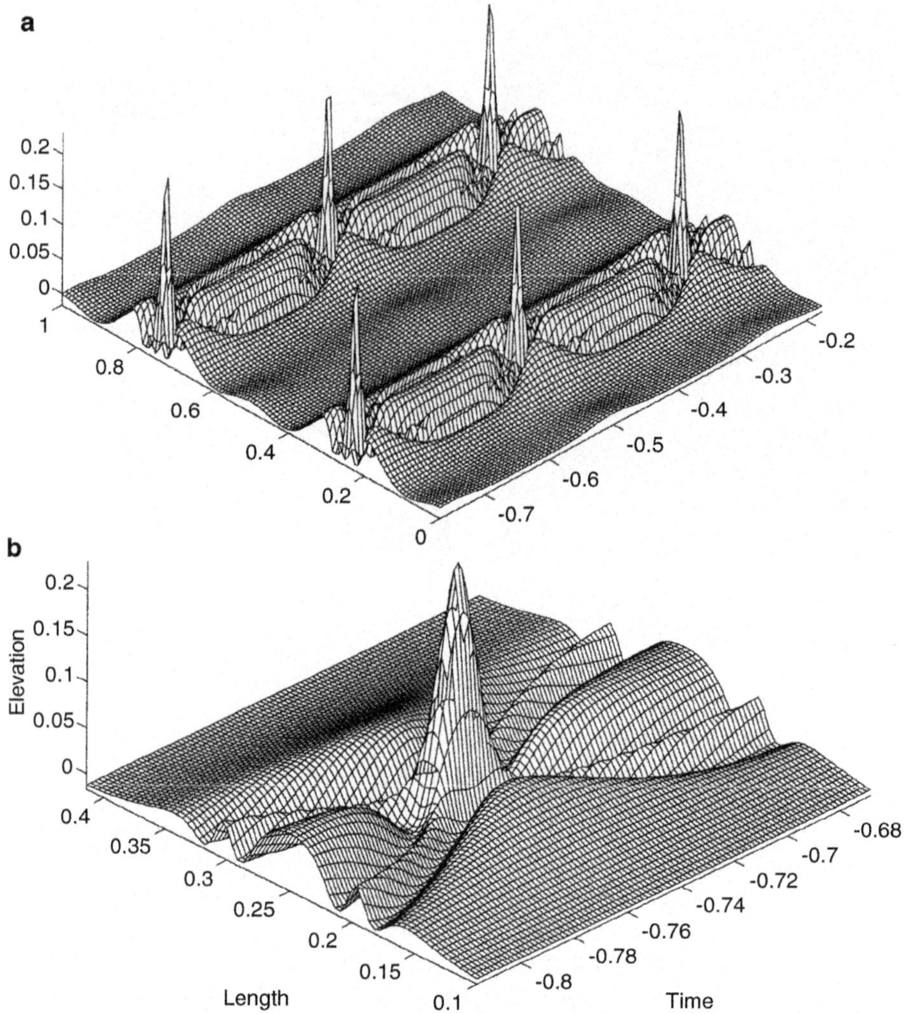

Fig. 5.45 Periodic eruptions of material and formation of surface craters (**a**). Detailed dynamics of the crater and the eruption (the formation of the oscillon) (**b**) [26]

arise and their amplitude sharply increases during the parametric resonances. We emphasize that the considered cases of oscillations do not describe all versions. For example, during very strong vertical excitation so-called 'arch' forms may be excited [62]. However, they become less and less steady during the process of increasing acceleration and the flight time. The same takes place for oscillations of compact masses of granules. It is interesting that oscillations of compact masses can be stable even for large vertical accelerations. For example, in [63] the flight time steadily exceeded 3 periods of the forcing oscillations.

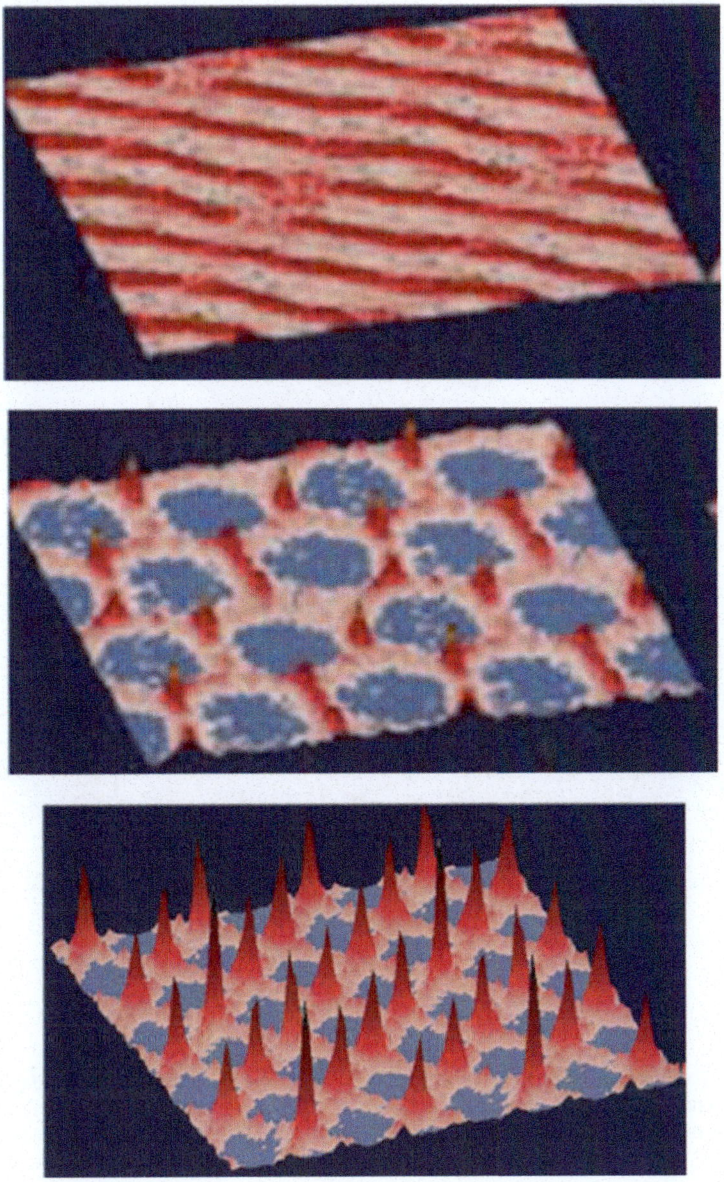

Fig. 5.46 Examples of the surface patterns [60]

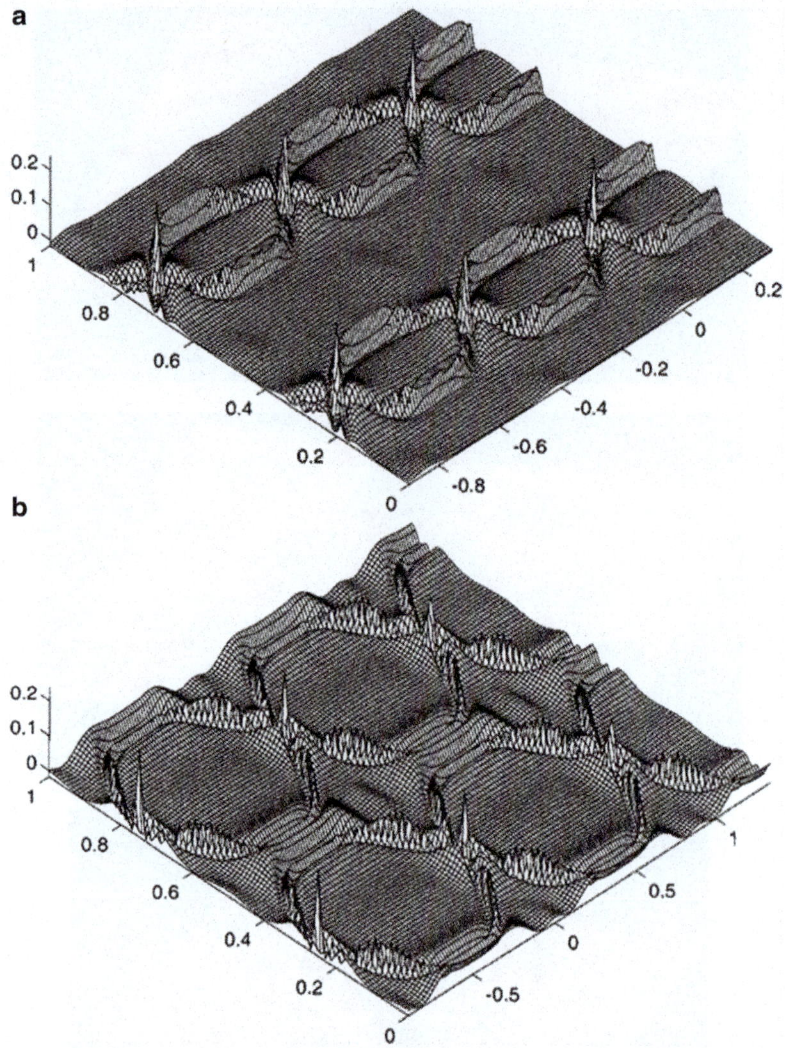

Fig. 5.47 Examples of surface patterns. The evolution of the oscillons into ellipsoidal, hexagonal and other complex periodic structures (**a**–**d**) [26, 27, 18]. Evolution of weakly-oscillating waves because of reduction of the forcing frequency: 15 Hz (a), 7 Hz (**b**) и 5 Hz (c) [26]

The results presented in this section are interesting as examples of extreme amplification of Faraday waves. At the same time they allow us to understand qualitatively the processes arising during great earthquakes and seaquakes. We underline that the localized oscillating waves (oscillons) began to use so that to describe wave processes in the very early Universe [18, 37, 38, 64–69] (see, also, Chap. 7).

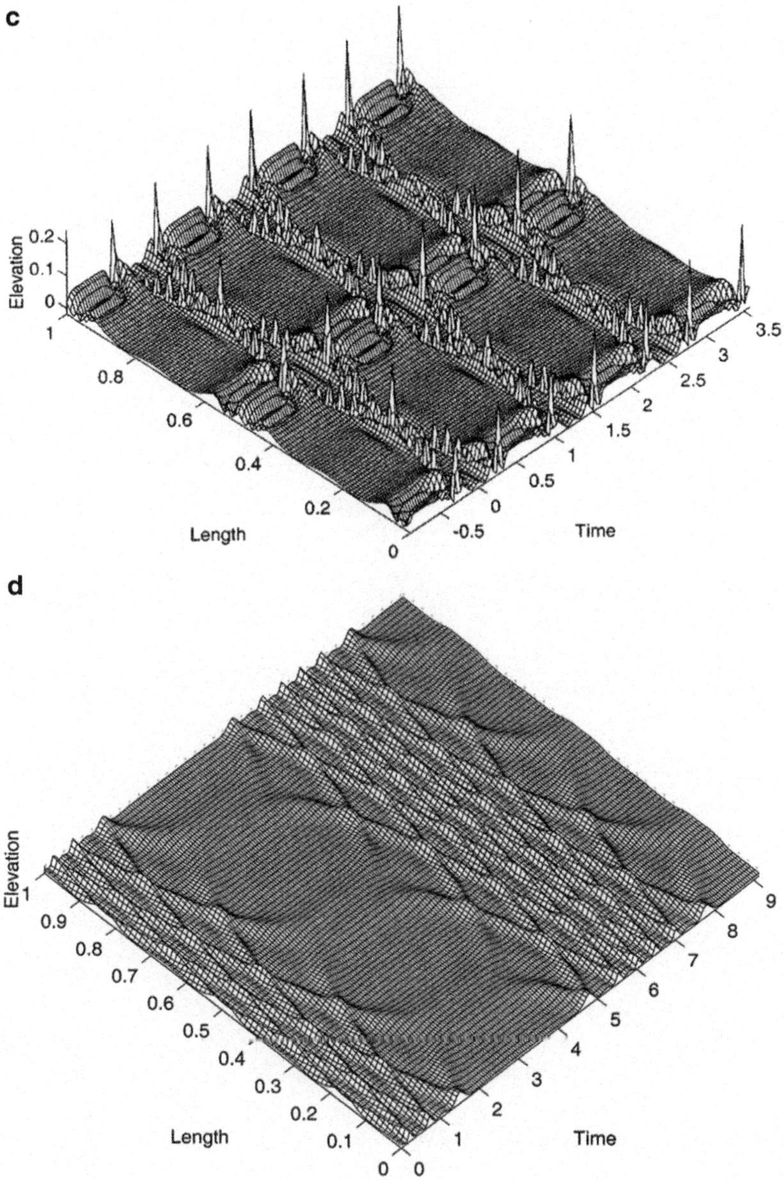

Fig. 5.47 (continued)

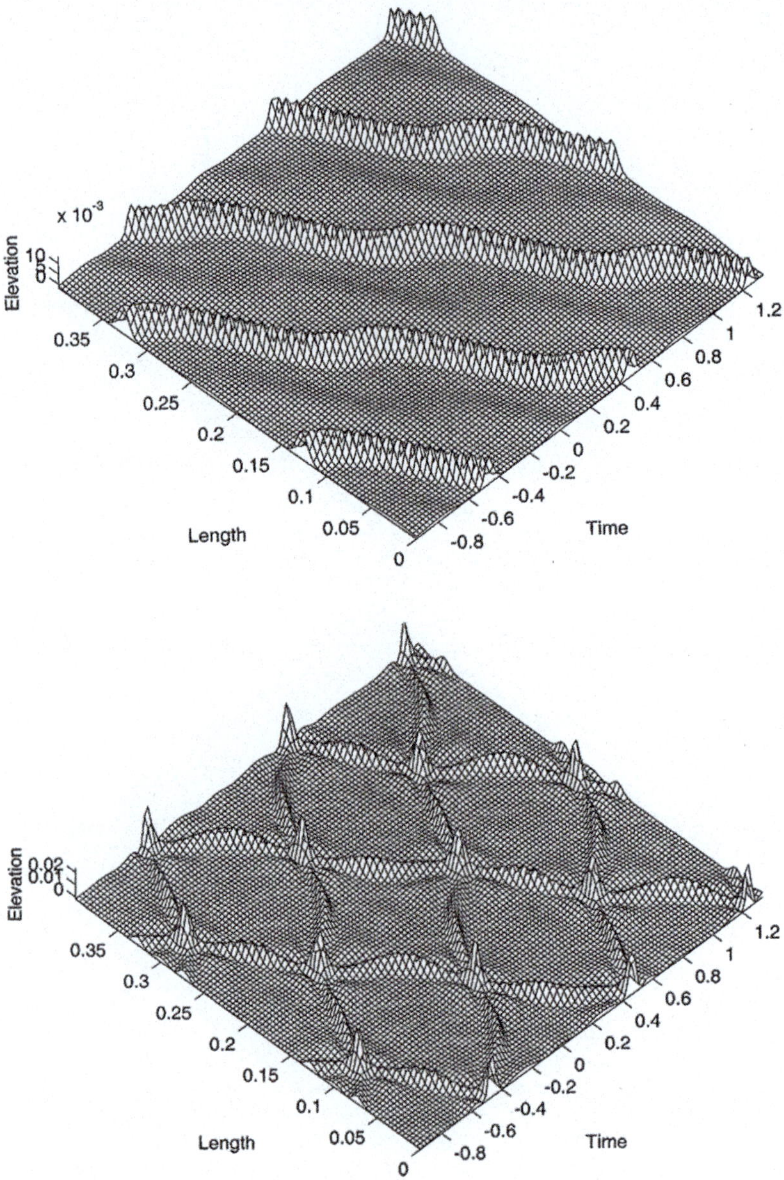

Fig. 5.47 (continued)

References

1. Darwin C (1890) Journal of researches into the natural history and geology of the countries visited during the voyage of H.M.S. Beagle around the world, under the command of Captain FitzRoy, R.N. T. Nelson and Sons, London
2. Darwin C (1839) Journal of researches into the geology and natural history of the various countries visited by H.M.S. Beagle, under the command of Captain FitzRoy, R.N. from 1832 to 1836. Henry Colburn, London
3. Darwin C (1840) On the connexion of certain volcanic phenomena in South America; and on the formation of mountain chains and volcanoes, as the effect of the same power by which continents are elevated. (Read 7 March 1838). In: Barrett PH (ed) (1977) The Collected Papers of Charles Darwin. The University of Chicago Press, Chicago
4. Stein S, Okal EA (2005) Speed and size of the Sumatra earthquake. Nature 434:581–582
5. Liu PL-F, Yeh HC, Synolakis C (eds) (2008) Advanced numerical models for simulation tsunami waves and runup. World Scientific, Hackensack
6. Mei CC, Stiassnie MS, Yue DK-P (2005) Theory and applications of ocean surface wave. Part 2: nonlinear aspects. World Scientific, London
7. Imamura F, Shuto N, Ide S, Yoshida Y, Abe K (1993) Estimate of the tsunami source of the 1992 Nicaraguan earthquake from tsunami data. Geophys Res Lett 20(14):1515–1518
8. Tadepalli S, Synolakis CE (1994) The run-up of N-waves on sloping beaches. Proc R Soc Lond A 445:99–112
9. Synolakis CE, Bernard EN (2006) Tsunami science: before and beyond Boxing Day 2004. Phil Trans R Soc A 364:2231–2265
10. Fritz H, Kalligeris N (2008) Ancestral heritage saves tribes during 1 April 2007 Solomon Island tsunami. Geophys Res Lett 35, L01607
11. Taylor FW, Briggs RW, Frohlich C, Brown A, Hornbach M, Parabatu AK, Meltzner AJ, Billy D (2008) Rupture across arc segment and plate boundaries in the 1 April 2007 Solomons earthquake. Nat Geosci 1:253–257
12. Satake K (2007) Tsunamis. In: Schubert G (ed) Treatise on geophysics, vol 4. Elsevier, Amsterdam/Boston, pp 483–511
13. Segur H (2007) Waves in shallow water, with emphasis on the tsunami of 2004. In: Kundu A (ed) Tsunami and nonlinear waves. Springer, Berlin/New York
14. Gisler GR (2008) Tsunami simulation. Annu Rev Fluid Mech 40:71–90
15. Weiss R, Wunnemann K (2007) Large waves caused by oceanic impacts of meteorites. In: Kundu A (ed) Tsunami and nonlinear waves. Springer, Berlin
16. Galiev SU (2008) Strongly-nonlinear wave phenomena in restricted liquids and their mathematical description. In: Perlidze T (ed) New nonlinear phenomena research. Nova Science Publishers, New York
17. Galiev ShU (2009) Modelling of Charles Darwin's earthquake reports as catastrophic wave phenomena. http://researchspace.auckland.ac.nz/handle/2292/4474
18. Galiev SU (2011) Геофизические Сообщения Чарльза Дарвина как Модели Теории Катастрофических Волн (Charles Darwin's geophysical reports as models of the theory of catastrophic waves). Centre of Modern Education, Moscow (in Russian)
19. Mader CL (2004) Numerical modelling of water waves. CRC Press, London
20. Monin AS, Korchagin NN (2008) Десять Открытий в Физике Океана (Ten discoveries in physics of ocean). Scientific World, Moscow (in Russian)
21. Levin B, Nosov M (2009) Physics of tsunamis. Springer, Dordrecht/London
22. Faraday M (1831) On a peculiar class of acoustical figures, and on certain forms assumed by group of particles on vibrating elastic surface. Phil Trans R Soc 121:299–340
23. Natanzon MS (1977) Продольные Автоколебания Жидкостной Ракеты (Longitudinal self-excited oscillations of a liquid-fuel rocket). Mashinostroenie, Moscow (in Russian)
24. Galiev SU, Iakovtsev AV, Zelenuk NI (1987) Effect of the breakdown of a fluid on its oscillations in a resonator. Strength Mat 12:1715–1720

25. Galiev SU (1998) Topographic amplification of vertical-induced resonant waves in basins. Adv Hydrosci Eng 3:179–182
26. Galiev SU (1999) Topographic effect in a Faraday experiment. J Phys A Math Gen 32:6963–7000
27. Galiev SU (2000) Unfamiliar vertically excited surface water waves. Phys Lett A 266:41–52
28. Zamankhan P (2012) Solid structures in a highly agitated bed of granular materials. Appl Math Model 36:414–429
29. Wright J, Yon S, Pozrikidis C (2007) Numerical studies of two-dimensional Faraday oscillations of inviscid fluids. J Fluid Mech 588:279–308
30. Niederhaus CE, Jacobs JW (2003) Experimental study of the Richtmyer-Meshkov instability of incompressible fluids. J Fluid Mech 485:243–277
31. Luo J-J (2011) Ocean dynamics not required. Nature 477:544–546
32. Longuet-Higgins MS (1983) Bubbles, breaking waves and hyperbolic jets at a free surface. J Fluid Mech 127:103–121
33. Bredmose H, Brocchini M, Peregrine DH, Thais L (2003) Experimental investigation and numerical modelling of steep forced water waves. J Fluid Mech 490:217–249
34. James AJ, Vukasinovich B, Smith MK, Glezer A (2003) Vibration-induced drop atomization and bursting. J Fluid Mech 476:1–28
35. Das SP, Hopfinger EJ (2008) Parametrical forced gravity waves in a circular cylinder and finite-time singularity. J Fluid Mech 599:205–228
36. Zeff BW, Kleber B, Fineberg J, Lathrop DL (2000) Singularity dynamics in curvature collapse and jet eruption on a fluid surface. Nature 403:401–404
37. Galiev SU, Galiyev TS (2001) Nonlinear transresonant waves, vortices and patterns: from microresonators to the early Universe. Chaos 11:686–704
38. Galiev ShU (2003) Transresonant evolution of waves and the Richtmyer-Meshkov instability: from microstructures to the Universe. In: Khovanskii AG, Nourgaliev DK, Petrova NK, eds. Proceedings of the International Conference. 'New Geometry of Nature.' Kazan, August 25 – September 5, 2003. Vol. 1, pp 88–91
39. Galiev SU (2003) The theory of non-linear transresonant wave phenomena and an examination of Charles Darwin's earthquake reports. Geophys J Inter 154:300–354
40. Galiev SU (2003) Transresonant evolution of wave singularities and vortices. Phys Lett A 311:192–199
41. Taylor GI (1969) Instability of jets, threads, and sheets of viscous fluid. In: Proceedings of the 12th International Congress of Applied Mechanics, Stanford (1968). Springer, Berlin
42. Love AEH (1944) The mathematical theory of elasticity. Cambridge University Press, Cambridge
43. Levien R (2008) The elastica: a mathematical theory. http://levien.com/phd/elastica_hist.pdf
44. Pocheau A, Roman B (2004) Uniqueness of solutions for constrained Elastica. Physica D192:161–186
45. Jones N (1989) Structural impact. Cambridge University Press, Cambridge
46. Karsten RH, Swaters GE (2000) Nonlinear effects in two-layer large-amplitude geostrophic dynamics. Part 2. The weak-beta case. J Fluid Mech 412:161–196
47. Sarpkaya T (2002) Experiments on the stability of sinusoidal flow over a circular cylinder. J Fluid Mech 457:157–180
48. Schowalter DG, van Atta CW, Lasheras JC (1994) A study of streamwise vortex structure in a stratified shear layer. J Fluid Mech 281:247–291
49. Arnol'd VI (1981) Теория Катастроф (Catastrophe theory). Znanie, Moscow
50. Jiang L, Perlin M, Schultz WW (1998) Period tripling and energy dissipation of breaking standing waves. J Fluid Mech 369:273–299
51. Kalinichenko VA (2009) Nonlinear effects in surface and internal waves of Faraday. Dissertation, Russian Academy of Sciences, Ishlinsky Institute for Problems in Mechanics
52. Russell JS (1844) Report on waves. Fourteenth meeting of the British Association for the Advancement of Science

53. Wang X, Wei R (1997) Dynamics of multisoliton interactions in parametrically resonant systems. Phys Rev Lett 78(14):2744–2747
54. Tu J, Lin H, Lu L, Chen W, Wei R (2002) Spatiotemporal evolution form a pair of like polarity solitons to chaos. Phys Lett A 304:79–84
55. Umbanhowar PB, Melo F, Swinney HL (1996) Localized excitations in a vertically vibrated granular layer. Nature 382:793–796
56. Lioubashevski O, Hamiel Y, Agnon A, Reches Z, Fineberg J (1999) Oscillons and propagating solitary waves in a vertically vibrated colloidal suspension. Phys Rev Lett 83(16):3190–3193
57. Suplee C (1999) Physics in the 20th century. Harry N. Abrams, New York
58. Cerda E, Melo F, Rica S (1997) Model for subharmonic waves in granular materials. Phys Rev Lett 79(23):4570–4573
59. Clement E, Labous L (2000) Pattern formation in a vibrated granular layer: the pattern selection issue. Phys Rev E 62(6):8314–8323
60. Shinbrot T, Muzzio FJ (2001) Noise to order. Nature 410:251–258
61. Cross MC, Hohenberg PC (1993) Pattern formation outside of equilibrium. Rev Mod Phys 65:851–1112
62. Douady S, Fauve S, Laroche C (1989) Subharmonic instability and defects in a granular layer under vertical vibrations. EuroPhys Lett 8(7):621–627
63. Wassgren CR, Brennen CE, Hunt ML (1996) Vertical vibration of a deep bed of granular material in a container. Trans ASME J Appl Mech 63:712–719
64. Belova TI, Kudryavtsev AE (1997) Solitons and their interactions in classical field theory. Physics Uspekhi 40(4):359–386
65. Easther R, Parry M (2000) Gravity, parametric resonance, and chaotic inflation. Phys Rev D62:103503
66. Graham N, Stamatopoulos N (2006) Unnatural oscillons lifetimes in an expanding background. arXiv:hep-th/0604134v2
67. Graham NN (2007) An electroweak oscillon. Phys Rev Lett 98(10):101801
68. Farhi E, Graham N, Guth AH, Iqbal N, Rosales RR, Stamatopoulos N (2008) Emergence of oscillons in an expanding background. Phys Rev D77:085019
69. Amin MA, Easther R, Finkel H, Flauger R, Hertzberg MP (2012) Oscillons after inflation. Phys Rev Lett 108:241302

Chapter 6
Extreme Wave/Ship Interaction

.....a great sea broke over us, ...The poor Beagle trembled at the shock,... (***Darwin***)

6.1 Extreme (Catastrophic) Ocean Waves

Catastrophic (extreme) ocean waves differ from tsunamis. Tsunamis become dangerous only when they come to the coastal zone. On the contrary, catastrophic ocean waves most often arise far from the coast. The *Beagle* met an extreme wave near Cape Horn. Darwin wrote ... *On the 13th the storm raged with its full fury; our horizon was narrowly limited by the sheets of spray borne by the wind ... At noon a great sea broke over us, and filled one of the whale-boats, ... The poor Beagle trembled at the shock, and for a few minutes would not obey her helm; but soon, like a good ship that she was, she righted and came up to the wind again. Had another sea followed the first, our fate would have been decided soon and for ever* ... [1, pp. 264–265].

This description indicates that the *Beagle* met a storm transforming into a hurricane. According to the standard scale of Sir Francis Beaufort, during a storm there are extremely high waves whose crests are blown off in foam. The maximum height of these waves is up to 16 m, an average height is 11.5, the wind is 28.5–32.6 m/s. During a hurricane the wind speed may be larger, the air is filled with foam and splashes, and the visibility is very bad. The situation is extremely dangerous for small vessels, such as the *Beagle*. Thus, according to the scale, the storm described by Darwin was a hurricane, for which the ocean around Cape Horn is famous. Huge waves reaching 16 m kept the crew under extreme pressure. However, during such an event which the crew had probably got used to, the vessel was shaken by a single huge wave. From Darwin's description, it appears that this wave may be described as catastrophic. Only during recent years have these waves drawn the attention of the public and of scientists [2–18]. Indeed, over the last 20 years more than 200 super-carrier-cargo ships over 200 m long have been lost at sea. Eyewitness reports suggest that many were sunk by high and violent walls of water that rose up out of calm seas [2, 8] (see, also, Sect. 3.3.6).

Sh.U. Galiev, *Darwin, Geodynamics and Extreme Waves*,
DOI 10.1007/978-3-319-16994-1_6

For many years these ocean monsters were considered as fantasy; and many marine scientists clung to statistical models stating that monstrous deviations from the normal sea state occur once every 1,000 years.

An extreme wave is often defined as a wave whose height exceeds twice the significant height H_s, which is defined as the average height of one third of the highest waves.

One of the consequences of the possible presence of these large extreme (rogue) waves is that they could lead to unprecedentedly large ratios of H_{max}/H_s. How high can this ratio be expected to be, or is there an upper limit for this ratio?

Based on the Rayleigh distribution, for ratios of H_{max}/H_s to be equal to 3, 4, 5, and 6, for instance, it simply requires the number of waves to be 6.5×10^7, 7.9×10^{13}, 7.9×10^{21} and 1.9×10^{31} respectively. These are extremely large numbers, which translate into millions of years for the ratios to occur. Basically, the Rayleigh approach considers large ratio of H_{max}/H_s as extremely rare occurrences [3].

Therefore, until recent years, researchers have considered these waves to be legends and myths of sailors. However, these monsters have now documented convincingly. It has been confirmed, that such monsters are much more common than mathematical probability theory had predicted.

It is common for ocean storm waves to reach 7 m in height, and in extreme conditions such waves can reach heights of 15 m. However, in recent years researchers have observed the existence of vastly more massive waves — veritable monsters up to 30 m high (approximately the height of a 10-storey building). They can appear without warning in the open ocean, against the prevailing current and the wave direction, and often in perfectly clear weather. Such waves were said to consist of an almost vertical wall of water preceded by a trough so deep that it was referred to as ‘a hole in the sea’. A ship encountering this wave would almost certainly be sunk in a matter of seconds.

These waves were named extreme waves (also giant or rogue, or murderers). We also call these waves catastrophic, because they represent huge danger to ships. On the other hand, curved form of these waves have often folds. Similar forms are being studied by catastrophe theory. These waves differ essentially from tsunamis, which get amplified only near a coast. Therefore, as we repeatedly mentioned, tsunamis are dangerous for coastal structures and for ships close to a coast, while extreme waves represent catastrophic danger to ships and sea structures both near a coast, and in the open ocean.

However, the absolute height of a wave is less important to sailors than is its steepness. Most wind-induced waves have steepness of the form kA, where A is the wave amplitude and k is the wavenumber. The steepness is of the order from 0.03 to 0.06. Steeper waves can present problems for ships, but fortunately the wave steepness very rarely exceeds 0.1. However, in opposing currents the steepness of storm waves may be up to 1/7. In general, the wave steepness diminishes with increasing wavelength. That is true, but not true for extreme waves which may be very steep.

In about 2003 the European Space Agency (ESA) found that catastrophic waves do occur in the ocean much more often than had been supposed earlier. Around 30,000 separate images were made by two satellites during 3 weeks in 2001. Radars monitored the sea surface in a rectangle measuring 10 by 5 km (6 by 2.5 miles). ESA reports that the survey revealed 10 massive waves – some nearly 30 m (100 ft) high. This conclusion, which has been confirmed by independent measurements of waves in the South Atlantic, could change completely the approach to standards of safety for ships and sea structures. It shows that existing criteria of safety must be reconsidered. Such criteria cannot rely only upon observations of catastrophic waves, since those are rare events. Theoretical models can serve as the basis, together with unconventional methods of observation, first of all from oil platforms or from satellites.

For example, the Draupner wave was a single giant wave which struck an oil platform in the North Sea on New Year's Day 1995 (Fig. 6.1).

The first accurate measurements of a catastrophic ocean wave were recorded on the offshore oil platform 'Draupner', for that well-known 'New Year Wave' (Fig. 6.1). The wave measurements were made in the North Sea by means of a laser device, and a 20-min fragment of the record is shown in Fig. 6.1. The depth of the sea around the platform is 70 m. The significant height of the waves, running on the platform that day, was 11.9 m, and the period of the waves was 10.8 s. The observed catastrophic wave was vertically nonsymmetrical. The crest height (18.6 m) much exceeded the depth of the wave hollow (near 7 m). The wave was steeper and slightly shorter (about 220 m in length) than the other waves. The dispersion parameter was $kh \approx 2$, with $h/\lambda \approx 0.3$. The wave steepness was $ka \approx 0.37$, if $A \approx 13$ m. Here h is the depth, λ is the wave length. Thus, this wave was enough long for it to be within the limits of the theory of long waves [18].

The front of such waves often forms an almost vertical wall of water. In a collision between the wave and a ship, the outcome depends mainly on the dimensions and mass of the ship, the wave parameters, and, especially on the steepness of the wave front. The steepness of the wave determines the collision's pressure on the hull [19–28].

Some waves might be very long. The ship has often time to avoid similar waves. But short waves appear sometimes from nowhere, break and disappear. In 1980 the Russian tanker Taganrogsky Zaliv met an extreme wave (Fig. 6.2). It is interesting that in this case, the wave width was about the same as the ship width (generally speaking, the width of extreme waves can reach several hundred metres) [27].

Extreme waves become especially dangerous during the breaking [21, 26]. Engineers, designing ships and building offshore platforms that are farther and farther from land, obviously require detailed knowledge about impact loads from these extreme waves.

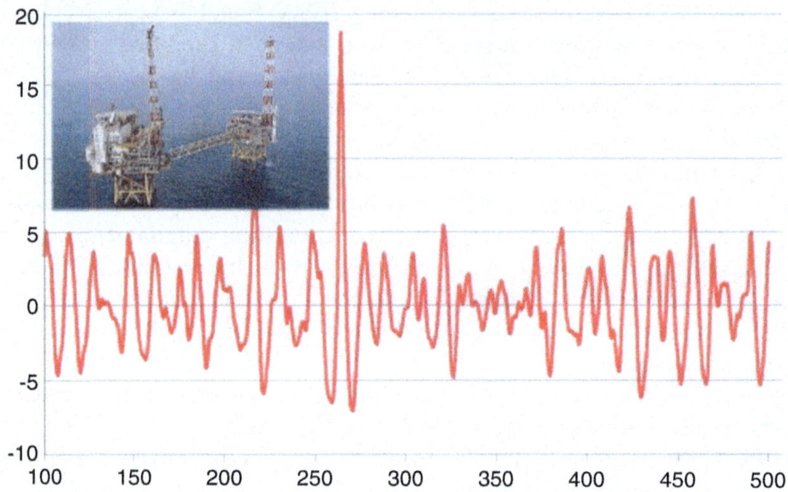

Fig. 6.1 The Draupner wave record. The wave height is in metres, time is in seconds. Internet

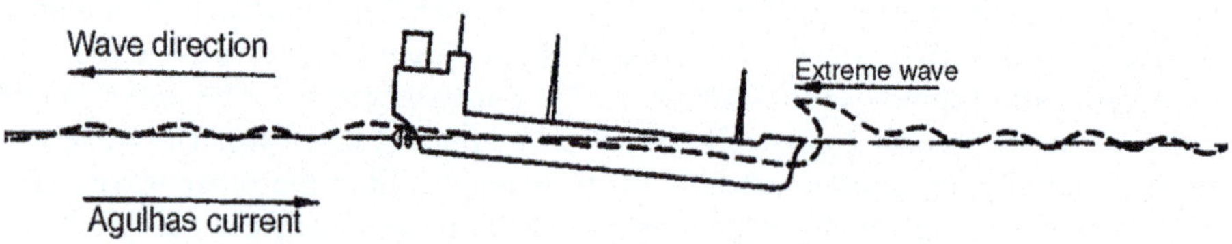

Fig. 6.2 Extreme wave collision with the *Taganrogsky Zaliv* described by a witness [27]

6.2 Reasons for Catastrophic Ocean Wave Generation

There are many kinds of waves on the ocean. Although each wave is characterized by general elements, such as the wave crest and the trough preceding or following the crest, they do have strong differences in the form, speed and type of origin. For example, very short waves (ripples) may be generated by a breeze. At the same time very long waves (tides and tsunami) may be on the water surface. The long waves may be generated by ocean currents, or by atmospheric pressure (hurricanes), or by internal waves. It is known that the length of internal waves can be of the order of 10 km, and their amplitudes may be very large. Therefore, the internal wave crests can interact with the ocean surface. But most waves are raised by the wind. Such waves depend on the wind velocity and the distance over which the wind blows. Typical ocean waves can have length from 100 to 300 m. Thus, usually the length of the ocean waves is much smaller than for internal waves.

If the atmospheric pressure decreases while the wind speed increases, two wave systems may be generated: wind-induced waves and long waves generated by a cyclone. On the other hand, a short wind gust can generate a short wave on the surface of a long swell. The short wave can begin to 'soak' energy from the long waves, especially when the wave speeds and the phases are similar or coincide. In this case, resonant amplification of the short wave can take place.

In an initial stage of development the wind waves can propagate as parallel lines, which then break up to the isolated ridges (three-dimensional excitation). The water surface excited by the wind gets a complex relief, continuously changing in time. If the wind which has caused the excitement abates, the wind waves will gradually be transformed into the free waves called swells, having more regular form than the wind waves, and greater length. As these waves propagate almost without loss of energy they can run huge distances. So the swells formed near the Cape Horn can run a distance up to 10,000 km. Most often there is a commixed excitation at which the swells and the wind waves are simultaneously observed. On the ocean surface always there are wind waves and swells of different dimensions (sometimes reaching length of 800–1,500 m, height of 12–13 m and speeds of 14–15 m/s).

In a windy area, waves of different heights and periods can appear. These waves have different directions of propagation. However, a group of waves possessing the same direction and the speed as the wind can begin to grow (Fig. 6.3). If the wind duration is long enough, then the group of waves may be amplified up to extreme amplitudes within the resonant band. Generally speaking, the length of these bands may be close to the wavelength. The band length can depend on wave stability. This band can disappear after the breaking of the wave.

Other waves will not increase in size. In this process the wind acts as a filter. As a result, only the group of waves which resonates with the wind is amplified. Thus, only if the resonant conditions are fulfilled (strong temporary and spatial correlations of the wind and the waves) can strong amplification of the waves take place. The amplification is limited by the dissipation of wind energy within the waves.

A British research team has observed some of the biggest sea waves ever measured. A whole series of giant waves were so big, that according to computer models, they should not even exist [29]. The data were collected on cruise 245 on the RRS Discovery (January–February 2000). Meteorologists had predicted a violent storm. The scientists – a team from Britain's National Oceanography Centre – wanted to observe it closer. What they ended up experiencing went far beyond anything they could have imagined – and could have cost them their lives.

Near the isolated rock called Rockall, 250 km west of Scotland, enormous waves came racing toward the vessel. When they checked their measuring instruments later, the scientists discovered that the tallest of these monster waves were nearly 30 m! 'We were shaken up by these waves for 12 h,' said Naomi Holliday, the leader of the expedition.

For our research it is important that the scientists observed the waves caused by wind. Groups of waves consisting of a wave of the maximum height and two accompanying waves of slightly smaller amplitude have been noted (Fig. 6.4). These waves can be related to the so-called 'three sisters'. These waves are without very steep fronts. We notice also that the waves presented on Fig. 6.4 are almost symmetrical relatively the undisturbed smooth surface of ocean. Thus, they differ essentially from the single 'New Year' wave shown in Fig. 6.1.

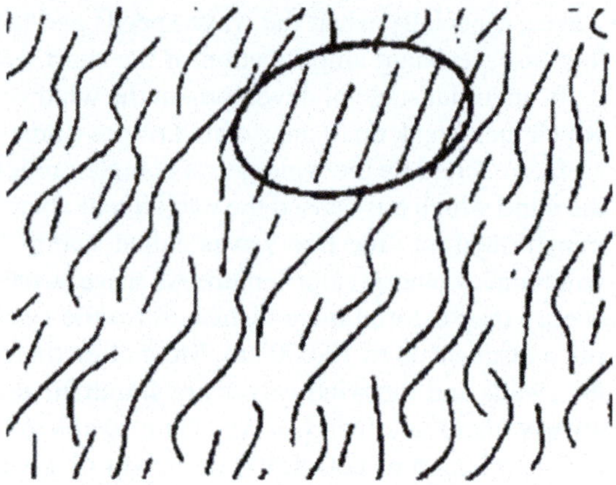

Fig. 6.3 The area (the resonant band) of local resonant amplification of wind-induced ocean waves [28]

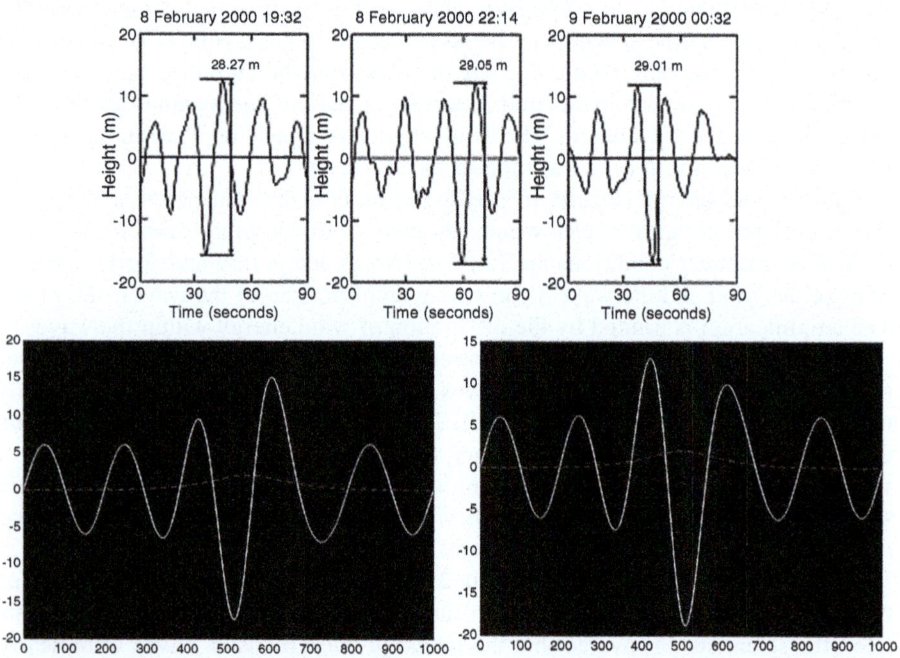

Fig. 6.4 Waves with height more than 29 m were measured by a British oceanographic research vessel near the isolated rock of Rockall (Scotland). Records of three biggest catastrophic waves are given [29] (*upper pictures*). Results of calculations [28] (*low pictures*). *Dashed lines* denote resonant bands

For thousands of years, people have learned to struggle with dangers of the ocean. Now, sailing directions specify a safe route, weather forecasters warn about storms, satellites observe icebergs and other dangerous objects. However, it is still not clear how to be saved from waves about 30 m high, which unexpectedly arise without any visible reasons.

Whence are the catastrophic waves? Universal and satisfactory answers to this question do not exist. The frequency of their occurrence differs markedly from theoretical predictions. Some theories which attempt to explain this phenomenon have been developed. A few of them have been successful in some aspects, but none has yet gained full acceptance [2, 6–8, 28].

Nevertheless, some empirical data about conditions which increase the possibility of a huge wave are known. For instance, if the wind drives the waves against a strong current, it can lead to an appearance of high steep waves. For example, Cape Agulhas (the southernmost point of Africa) is notorious. Other zones of large danger are the Kuroshio Current, the Gulf Stream, the North Sea and adjoining areas.

In addition, experts name the following preconditions for the occurrence of catastrophic waves: (1) low-pressure areas; (2) wind blowing in one direction for 12 or more hours; (3) waves, moving with the same speed as the pressure area; (4) fast waves catching up some slower waves and merging with them. According to some hypotheses, an occurrence of such waves may be explained by a steep slope of the seabed, by a strong wind and by large surges radiating from areas of the Antarctic Ocean.

Reasons for appearance of catastrophic waves may be most varied. In particular, Darwin wrote about an extreme wave which had arisen in the case of wind speed about 27–28 m/s. We do not exclude another possibility, the resonant mechanisms of amplification [28, 29].

6.2.1 Underwater Topographies and Topographic Resonance

Now we will consider an example, which explains ideas about nonlinear resonant self-excitation of the surface wave. On the ocean surface there can simultaneously be waves having different lengths and directions. They propagate with different speeds depending on the depth, the wind, the current and other factors (Figs. 6.3 and 6.5). Here we demonstrate the depth influence. We consider the surface waves crossing an underwater topography (mountain or bank) (Fig. 6.5).

The wave may be strongly amplified over the top of the underwater topography. A resonant band exists around the top, where the surface waves are amplified. We will use the Lamb's result [30] to illustrate this amplification

$$\eta = \frac{h p_0}{\rho(U^2 - gh)}. \tag{6.1}$$

Here η is the water elevation, h is the depth, ρ is a constant, p_0 is the wave and U is the wave speed. The resonance (topographic resonance) takes place, if $U^2 \to gh$.

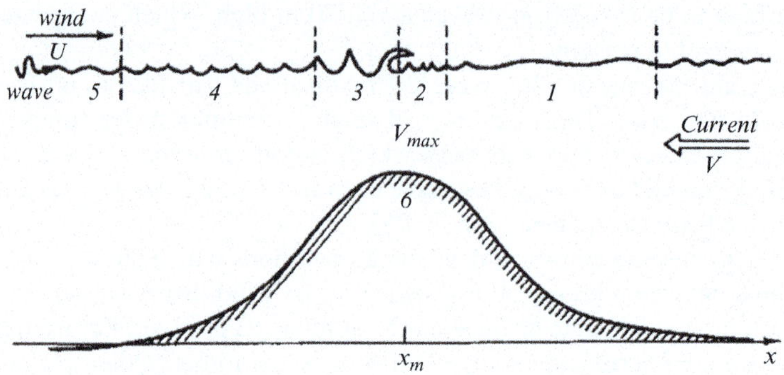

Fig. 6.5 Scheme of zones where different type of waves are generated: *1* – calm water; *2* – transition zone; *3* – breaking zone; *4* – whirlpool; *5* – background waves; *6* – submarine bank; x_m – point of the maximum speed [27, 28]

According to (6.1) the wave amplitude may be maximal when the wave crosses the top of the underwater topography. The amplitude is infinite, if $U^2 = gh$. For the real situation the elevation grows up only during finite time. After some time the wave breaks as a result of nonlinear effects.

It is obvious that within the resonant band, where $U^2 - gh$ is small, various waves can be generated. For example, near the band border (near the plane seabed) the quadratic effects can prevail. In this case, waves having a steep front may be generated. However, near the top (Fig. 6.5) cubic nonlinear effects can prevail. As a result the wave can break.

There are areas of fast variation of the depth in seas and the ocean, for example, coastal areas. In the open ocean there are regions where the seabed almost reaches the surface. The catastrophic waves can form above the noted underwater topographies. In particular, the area of the Bermuda region is notorious for the many ships which have perished there. It is interesting to note that the seabed of the Bermuda region differs strongly from the bottom relief in other places of the ocean.

Various types of seabed are there found close together: abyssal plains (35 %), underwater shelves with shallow banks (25 %), a continental slope (18 %), edge and median plateaus (15 %), deep-water trenches (5 %), deep water straits (2 %). Here we specify the relevant areas of the formations.

They tower above the bottom plane by 150–200 m or more. Some underwater mountains are deep under water, others appear above the surface in the form of islands. In the same region the deep-water Puerto Rico Trench is located with depth of 8,742 m and length of 1,550 km. That is the maximum depth of the Atlantic Ocean. But that depth is in close proximity to the Bahama Banks, on which the depth is only a few metres.

We will present results of calculations for the catastrophic resonant waves. The wave will be regarded as extreme if its amplitude is doubled (at least) during the resonant process.

6.2.2 Discussion of Some Ocean Resonances

In particular, very long waves can completely cover some underwater topographic features. In other cases, the wave profile lies within various zones of the resonant band. In these situations, the elevation of the various points of the wave depends on the various points of the resonant band and the initial wave shape. The waves arising under such circumstances we called transresonant waves [28, 31].

We already mentioned above the transresonant waves. The evolution of similar ocean waves may be very complex. In particular, we assume that the formation and the evolution of the extreme ocean waves are defined by the transresonant development of ocean waves, which are quite small far from the resonant band. Let us notice that much of Chaps. 5 and 7 of this book are devoted to research on resonant and transresonant transformations of different waves.

These concepts are illustrated by Fig. 6.6. We modelled the experimental and numerical data described in [32]. In the experiments, nonlinear waves propagating along a channel with underwater trapezium-like topography were studied. The source of the waves was located at the beginning of the channel. The wave curves in Fig. 6.6 were calculated in [28, 33, 34] using the analytic solution. The results of our calculations are presented in Fig. 6.6 for three different waves.

The wave profile changes strongly during the propagation above the forward slope of the topography, and the wave amplitude increases. This is a consequence of entry of the wave into a resonance band located above the topography top. As a result, the wave front starts to resemble overturning waves. We emphasize that the waves above the top and exactly behind the top tend to overturn in different directions, according to Fig. 6.6C, B. In the beginning of the propagation there is the tendency to overturning in the direction of the movement of the wave. However, the situation varies directly behind the top. There is the tendency of overturning backwards contrary to the direction of the wave movement.

The reader can compare the results of the analytic research (Fig. 6.6) with experimental and numerical data [32]. These data are shown in 6.7. The wave profiles are shown at various points of the channel. Continuous lines show the calculations, and dashed lines show the experiments. The wave period is 2.02, initial amplitude is 0.02 m, and the vertical scale is 0.02 m at one point.

On the whole, the analytical solution (Fig. 6.6B) describes well the experiment (Fig. 6.7).

The initially harmonious waves were studied above. Now we consider a solitary wave propagating in the channel of the variable depth. The wave profile is described at $a=0$, as $\eta = l\text{sech}^2\omega t$. It is found that the evolution strongly depends on the depth over the topography top. If the depth is big enough the wave does not actually change during propagation (Fig. 6.8a). The amplitude is increasing and the wave profile becomes steeper, when the depth decreases (Fig. 6.8b). If the depth decreases further, the amplitude strongly increases and the wave transforms into the complex wave structure with steep back side (Fig. 6.8c, d).

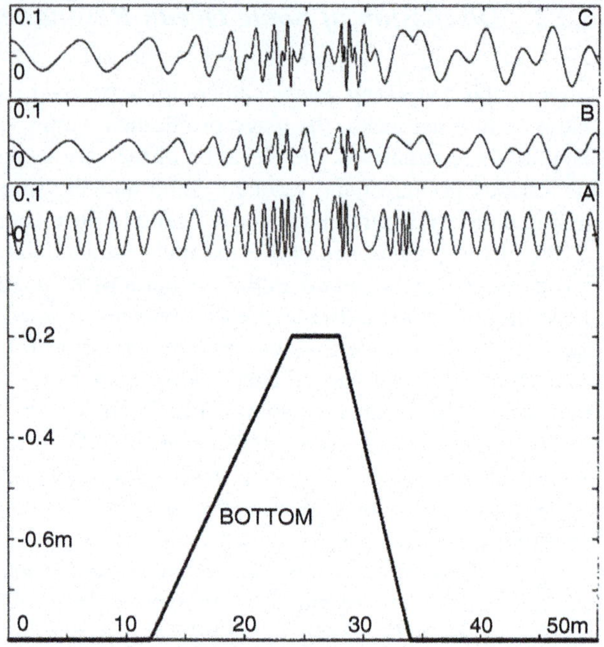

Fig. 6.6 Example of wave evolution above the trapezium-like topography: (*A*) – the incident wave period is 1.01 and the amplitude is 0.041 m; (*B*) – the incident wave period is 2.02 and the amplitude is 0.02 m; (*C*) – the incident wave period is 2.525 and the amplitude is 0.029 m [28, 33, 34]

Thus, we see a tendency to overturning the front of the wave, which is passing above the topography. Initially smooth waves may strongly increase the amplitude and reduce the length, when they cross an underwater topography. It is typically a transresonant phenomenon. We have already observed that such resonance can explain many catastrophic wave phenomena, both on the ocean surface, and on the earth surface.

The resonant mechanisms may be different. In particular, the resonant mechanism (band) may be formed by thermoclines on ocean surface (the thermocline is the transition layer between the surface light (warm) water and the heavy (cool) deep water).

Vertical motion of the trapezium-like topography shown in Fig. 6.8 we consider as a model describing a current of cool water or the vertical displacement of ocean animals (for example, sardines near the Agulhas Band, the east coast of South Africa). Similar currents and displacements can form resonant situations in different points of the world ocean, for a short time.

I think that sardines near the east coast of South Africa can form a very sharp border which is similar to a thermocline sheet or an ocean bottom. These structures can also form resonant bands where the extreme waves may be generated. The last situation is illustrated by Fig. 6.9.

The occurrence of small marine animals and fishes can be explained, and also the occurrence of a resonant upper layer. In particular, small marine crustaceans, which serve as food for whales, can form large schools. Sometimes, these small animals live under a surface water layer, which has a very sharp border. Such cases are

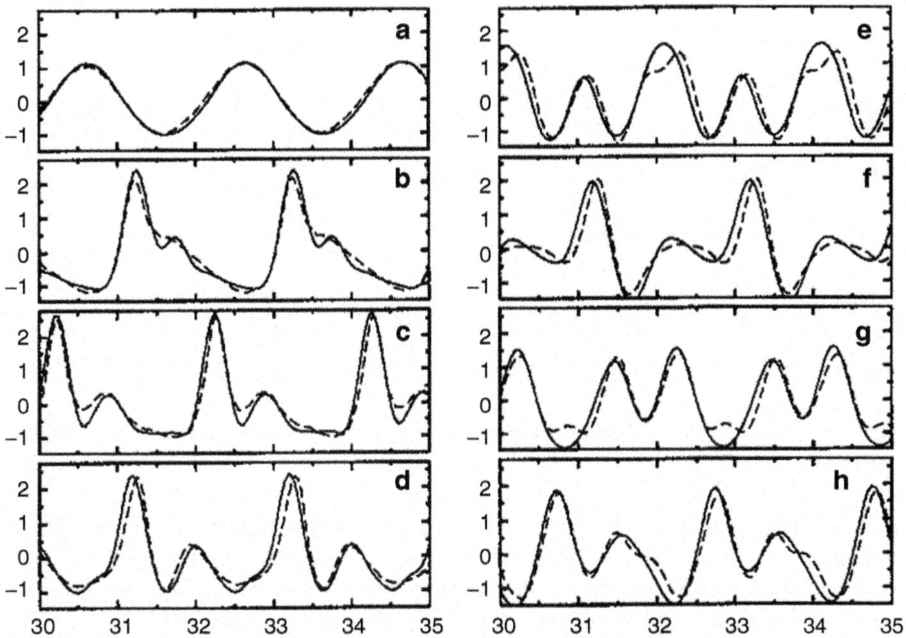

Fig. 6.7 Curves are calculated for various distances from the wave source: (**a**) $a = 4$ m; (**b**) $a = 25$ m; (**c**) $a = 27$ m; (**d**) $a = 29$ m; (**e**) $a = 31.4$ m; (**f**) $a = 34.6$ m; (**g**) $a = 38$ m; (**h**) $a = 42$ m [32]

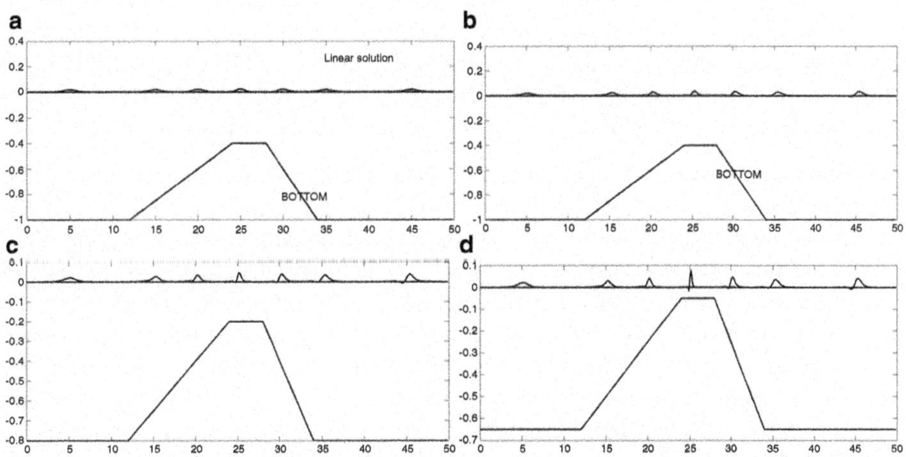

Fig. 6.8 The effect of the variation of the depth on solitary-wave evolution above the trapezium-like topography. The wave propagates from *left* to *right*. The curves **a**, **b**, **c** and **d** are calculated for different distances between the height top and the free surface. The reduction of the distance (from **a** to **d**) determines the local wave resonance. This resonance is shown clearly in **c** and **d** [28, 34]

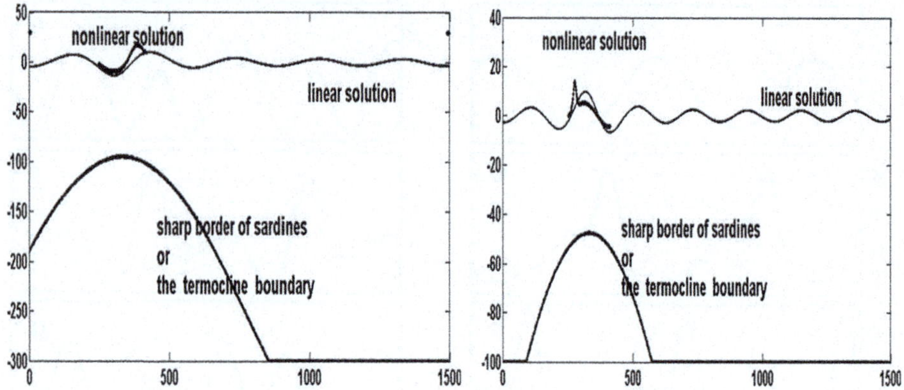

Fig. 6.9 Resonant amplification of ocean wave because of vertical displacement of a school of sardines

normally created by meeting of cold and warm ocean waters. Those are places where ocean catastrophic ocean waves are mostly encountered. For example, cold water of the Labrador Current is denser than warm water of the Gulf Stream. The combination of these two currents can create local resonant bands [28].

Similar complex structures, together with wind, surface and vertical currents, internal waves and vertical oceanic structures (temperature, salinity, density) promote the formation of multi-layer resonant systems in the ocean. In particular, in different areas of the ocean there are quasihomogeneous layers (sheets) of water differing by one or several characteristics. They are often at depth down to 300 m. Their thickness changes from several metres to hundreds of metres, and their horizontal dimensions are in the range from 100 to 10,000 m. The lifetime of these non-uniformities can reach up to months. We do not exclude the possibility that catastrophic amplification of waves may occur in these non-uniformities.

Resonant 'dead water' In particular, a layer of fresh or warm water may be on the ocean surface, and as a result a local two-layer system is formed. This system has some specific properties and its own wave speed (resonant speed). The upper fluid usually reduces the wave speed. Resonant effects can take place if some vessel (or some wave) crosses such a system with the resonant speed. In particular, the resistance to movement of the vessel is amplified. The famous researcher Horace Lamb wrote in his fundamental book 'Hydrodynamics' [30, p. 371] 'near the mouths of some of the Norwegian fiords there is a layer of fresh over salt water....waves of considerable height in this boundary are easily produced. To this cause is ascribed the abnormal resistance occasionally experienced by ships in those waters'.

The effect of the upper fresh water, so-called 'dead water', was described by Pliny the Elder at the beginning of our era, and it was well-known to Vikings. One of their outstanding descendants Fritjof Nansen wrote 'we did not move almost due to dead water; the vessel was carrying away all surface layer of water. Dead water forms a swell or even waves of big or smaller sizes'. The upper layer slides on the salty water, as on a solid surface, and moves together with the vessel. We think that waves arising in this manner, even in the deep ocean, can be described by the theory of long waves developed in [28].

Nansen described this phenomenon to Professor Vilhelm Bjerknes. Special experiments showed that if the vessel speed is far from the resonant speed, the specified effect does not arise. Of course, the effect takes place while a two-layer system exists. Such bands of surface fresh water arise near mouths of the big rivers, near ice fields, and in fiords, notably at Fiordland in New Zealand.

It is possible to assume that similar effects can occur in places of fast rising of highly-salty and cold ocean waters to the surface. It can take place in zones of interaction of ocean currents (for example, the Agulhas leakage around the southern tip of Africa), in places of uneven seabed (for example, the Bermuda region), in areas of meetings of warm and cold waters (for example, the Cape Horn or the North Sea). Such two-layer systems are unstable, but the period of their existence can be sufficient to form a catastrophic wave. In the case of considerable amplification and overturning, such a wave can increase the instability of the system.

Analogy with tsunami According to Luo [35], in some ocean areas the surface separating upper light water from lower dense water lies close to the free surface, at a depth of about 50 m. In principle, a thermocline layer can occur in places with horizontal and vertical currents. As a result, some areas appear with the thermocline layer rising and the warm light water reducing its thickness. Probably, such areas cannot exist for long.

However, for a period the action of this surface may be similar to the influence of a shallow beach on ocean waves. In particular, the last leads to tsunami formation. We can assume a possibility of similar amplification of ocean waves in layers of warm water, decreasing in thickness, which are restricted below by a thermocline layer. This analogy is especially important for us, because in both cases the theory of long waves is applicable. The wave evolution from thermoclines might be very surprising in the open ocean. This evolution may be similar to the coastal evolution of tsunamis (cf. Sects. 3.3.6 and 5.1) or to the nonlinear coastal waves (cf. Sect. 7.5).

Different cases of the resonant wave evolution will be studied in the next section.

6.3 Results of Modelling of Catastrophic Ocean Waves

First we consider long waves.

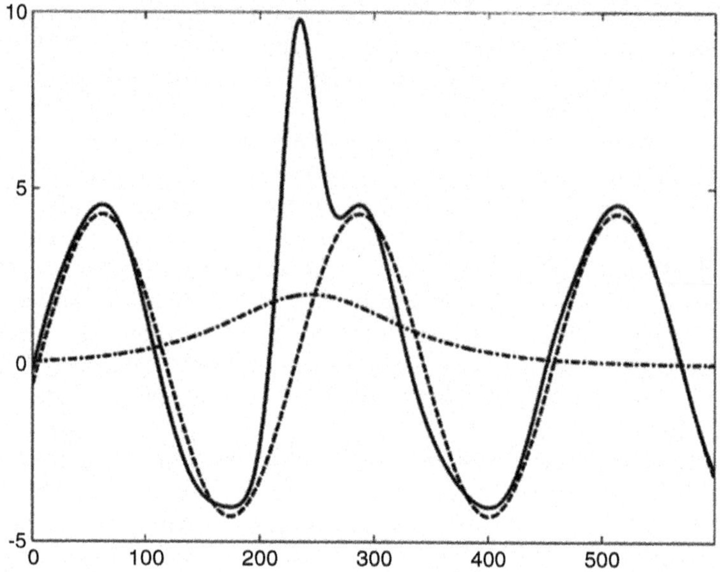

Fig. 6.10 The resonant wave 'water wall' (*solid line*). The *dash-dotted line* determines the resonant band. The wave travels from right to left. The harmonic wave (*dashed line*) determines the linear component of the solution and the wave coming from deep ocean

6.3.1 Long Waves

It is stressed again that although we use here the term the 'long' waves, our theory presented in [28, 33, 34] describes the typical storm waves and surges in shallow seas, for example, in the North Sea and the east coast of Southern Africa. The waves are modelled described in [8, Fig. 2.4] below.

Wave 'water wall' The transresonant evolution of these waves is shown in Fig. 6.10 (solid line).

For calculations (Fig. 6.10), we used parameters of extreme waves given in [8] (Fig. 2.4a). It has been found that Fig. 6.10 describes the observational data.

Extreme waves often have a steep front and smoother back slope. However, two profiles (from the 5 profiles of extreme waves, shown in Fig. 2.4 [8]), have a shelf on the back slope. Such shelves are often recorded in experiments with nonlinear water waves.

Wave 'front shelf' In this case, the shelf occurs on the wave front. The slope of the back is generally stronger than that of the front side. These wave particularities are described by Fig. 6.11.

In Fig. 6.11 (left) the resonant wave is the continuous line, the dashed line characterizes the linear component of the wave (the harmonious wave coming from deep ocean), dash-dotted line describes the nonlinear component of a wave. The

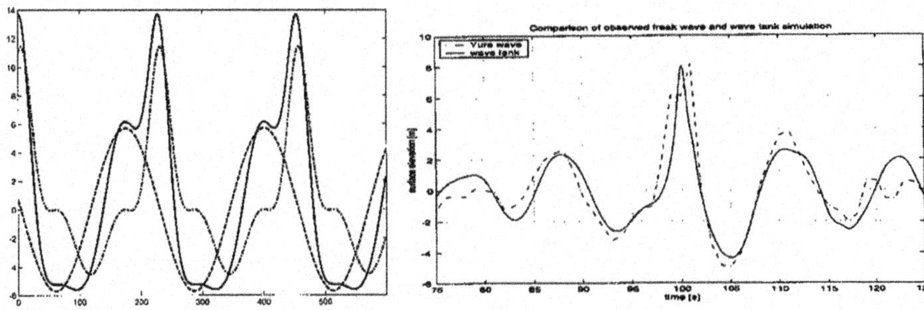

Fig. 6.11 Wave 'front shelf': theory (*left*) [28] and experiments (*right*) [19]

wave runs from right to left [28, 34]. There is a small shelf on the wave front [19]. We notice also that Fig. 6.11 describes the surprising wave [9] shown by the thin line in the top part of Fig. 6.12. Numerical methods sometimes cannot describe such complex waves.

Figure 6.12 shows a comparison of an observed wave (the thin line in the top part of drawing) and result of its numerical modelling (a thick harmonious curve in the same place) [19]. The wave runs from the right to the left. Our theory (Fig. 6.11 (left)) describes the experiment (Fig. 6.12) better than the numerical modelling.

'New Year's' wave Now we will address to a wave most often mentioned by experts. On the whole, the waves presented in Fig. 6.13, correspond to instrument readings on the Draupner oil platform (Fig. 6.1). In the increased scale these evidences are given in Fig. 6.13 (right). Results of calculations for a case of an exact resonance are shown in Fig. 6.13 (left), where the designations correspond to Fig. 6.10.

On the whole, Fig. 6.13 shows that our theory [28, 33, 34] describes the 'New Year's' wave.

'Three sisters' We have considered the most-often mentioned types of extreme (catastrophic) ocean waves. Now we will consider 'three sisters'. These extreme waves are groups of a small number of very steep waves. An example of these waves was described by the captain Frédéric-Moreau – the commander of the training cruiser of Naval Forces of France 'Joan of Arc'.

Results of a simulation of these waves [28] are presented in Fig. 6.14. The theory shows that a sequence of extreme (catastrophic) waves (Fig. 6.14) is possible. It qualitatively models the observations from aboard the 'Joan of Arc'.

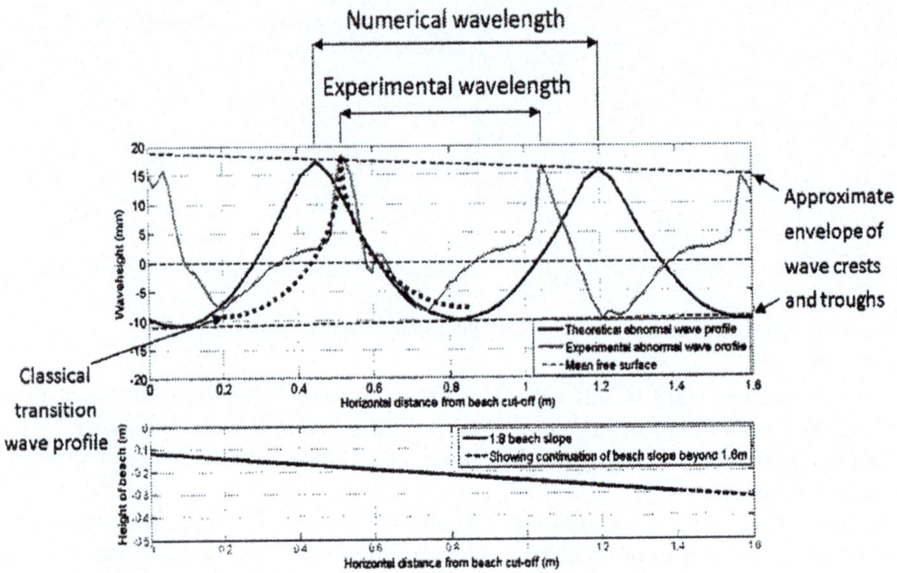

Fig. 6.12 Wave of the surprising form – 'step front' [9]

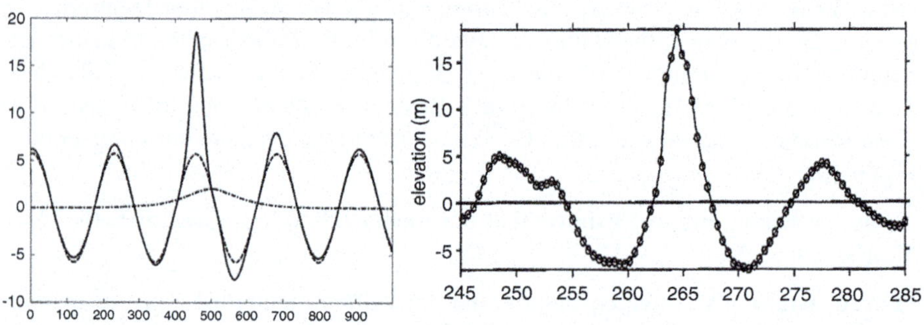

Fig. 6.13 Transresonant evolution of the 'New Year's' wave calculated in [28] (*left*). Instrument readings on the Draupner oil platform. Wave height – in metres, time – in seconds (*right*)

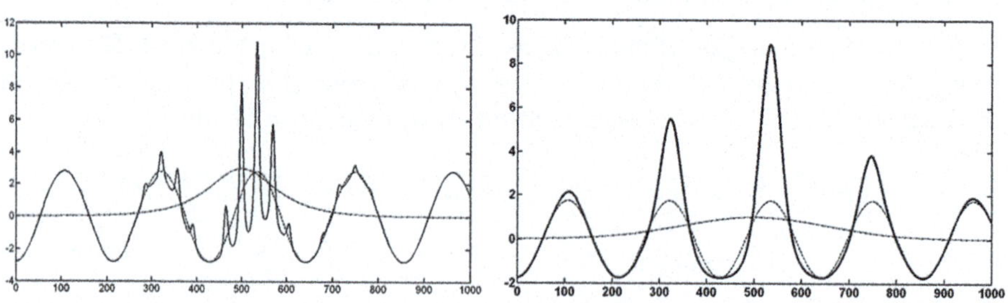

Fig. 6.14 Examples of transresonant evolution of a group of waves 'three sisters' [28]

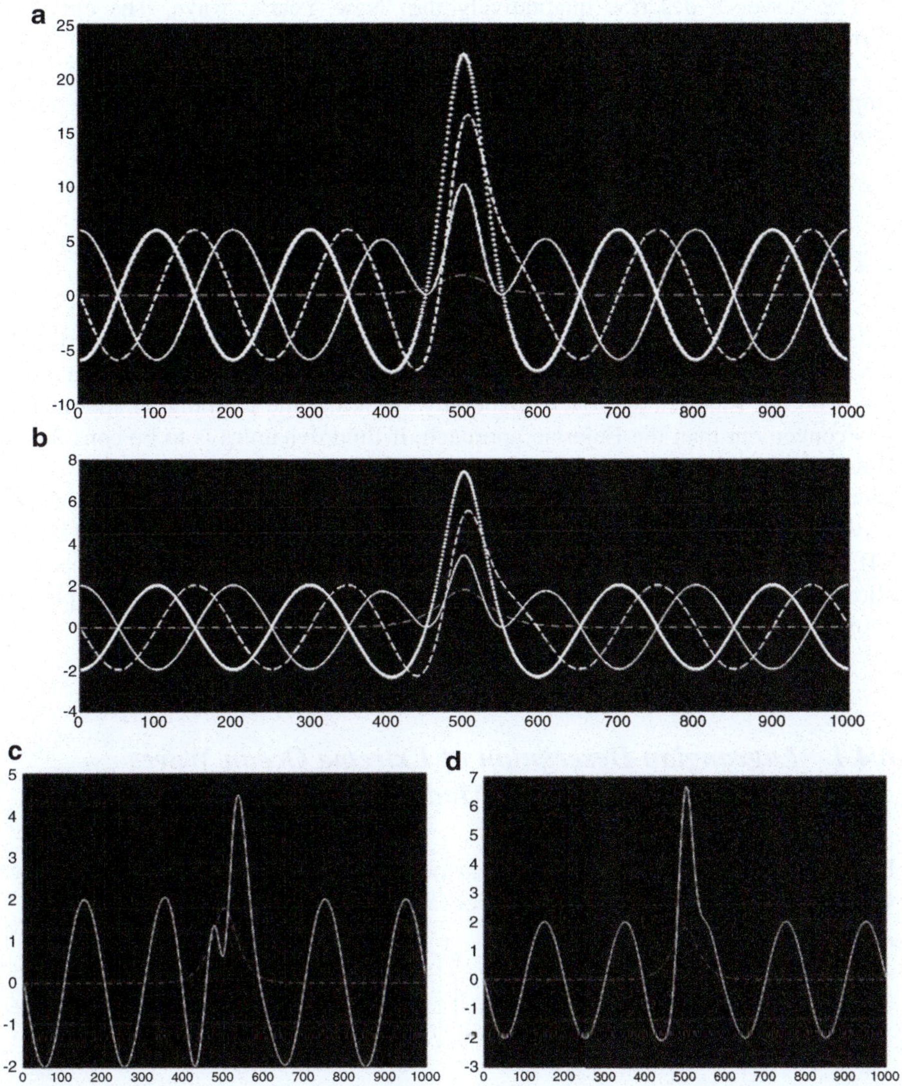

Fig. 6.15 Examples of transresonant evolution of ocean waves into catastrophic waves. The *dash-dotted line* determines the resonant band [28]

6.3.2 *Short Waves*

We will consider waves having length much smaller than the depth of water. These waves are strongly influenced by the dispersion. In particular, we cannot describe them as solitons. We assumed the they can be described by one or two harmonics. Results of calculations are shown in Fig. 6.15.

The curves a describe qualitatively the 'New Year's' wave. The curves b correspond to wave *e* from [8, Fig. 2.4]. The curves c describe the wave 'front shelf', while the curves d model the wave 'white wall'. It is clear that wave amplification in a resonant band can be very large. As a whole, this result corresponds to calculations presented for long waves.

6.4 The Generation of Catastrophic Ocean Waves

Eulerian coordinates are traditionally used to describe extreme waves. In particular, the nonlinear Schrödinger equation, which is often used to model ocean waves, is based on the Eulerian approach. Generally speaking, the Lagrangian approach is less convenient than the Eulerian approach, if fluid dynamics is to be considered. However, this is usually true only for three-dimensional problems, which will not be studied here. Here, we present a brief description of a theory of nonlinear ocean waves, based on the Lagrangian formulation. This theory was developed in [18, 28, 34]. The choice of the Lagrangian approach together with suitable assumptions, allows us to derive nonlinear wave equations for extreme waves. Examples of approximate solutions are presented.

6.4.1 *Lagrangian Description of Extreme Ocean Waves and a Depth-Averaged Model*

First, we derive equations for nonlinear ocean waves. Following [18, 28, 34] we write the depth-averaged equation for ocean waves

$$\int_h^0 \left[u_{tt}(1+u_a)+g\eta_a+w_{tt}w_a\right]dc=-\int_h^0\left[\int_c^0 u_{tt}u_c dc+\int_c^0 w_{tt}(1+w_c)dc\right]_a dc. \quad (6.2)$$

Here $u=u(a,c,t)$ and $w=w(a,c,t)$ are the horizontal and vertical displacements of water particles, a and c are the Lagrangian coordinates in the horizontal and vertical directions respectively, g is the acceleration due to gravity, h is the depth and η is the water surface elevation. Subscripts t, a and c indicate derivatives with respect to time and the Lagrangian coordinates. The incompressible model of water is used, therefore the equation of continuity is written in the form

$$1=(1+w_c)(1+u_a). \quad (6.3)$$

Now, using different approximations for the variation of w with c, we can derive further Eqs. from (6.2) to (6.3). Using the result of classical linear analysis we first assume that w varies with depth as a hyperbolic sine, i. e.

$$w = \eta\beta \sinh k(h - c). \tag{6.4}$$

Here $\beta = -[\sinh(kh)]^{-1}$, $k = 2\pi\lambda^{-1}$ and λ is the wavelength. Using (6.4), we find the following expressions for the terms in (6.2) which depend on w:

$$\int_h^0\int_c^0 w_{tta}dcdc = \beta k^{-1}\left(h\cosh kh - k^{-1}\sinh kh\right)\eta_{tta}, \tag{6.5}$$

$$\int_h^0 w_a w_{tt}dc = -\frac{1}{2}k^{-1}\beta^2\left(\frac{1}{2}\sinh 2kh - kh\right)\eta_a\eta_{tt}, \tag{6.6}$$

$$\int_h^0\left(\int_c^0 w_{tt}w_c dc\right)_a dc = \frac{1}{4}\beta^2\left(\frac{1}{2}k^{-1}\sinh 2kh - h\cosh 2kh\right)(\eta\eta_{tt})_a. \tag{6.7}$$

Since $1 = (1 + w_c)(1 + u_a)$ (6.3), we can approximately write

$$\begin{aligned} u = {} & k\beta\cosh k(h-c)\int\eta da + k^2\beta^2\cosh^2 k(h-c)\int\eta^2 da \\ & + k^3\beta^3\cosh^3 k(h-c)\int\eta^3 da \end{aligned} \tag{6.8}$$

and

$$\begin{aligned} \int_h^0 u_{tt}dc = {} & -\beta\sinh kh_0\int\eta_{tt}da - \frac{1}{2}k\beta^2\left(\frac{1}{2}\sinh 2kh + kh\right)\int\eta^2{}_{tt}da \\ & + k^3\beta^3\int_h^0\cosh^3 k(h-c)dc\int\eta^3{}_{tt}da, \end{aligned} \tag{6.9}$$

$$\begin{aligned} \int_h^0 u_{tt}u_a dc = {} & -\frac{1}{2}k\beta^2\left(\frac{1}{2}\sinh 2kh + kh\right)\int\eta_{tt}da\int\eta_a da \\ & + k^3\beta^3\int_h^0\cosh^3 k(h-c)dc\left(\int\eta^2{}_{tt}da\int\eta_a da + \int\eta_{tt}da\int\eta^2{}_a da\right), \end{aligned} \tag{6.10}$$

$$\int_h^0 \left(\int_c^0 u_{tt} u_c dc \right)_a dc = \frac{1}{4} k^2 \beta^2 \left(-h\cosh 2kh + \frac{1}{2} k^{-1} \sinh 2kh \right) \left(\int \eta_{tt} da \int \eta da \right)_a$$
$$+ \frac{1}{3} k^3 \beta^3 \int_h^0 \left[\cosh^3 kh - \cosh^3 k(h-c) \right] dc \left(\int \eta^2{}_{tt} da \int \eta da + 2 \int \eta_{tt} da \int \eta^2 da \right)_a . \tag{6.11}$$

In (6.9, 6.10 and 6.11)

$$\int_h^0 \cosh^3 k(h-c) dc = -\frac{1}{3} k^{-1} \sinh kh \left(\cosh^2 kh + 2 \right). \tag{6.12}$$

Substituting (6.8, 6.9, 6.10 and 6.11) into (6.2) it follows that

$$\int \eta_{tt} da - gh\eta_a + \eta_{tta} \beta k^{-2} (kh\cosh kh - \sinh kh) - \frac{1}{4} k\beta^2 (\sinh 2kh + 2kh) \left[\int (\eta^2)_{tt} da + \eta \int \eta_{tt} da \right]$$
$$- \frac{1}{9} k^2 \beta^3 \sinh kh \left(\cosh^2 kh + 2 \right) \left[2\eta \int (\eta^2)_{tt} da + \eta^2 \int \eta_{tt} da + 3 \int (\eta^3)_{tt} da \right.$$
$$\left. - (\eta^2)_{tt} \int \eta da - 2\eta_{tt} \int \eta^2 da \right] - \frac{1}{4} \beta^2 (k^{-1} \sinh 2kh - 2h) \eta_a \eta_{tt}$$
$$= \frac{1}{8} k\beta^2 (2kh\cosh 2kh - \sinh 2kh) \left(\int \eta_{tt} da \int \eta da \right)_a$$
$$+ \frac{1}{3} k^3 \beta^3 h\cosh^3 kh \left[\int (\eta^2)_{tt} da \int \eta da + 2 \int \eta_{tt} da \int \eta^2 da \right]_a$$
$$+ \frac{1}{8} \beta^2 (2h\cosh 2kh - k^{-1} \sinh 2kh) (\eta \eta_{tt})_a . \tag{6.13}$$

We recall that the solution (6.4) approximates the variation of w with depth according to the hyperbolic sine. Thus, the influence of nonlinearity on this variation is not taken into account.

Linear- dispersive relation Let us consider the following linearized version of the Eq. (6.13):

$$\int \eta_{tt} da - gh\eta_a + \eta_{tta} \beta k^{-1} \left(h\cosh kh - k^{-1} \sinh kh \right) = 0. \tag{6.14}$$

Let

$$\eta = \exp i(ka - \omega t). \tag{6.15}$$

In this case, (6.13) yields the classical expression

$$\omega^2 = gk\tanh(kh). \tag{6.16}$$

This relation describes the interaction between inertia and gravitational forces. (6.16) may be written in terms of the wave speed C as

$$C^2 = gk^{-1}\tanh(kh). \tag{6.17}$$

For a deep ocean $kh >> 1$ then relation (6.17) reduces to

$$C = \sqrt{gk^{-1}}. \tag{6.18}$$

This means that the phase speed of gravity waves in deep water is proportional to the square root of the wavelength.

6.4.2 *Quadratic and Cubic -Nonlinear Equations for Gravity Waves in Deep Ocean*

Let us consider the quadratic and cubic terms of (6.13) assuming that $kh >> 1$. It is easy to see that for this case

$$\begin{aligned}\frac{1}{9}\, k^2\beta^3 \sinh kh(\cosh^2 kh + 2)\Big(2\eta\int {\eta^2}_{tt} da + \eta^2 \int \eta_{tt} da + 3\int {\eta^3}_{tt} da \\ - {\eta^2}_{tt}\int \eta da - 2\eta_{tt}\int \eta^2 da\Big) << \tfrac{1}{3}k^3\beta^3 h\cosh^3 kh\left(\int {\eta^2}_{tt} da \int \eta da + 2\int \eta_{tt} da \int \eta^2 da\right)_a.\end{aligned} \tag{6.19}$$

Consider first the quadratic terms. It is convenient to introduce dimensionless variables

$$a^* = a/h,\ \ t^* = \sqrt{gh}t/h = Ct/h,\ \ \eta^* = \eta/A. \tag{6.20}$$

Now dimensionless expressions for the terms in (6.13) may be introduced. For example,

$$\eta_a = Ah^{-1}\frac{\partial \eta^*}{\partial a^*},\quad \eta_t = Ah^{-1}C\frac{\partial \eta^*}{\partial t^*},\quad \eta_{tt} = Ah^{-2}C^2\frac{\partial^2 \eta^*}{\partial (a^*)^2}. \tag{6.21}$$

Using the dimensionless variables we rewrite correspondingly the quadratic-nonlinear terms of (6.13) in the form

$$-\frac{1}{4}kh^{-1}\beta^2A^2C^2(\sinh 2kh + 2kh)\left(\int \eta^2{}_{tt}da + \eta\int \eta_{tt}da\right), \tag{6.22}$$

$$-\frac{1}{4}k^{-1}h^{-3}\beta^2A^2C^2(\sinh 2kh - 2kh)\eta_a\eta_{tt}, \tag{6.23}$$

$$\frac{1}{4}kh^{-1}\beta^2A^2C^2(kh\cosh 2kh - 0.5\sinh 2kh)\left(\int \eta_{tt}da\int \eta da\right)_a, \tag{6.24}$$

$$\frac{1}{8}k^{-1}h^{-3}A^2C^2\beta^2(2kh\cosh 2kh - \sinh 2kh)(\eta\eta_{tt})_a. \tag{6.25}$$

For clarity we drop the asterisks in (6.22, 6.23, 6.24 and 6.25). It is easy to see that for deep water waves the expression (6.24) is much larger than (6.22, 6.23 and 6.25). Now taking into account only the most important quadratic and cubic-nonlinear terms, we rewrite the Eq. (6.13) in the form

$$\begin{aligned}\int \eta_{tt}da - gh\eta_a + \eta_{tta}\beta k^{-1}(h\cosh kh - k^{-1}\sinh kh) &= \frac{1}{4}k^2\beta^2 h\cos 2kh\left(\int \eta_{tt}da\int \eta da\right)_a\\ &+ \frac{1}{3}k^3\beta^3 h\cosh^3 kh\left(\int \eta^2{}_{tt}da\int \eta da + 2\int \eta_{tt}da\int \eta^2 da\right)_a.\end{aligned} \tag{6.26}$$

Since $kh >> 1$, we assume that $\tanh kh \approx 1$. In this case, (6.26) yields

$$\begin{aligned}\int \eta_{tt}da - gh\eta_a - hk^{-1}\eta_{tta} &= \frac{1}{2}hk^2\left(\int \eta_{tt}da\int \eta da\right)_a\\ &- \frac{1}{3}hk^3\left(\int \eta^2{}_{tt}da\int \eta da + 2\int \eta_{tt}da\int \eta^2 da\right)_a.\end{aligned} \tag{6.27}$$

Finally, we introduce the elevation potential Φ, defined such that

$$\eta = \Phi_a \tag{6.28}$$

and rewrite (6.27) in the form

$$\Phi_{tt} - gh\Phi_{aa} - hk^{-1}\Phi_{ttaa} = \frac{1}{2}hk^2(\Phi_{tt}\Phi)_a - \frac{1}{3}hk^3\left[\Phi\int\left(\Phi_a{}^2\right)_{tt}da + 2\Phi_{tt}\int\Phi_a{}^2da\right]_a. \tag{6.29}$$

The quadratic-nonlinear version of this equation is

$$\Phi_{tt} - gh\Phi_{aa} - hk^{-1}\Phi_{ttaa} = \frac{1}{2}hk^2(\Phi_{tt}\Phi)_a. \tag{6.30}$$

We shall use the nonlinear Eqs. (6.29) and (6.30) to study the possibility of extreme wave generation. Recent observations show that the probability of generation of extreme large waves in the open ocean is much larger that was expected from ordinary wave-amplitude statistics [6, 8]. Although considerable efforts have been directed towards understanding this phenomenon, the complete picture remains unclear. However, it is known that the generation of large ocean waves may be explained by the resonant interaction of several surface waves [8, 14, 28, 29].

6.4.3 *Solitary Ocean Waves*

Here the extreme wave problem is considered as a part of the problem of generation and evolution of nonlinear ocean waves. Strongly localised extreme waves are described thus in [8, pp. 17–18]: 'Russian kayakers lucky to have observed and taken photos of strange waves 25 km from Cape Olga, Kronotsky Peninsula, about 1–1.5 km offshore. They reported that the weather was calm with only very gently sloping surge waves coming from the open ocean every 15–20 s. About 10 strange waves were observed in the same area with irregular lengths. Freak waves arose, propagated, and collapsed during tens of second.... Wave heights were about 2–4 m...They seem to be a solitary wave variety'. We underline that the length of the wave in Fig. 6.16 is approximately similar to its height.

We assume that the elevation of water has a form of a solitary wave

$$\eta = \Phi_a = A\,\mathrm{sech}^2(\omega t - ka) \tag{6.31}$$

Substituting (6.31) in (6.30) we obtain

$$\begin{aligned}&\left(\omega^2k^{-2} - gh\right)A\,\mathrm{sech}^2(\omega t - ka) - h\omega^2k^{-3}\left[6k^2A\,\mathrm{sech}^2(\omega t - ka)\tanh^2(\omega t - ka)\right.\\&\left.-2k^2A\,\mathrm{sech}^2(\omega t - ka)\right] = -h\omega^2A^2\,\mathrm{sech}^2(\omega t - ka)\tanh^2(\omega t - ka) + D.\end{aligned} \tag{6.32}$$

Fig. 6.16 Photo (A.V. Sokolovsky) of strange wave [8]

Here D is a constant of integration. We assume that

$$D = 0. \tag{6.33}$$

In this case Eq. (6.32) determines the dispersion equation in the form

$$\omega^2 k^{-2} - gh + 2h\omega^2 k^{-1} = 0 \tag{6.34}$$

and the wave amplitude

$$A = 6k^{-1} = 3\lambda\pi^{-1}. \tag{6.35}$$

Here λ is the wavelength. Expressions (6.31, 6.34 and 6.35) describe a solitary wave which exist in deep ocean according to the Eq. (6.30). According to (6.35) – the longer the wave length, the higher the amplitude. We underline that according to the presented theory the length of the wave may be comparable to its height ($A \approx \lambda$). On the whole, this conclusion is supported by Fig. 6.16.

However, the last conclusion is not correct if the amplitude is very large. In that case we should take into account cubic nonlinear effects.

6.4.4 Catastrophic Amplification of Harmonic Ocean Waves

We consider in this subsection the problem of the determination of harmonic travelling waves using the Eq. (6.29). If we neglect the nonlinear term in it, we obtain the linear solution

$$\Phi = A\cos(ka - \omega t), \;\; \omega^2 - ghk^2 + \omega^2 hk = 0. \tag{6.36}$$

The phase speed for this wave is ωk^{-1} which is independent of the amplitude A. In the nonlinear problem the phase speed is in general a function of the amplitude. To determine qualitatively the dependence of the phase speed C on the amplitude, we assume that

$$\Phi = \Phi(r), \quad r = a - Ct \tag{6.37}$$

so that (6.29) becomes

$$\begin{aligned}(1 - ghC^{-2})\Phi_r - hk^{-1}\Phi_{rrr} &= \frac{1}{2}hk^2\Phi_{rr}\Phi \\ &- \frac{1}{3}hk^3\left[\Phi\left(\Phi_r{}^2\right)_r + 2\Phi_{rr}\int\Phi_r{}^2dr\right] + D,\end{aligned} \tag{6.38}$$

where the subscript r denote differentiation with respect to r. We write both Φ and C as sums

$$\Phi_r = \Phi_r{}^{(1)} + \overline{\Phi}_r{}^{(1)} \tag{6.39}$$

and

$$C = C_0 + \overline{C}. \tag{6.40}$$

1. We first assume that

$$\Phi_r{}^{(1)} >> \overline{\Phi}_r{}^{(1)}, \quad C_0 >> \overline{C}. \tag{6.41}$$

Substituting (6.39) and (6.40) into (6.38) and equating terms having similar order, we obtain

$$(1 - ghC_0^{-2})\Phi_r{}^{(1)} - hk^{-1}\Phi_{rrr}{}^{(1)} = 0, \tag{6.42}$$

$$\begin{aligned}(1 - ghC_0^{-2})\overline{\Phi}_r{}^{(1)} - hk^{-1}\overline{\Phi}_{rrr}{}^{(1)} &= -2gh\overline{C}C_0^{-3}\Phi_r{}^{(1)} + \frac{1}{2}hk^2\Phi_{rr}{}^{(1)}\Phi^{(1)} \\ &+ D.\end{aligned} \tag{6.43}$$

We take the solution of (6.42) to be

$$\Phi_r{}^{(1)} = A\cos kr, \quad C_0^2 = gh - hkC_0^2. \tag{6.44}$$

In this case (6.44) coincides with (6.36) to order of the wave amplitude. Then (6.43) becomes

$$\left(1-ghC_0^{-2}\right)\overline{\Phi}_r{}^{(1)}-hk^{-1}\overline{\Phi}_{rrr}{}^{(1)}=-2gh\overline{C}C_0^{-3}A\cos kr+\tfrac{1}{2}hk^2\Phi_{rr}{}^{(1)}\Phi^{(1)}+D. \tag{6.45}$$

There is the singular term in (6.45). According to (6.45) an infinite amplitude wave may be generated on the ocean surface if the dispersive equation (6.44) occurs. Effect of the singularity may be eliminated if we take into account the strongly nonlinear effects.

2. We assume in (6.39) that

$$\Phi_r{}^{(1)}\approx\overline{\Phi}_r{}^{(1)}\ \text{ or }\ \Phi_r{}^{(1)}<<\overline{\Phi}_r{}^{(1)}. \tag{6.46}$$

Here $\Phi_r=\Phi_r{}^{(1)}+\overline{\Phi}_r{}^{(1)}$ (6.39). Let

$$\overline{\Phi}_r{}^{(1)}=B\cos\left(Kr+\phi\right), \tag{6.47}$$

where

$$K=k+\overline{k}. \tag{6.48}$$

Here ϕ is the wave phase and $k>>\overline{k}$. In this case the Eq. (6.38) is presented as

$$\begin{aligned}&\left(1-ghC_0{}^{-2}+hk\right)\left[A\cos kr+B\cos\left(Kr+\phi\right)\right]\\&\quad=-2ghC_0{}^{-3}\overline{C}\left[A\cos kr+B\cos\left(Kr+\phi\right)\right]+2hB\overline{k}\cos\left(Kr+\phi\right)\\&\quad+\frac{1}{2}hk^2\Phi_{rr}\Phi-\frac{1}{3}hk^3\left[\Phi\left(\Phi_r{}^2\right)_r+2\Phi_{rr}\int\Phi_r{}^2dr\right]+D.\end{aligned} \tag{6.49}$$

The Eq. (6.49) is rewritten using (6.39, 6.44 and 6.47). We obtain the equation containing different harmonics and constants after long but simple calculations. Now we remind that the Eq. (6.29) was derived for waves having the wave number k. Therefore we will take into account only the terms containing the corresponding harmonics in (6.49). In particular, the harmonics of the doubled and trebled frequencies will be ignored. In this case the Eq. (6.49) yields

$$\begin{aligned}&2ghC_0{}^{-3}\overline{C}\left[A\cos kr+B\cos\left(Kr+\phi\right)\right]-2h\overline{k}B\cos\left(Kr+\phi\right)\\&\quad\approx\frac{1}{4}hk^3A_2\left(B^2{}_2+A^2{}_2\right)\cos kr-\frac{1}{4}hk^3B_2\left(B^2{}_2+A^2{}_2\right)\sin kr.\end{aligned} \tag{6.50}$$

Here

$$A_2=A+B\cos\phi,\ \ B_2=B\sin\phi. \tag{6.51}$$

Then the Eq. (6.50) is rewritten in the form

$$\begin{aligned} &2ghC_0{}^{-3}\overline{C}A\cos kr + 2B\left(ghC_0{}^{-3}\overline{C} - h\overline{k}\right)\left(\cos Kr\cos\phi - \sin Kr\sin\phi\right) \\ &\quad\approx \frac{1}{4}hk^3\left[(A + B\cos\phi)\cos kr - B\sin\phi\sin kr\right]\left(B^2 + A^2 + 2AB\cos\phi\right). \end{aligned} \tag{6.52}$$

We group the terms with $\cos kr$. As a result we obtain the equation

$$\begin{aligned} &\left(2ghC_0{}^{-3}\overline{C} - \frac{1}{4}hk^3A^2\right)A + \left[2ghC_0{}^{-3}\overline{C} - 2h\overline{k} - \frac{3}{4}hk^3A^2\right]B\cos\phi \\ &\quad\approx \frac{1}{4}hk^3\left[AB^2(1 + 2\cos^2\phi) + B^3\cos\phi\right]. \end{aligned} \tag{6.53}$$

Resonance and resonant waves Let in (6.53),

$$gC_0{}^{-3}\overline{C} - \overline{k} = \frac{3}{8}k^3A^2. \tag{6.54}$$

In this case the Eq. (6.53) yields if $\cos\phi = 1$,

$$B^3 + 3AB^2 + A\left(A^2 - 8gk^{-3}C_0{}^{-3}\overline{C}\right) = 0. \tag{6.55}$$

First we produced series of calculations of the amplitude B using the Eq. (6.55). The coefficient $A^2 - 8gk^{-3}C_0{}^{-3}\overline{C}$ was varied near zero. We assumed that $10 > A > 1$. Some results are shown in Fig. 6.17.

Near the resonance (6.54) the Eq. (6.55) defines a complex picture of the curves consisting of real and imaginary parts. The real parts correspond to real roots of the equation (6.55). The imaginary parts correspond to imaginary roots of the equation (6.55). Real parts are shown in Fig. 6.17 as 'real'. Imaginary parts are defined by almost horizontal straight lines. Real parts of the curves form the fold where there are only real roots of the equation (6.55). There are also imaginary roots outside of the fold.

Our theory is valid for small and not too large amplitudes. Therefore, apparently, only parts of the curves forming the fold have a physical meaning. Let us consider the upper portion folds. According to it, the amplitude B varies from zero to about 2 (or 6). Thus, according to this analysis the two- time amplification of the amplitude of the wave is possible if the condition (6.54) takes place.

Let us assume in addition to (6.54) that

$$8gk^{-3}C_0{}^{-3}\overline{C} = A^2. \tag{6.56}$$

In this case the Eq. (6.55) yields $B = -3A$.

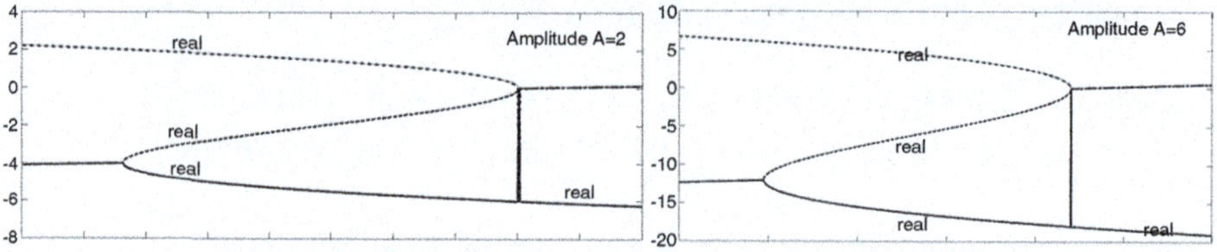

Fig. 6.17 Folded curve calculated for $A = 2$ describing the transresonant evolution of the amplitude B (*left*). Folded curve for B calculated for $A = 6$ (*right*)

In the case (6.54) and $\cos\phi = -1$, we have $B = 3A$. For both cases the water elevation is determined as

$$\eta = \Phi_a = \Phi_r = -2A\cos kr. \tag{6.57}$$

On the whole the analytical result agrees with the calculations. On the other hand it agrees with well-known lowest amplification of the amplitude of extreme waves [8]. According to [3, 8] 'tipical' rogue waves achieve amplification approximately from $2A$ to $4A$.

We consider the terms of the Eq. (6.52) with $\cos kr$ when $\cos\phi = \pm 1$. In this case $\sin\phi = 0$. Therefore the terms of the Eq. (6.52) with $\sin kr$ equal zero.We considered the most simple case of the interaction of waves $\Phi_r^{(1)}$ and $\overline{\Phi}_r^{(1)}$ (6.39) when the phase $\phi = 0$.

6.4.5 *Nonlinear Dispersive Relation and Extreme Waves*

Let

$$\Phi_r = A\cos kr, \tag{6.58}$$

where $r = a - Ct$ (6.37). In this case the Eq. (6.29) approximately yields

$$\left(1 - ghC^{-2} + hk\right)A\cos kr \approx \frac{5}{24}hk^3A^3\cos kr. \tag{6.59}$$

As a result we have the nonlinear dispersive relation of the problem

$$A^2 = 4\frac{4}{5}\left(h^{-1} - gC^{-2} + k\right)k^{-3}. \tag{6.60}$$

Let us assume a resonance condition $h^{-1} - gC^{-2} = -\frac{1}{2}k$ in order to appreciate qualitatively the result (6.60). In this resonance case

$$A \approx \frac{1}{4}\lambda. \tag{6.61}$$

In [4] results of an observational study of rogue waves were presented based on a measurement of wave height in South Indian Ocean. The depth was about 100 m. There were waves having lengths from 200 to 1,500 m coming from the deep ocean. At the same time short waves were with wave length of 200 m or less. It was found that the maximal wave height changed between 23.2 and 48 m. According to (6.61) the maximum height of these waves may be about 50 m. On the whole the last result correspond to the observations.

'Rogue waves', 'freak waves', 'killer waves' and similar names have been the topic of several recent publications related to giant single waves appearing in the ocean 'from nowhere'. Hitherto, we do not have a complete understanding of this phenomenon due to the difficult and risky observational conditions. It is difficult to explain the high amplitudes that can occur in the open ocean using linear theories based on the superposition principles. Nonlinear theories of ocean waves are more likely than linear theories to explain why the waves can "appear from nowhere". In this section, we considered certain cases of the amplification of ocean waves.

The remark by Darwin about the huge wave met at Cape Horn and the publications of the last 10 years devoted to extreme waves have attracted our attention to the problem of catastrophic (extreme) waves. Already the first acquaintance showed us the great importance of this problem, for which there are no settled concepts and decisions. It was decided to use for the description of catastrophic ocean waves the results of our publications during 1970–2000, in which the resonant amplification of nonlinear waves was studied (cf. Chap. 7).

We accepted as fact some reasons for the emergence of catastrophic waves. The reasons are the resonance effect as a mechanism for amplification of waves, and the effect of nonlinearity as the mechanism limiting the growth of waves. The accounts concern coastal waves, waves in shallow seas and waves over underwater topographies where conditions for wave resonance can occur often.

The occurrence of catastrophic ocean waves having abrupt, even folded fronts, is surprising, since the influence of wave dispersion in deep ocean is large. Generally speaking, the extreme ocean waves can arise there, as a result of random composition of harmonic waves. However, according to the theory developed in [28], they can arise quite naturally in following cases. First, if parameters of a wave, a wind and an interaction wind-wave satisfy some resonant condition.

Secondly, when on an ocean surface there are water layers of different mechanical properties. Such layers reaching one kilometre and more in length, have been discovered in recent decades. Those layers can form resonant bands. As well as in the first case, surface waves arising there can be described as nonlinear, long, weakly-dispersive waves. That does not exclude the possibility of catastrophic waves occurring on the surface of deep ocean. However the front of such ocean waves cannot be very abrupt. These waves are described by one or two harmonics. The influences of wind, currents and nonlinearity on the occurrence of these waves, probably varies significantly. In particular, the theory [28] explained wave amplification by a cubic nonlinearity in the deep-water equation.

When a catastrophic wave happens, it changes the environment (air streams, surface and wave characteristics of ocean). It leads to infringement of conditions for resonance and instant disappearance of a wave. Thus the occurrence of resonant conditions and their realization at open ocean is a rare and short-time event.

6.5 Surface-Breaking Waves, Underwater Explosions and Hull Cavitation

The fronts of some catastrophic waves are almost-vertical walls of water. This photograph of Wilstar (Fig. 6.18) allows to estimate the power of them.

Engineers, who now are designing ships and building offshore platforms that are further and further from land, obviously require detailed knowledge of structural loads imposed by these extreme waves. A typical wave having 12 m height impacts with 6 t weight per square metre. The majority of modern ships can sustain up to 15 t weight per square metre. In the case of the extreme (catastrophic waves) the pressure on the hull may be much more. The pressure can reach 100 t weight on per square metre (near 980,000 Pa).

It is very important to know the strength required for vessels to resist the loading caused by impacting waves, so that appropriate design can be undertaken to withstand these loads. The action of overturning (breaking) waves is especially dangerous, because in certain conditions they can produce very short and intense loadings that are similar in character to the pressure time-histories caused from underwater explosions. Furthermore, it is well-known that the action of an underwater explosion near a submarine hull can cause hull cavitation to occur. Such cavitation can completely change the characteristics of an underwater shock-wave loading signature compared to that which could occur if no cavitation were present. Because very high pressures can be developed in the case of breaking waves impacting on ship panels, it is reasonable to assume that cavitation can arise in a manner similar to that which arises from underwater explosions. The remaining sections of this chapter are devoted to research on this specific question: How significantly can hull cavitation change the pressure loading on a vessel hull from the impact of a breaking wave, compared to the pressure when there is no cavitation present?

I published at 1977 a book (in Russian) devoted to hull cavitation [36]. This book was translated into English (1980) by The Department of the Navy, Office of Naval Research, Arlington, USA [36]. The action of underwater explosions on ships was studied in this book. The different aspects of the cavitation interaction of submarines with underwater explosions were studied in [36–52]. However, the cavitation interaction of surface waves with ships was not considered, and was not mentioned at all until very recently. The first results devoted to that problem

Fig. 6.18 The damage received by Norwegian tanker Wilstar in 1974. The combination of ship motion and a steep incoming wave can cause extreme local structural damage. Internet. https://www.flickr.com/photos/40269779@N08/3928691965/

appeared in 2009 [21]. We used the similarity between profiles of the pressures on a solid surface arising during underwater explosions and the actions of surface breaking waves. There is one more important circumstance. The pressure of the breaking wave depends on the trapped air. The pressure of the underwater explosions also can depend very strongly on the dissolved gas in the water and the hull cavitation. In both considered cases the hull is affected not by ideally pure water, but by gas-water or bubbly-water mixture.

6.5.1 Effects of the Breaking

The most dangerous waves were found to be asymmetric in the wind direction, with steep fronts and less steep rear sides. The researches show that the wave impact from non-breaking waves on structures increases with wave height, however it was not the highest waves but the breaking waves that produced the most severe wave impact.

There is a vast difference in the destructive power of smooth and steep (breaking) waves [53–55].

Objects floating in water, such as ships, tend to make the same motion as the water they displace. A ship at sea in large waves will describe orbital circles, that are approximately the same size as the orbital circles of the water in that part of the wave (Fig. 6.19). There is thus little relative motion between the bulk of the ship and the surrounding water. The motion of a ship may be uncomfortable, but it is safe.

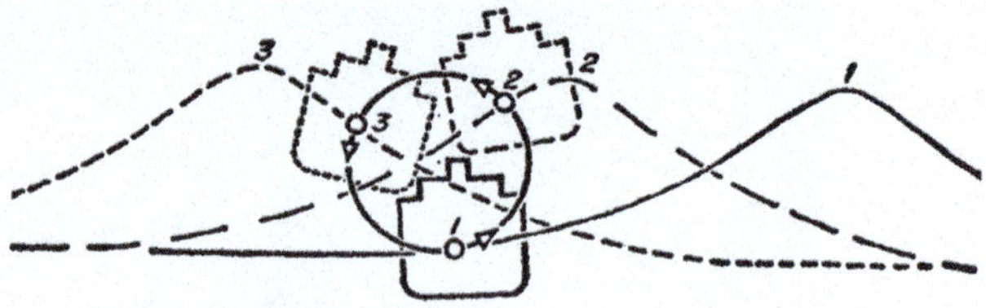

Fig. 6.19 Circular motion of a ship on a surface of a long wave, similar to the motion of water particles on a wave surface [53]

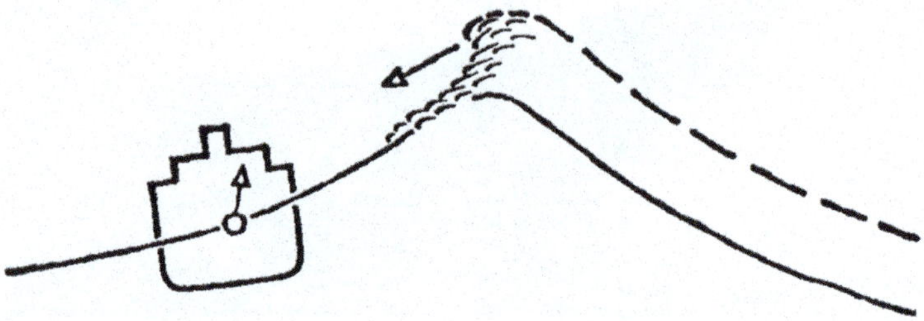

Fig. 6.20 Ship in breaking waves. Water in the crest of a large wave has broken free of the orbit and will collide violently with the ship [53]

In the case where a wave breaks, the motion of a ship is practically independent of the motion of the water particles in the wave. Moving in different directions the ship and the crest water may collide with disastrous results. A 1,000 t of violently moving water can sink even a large ship. The collision is especially dangerous for ships having a mass comparable with the mass of the crest wave. It is these breaking storm waves, that do the serious damage to ships which are unlucky enough to be hit (Fig. 6.20).

The force exerted by sea waves on a vertical wall depends, for any given shape of approaching wave, on the depth of water in front of the wall [54]. The force is quite small if the water is very deep, and is again quite small if the still water level is well below the foot of the wall; but there is a certain depth for which the force on the wall is a maximum. If the wave breaks somewhere near the wall, or should the wave be just about to break when it meets the wall, pressures from the impact lasting a few seconds will occur. In these cases the pressure amplitudes will not be that large. Should, however, the impact occur at one particular and very precise stage of the break (Fig. 6.21d), then higher shock pressures may result lasting for only approximately 0.01 s (see pressure–time histories in Figs. 6.22 and 6.23).

In the experiments described in [54] the pressure was measured by piezo-electric gauges at points located 0.64 m and 1 m below mean sea level, and at a point located 0.33 m above mean sea level. A few waves (less than 5 in a 100) were found to generate high pressures. The pressures reached their maxima in as short a time as 0.005 s or as long as 0.05 s. The maximum pressure at the lowest level was 0.68 MPa (14,300 lb per square foot) (see Fig. 6.22); at the intermediate level it was 0.29 MPa; while at the highest level pressure shocks were very seldom recorded.

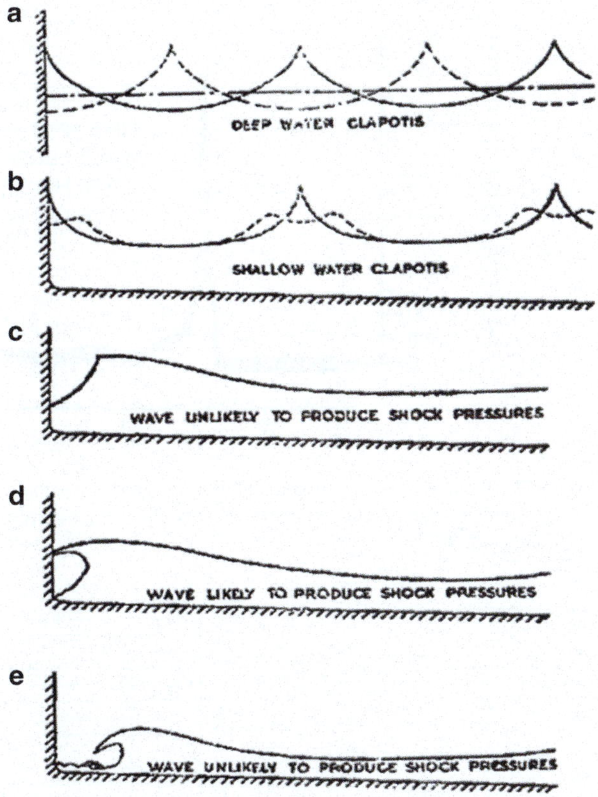

Fig. 6.21 Types of wave action to which a wall may be subjected, depend on many variables such as shape of the incident wave immediately prior to impact and depth of water (**a**–**e**). The situation depicted in (**d**) gives rise to the highest pressure [54]

During the breaking the wave traps a certain volume of air adjacent to the wall. This volume is often called an air 'pocket' [55] (Fig. 6.24).

The model of the hull loading is simulated as the action of a rectangular water mass (water wall) advancing at a speed of U_1 towards a vertical hull. The hull is modelled as the vertical wall. As a result of the loading the pressure impact that is formed can be similar to that presented in Figs. 6.22 and 6.23.

The 'pocket' effect can be very important [55]. It can significantly reduce the amplitude of the pressure. This effect, probably, explains the rarity of the sinking of ships because of the shock of ocean waves. Really, because of the air trapped in the pocket, a hull is loaded by the water-air mixture, instead of action of an incompressible water (the water hammer). At the same time it is found that the biggest pressure arises when the amount of the air in the pocket is not zero. Apparently, a small layer of air allows the wave front to accelerate additionally. At the same time, this amount of air does not change the water density on the wave front.

Thus, the duration of the shock front loading may be of the order of 0.005 s. In particular, Peregrine [55] noted that 'In the laboratory violent peak pressures act for times of approximately 1 millisecond (ms), and peak durations range from 10 to 100 ms in the field' (Fig. 6.23).

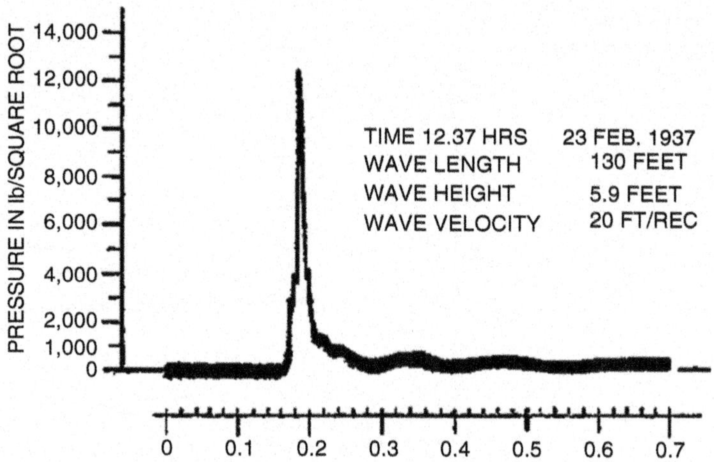

Fig. 6.22 Time history of pressures at a point on a vertical solid wall subjected to wave motion. The greatest pressure occurs when the crest of the wave overturns, generating a jet of water as the wave breaks [54]

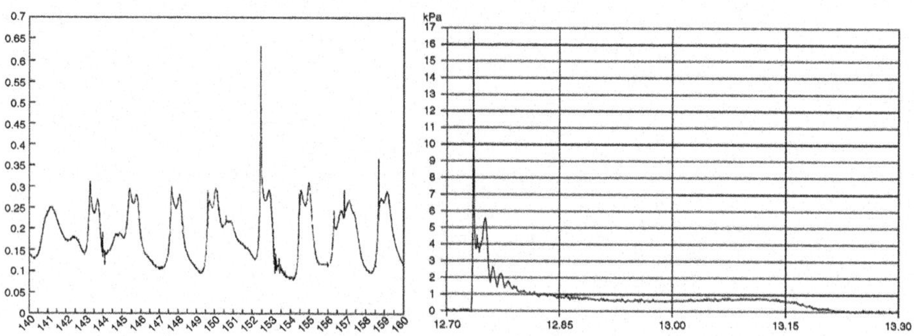

Fig. 6.23 Pressure on a vertical wall caused by breaking waves as a function of time in seconds [55]

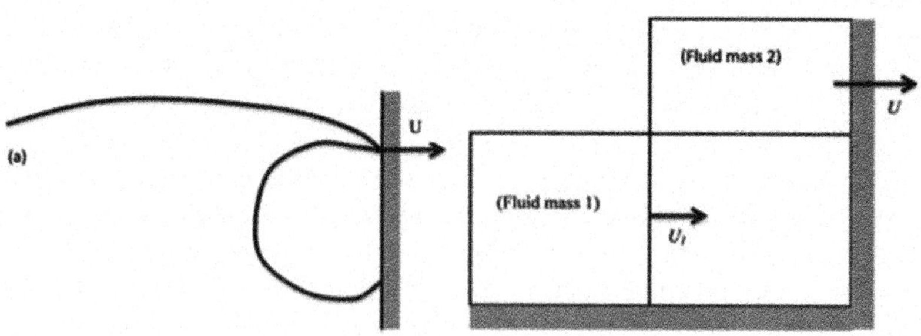

Fig. 6.24 The loading mechanism of the breaking wave (*a*) (*left*) and its simplified model (*right*) [26, 28]

It is important for us that the profile and the duration of the pressure time histories from violent surface waves are approximately the same as produced by underwater shock waves from explosions acting on immersed surfaces [37, 38].

6.5.2 *Underwater Explosions and Hull Cavitation: Descriptions*

A study of the dynamic response of ships and offshore structures to the action of explosions and shock waves is required for safety and serviceability assessments, including habitability on ships. For merchant ships accidental explosions have to be considered as possible events, as they could be severe enough to cause serious damage to structures and injuries to people. They are also becoming of great interest due to increasing public concern about ship safety and potential harm to the environment [56].

Basically, design methods adopted for naval and merchant ships are the same, but naval ships must fulfil some specific requirements [48–52, 57]. In particular, they are required to retain a high standard of operational effectiveness when under attack.

Figure 6.25 shows the results of an underwater explosion of a torpedo. It was found that the results of the attacks are connected with the transient cavitation (the negative pressure) on the ship girder. Above-water weapon attacks include air blast loads. The negative pressure and the impulse caused by the expanding bubble are main reasons of the final anomalous sag distortion of the ship girder [52].

Underwater weapons usually also explode near to the hull (Fig. 6.26).

The detonation of a high explosive charge underwater transforms the solid explosive material into a gaseous reaction that produces an extremely high temperature and high pressure. This pressure is transmitted to the surrounding water and propagates as a spherical wave disturbance (shock wave) moving approximately at the speed of sound in water, and it produces damage by shock and whipping. With the arrival of the shock wave, the pressure rises to a peak value and then the peak is followed by a decay that in its initial portion can be approximated by an exponential decay function. The time for the pressure to decay to e^{-1} (the exponential law) of its maximum value is of the order of a millisecond (0.001 s), after which the pressure decay rate becomes slower than the exponential rate. The results of underwater explosions depend on the amplitude and duration of the loading by the shock wave, and also on the character of the shock wave-structure interaction. Generally speaking, the shock may cause distortion or rupture of the hull if it is sufficiently severe.

This is why submarine surfaces are formed by stiffened and unstiffened hull panels. Namely, if the high peak of initial pressure of water is very large, an element of the hull can begin to move very rapidly and separates from the water. Thus, the hull cavitation can occur in the water due to reflection of the shock wave from the hull surface. The reflected shock wave becomes a rarefaction wave if the hull is

Fig. 6.25 A few stages of the sinking of a ship after an explosion [52]

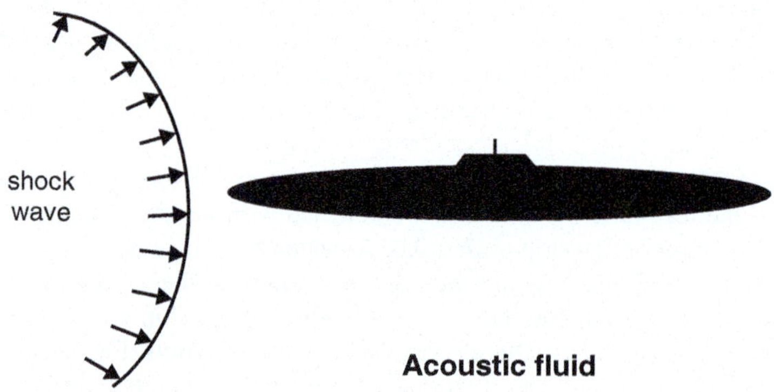

Fig. 6.26 Sketch of shock wave induced by an underwater explosion impinging on a submarine [50]

flexible enough and the depth is not very large. As a result so-called ‘hull cavitation’ (near the submarine) or so-called ‘bulk cavitation’ (far from the submarine) can occur at points where the local pressure (from the sum shock-wave pressures, and the ambient static pressure at that depth) drops below the saturation vapour pressure. In the case of the waves of sufficiently long duration, cavitation appears if the pressure in the water drops to the order of −0.2 to −0.35 MPa (Fig. 6.27).

The hull cavitation, owing to its specifics, is almost unknown in hydrodynamics, though some books and a number of articles are devoted to it [36–52]. In particular, a gas-water mixture may be generated due to the hull cavitation. The cavitation limits the reduction of the water pressure. As a result, cavitation can increase the pressure on the deforming surface and may determine the result of the ship hull/ extreme surface wave interaction.

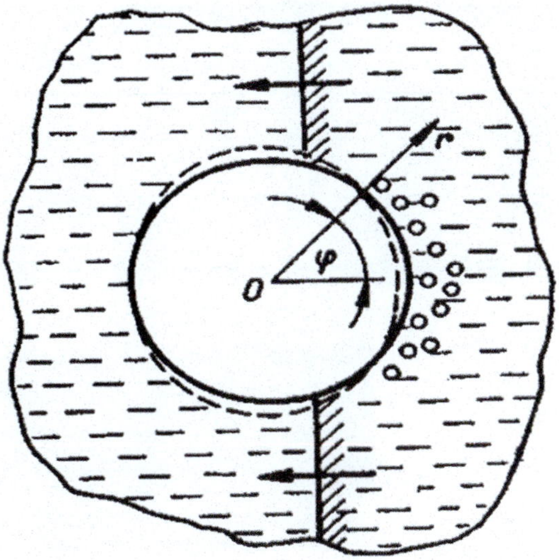

Fig. 6.27 A qualitative picture of the cavitation interaction of underwater shock waves with the hull of a submarine. The *solid line* (*circle*) corresponds to the undisturbed hull. The *dashed line* corresponds to the hull loaded by the underwater shock wave. The *bubbles* determine the cavitation zone [43]

Generally speaking, transient cavitation can take place due to different reasons. Darwin described its appearance in volcanic bombs because of the centrifugal force *...I think there can be little doubt,...relieving the pressure ...allowed the heated vapours to expand...* (cf. Sect. 4.5). In our case, the gas-water vapour appears because the water volume has locally emptied from the compression due to the hull element displacement [36, 41–52].

6.5.3 *Underwater Explosions and Hull Cavitation: Experiments*

Now we consider the influence of deformability of the water surface on the explosion pressure. First, as a limiting case, we consider the free water surface.

The photographs presented in Fig. 6.28 illustrate three instants following an underwater explosion near the water surface. The explosion product (a dark spherical spot) and the expanding spherical shock wave are shown in Fig. 6.28a. The initial stage of the reflection of this wave from the free water surface is shown in Fig. 6.28b. The dark zone located under the free surface and above the explosion product is the transient cavitation zone. This zone expands as shown in Fig. 6.28c.

Let us consider the hull cavitation on the thin flexible solid surfaces. In Fig. 6.29 these surfaces are modelled by a round plate. The plate was loaded by an underwater shock wave.

The circular closed pipe whose left-hand end is at the centre of Fig. 6.29 is surrounded by fluid and is orientated horizontally, extending to the right beyond the image. The experiments have been performed using a thin (0.020 in.) plate of

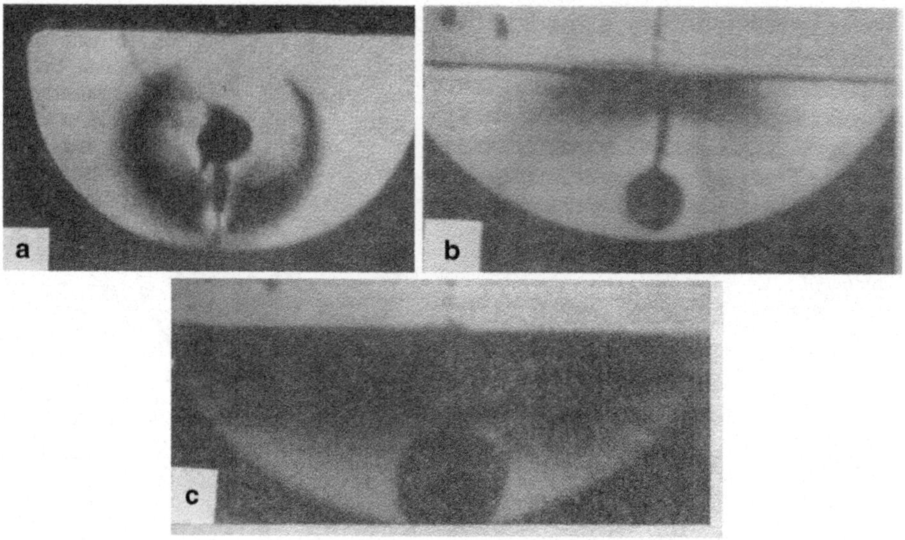

Fig. 6.28 The interaction of underwater explosion with a free water surface [58, 59]

Fig. 6.29 Cavitation beyond the end of a circular pipe closed by a plate after being subjected to a shock wave approaching from the left [37]

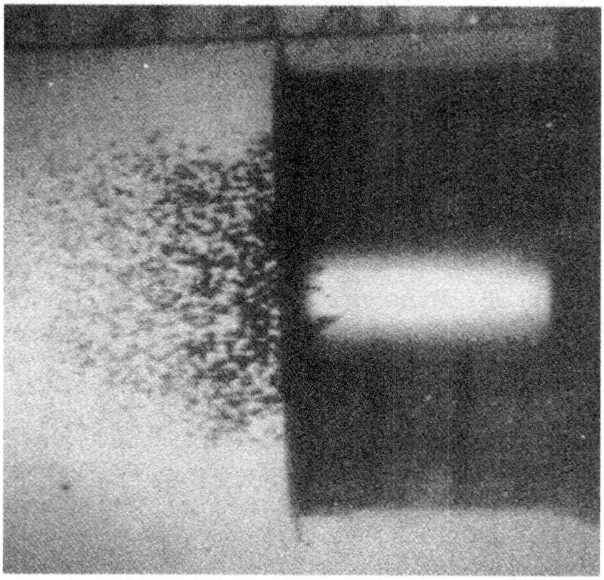

cellulose acetate 6 in. in diameter backed by air, and photographing the water in front of the plate after reflection of a shock wave. The cavitation zone is shown by a multitude of bubbles. This zone arose following the impact of an underwater shock wave onto a flexible plate which closed the end of a pipe surrounded by fluid, extending to the right of the image.

Fig. 6.30 A photograph of the cavitation interaction of a cylindrical shell/water/underwater explosion [60]

Thus, with sufficiently large amplitude the pressure will become negative at points in the fluid, and it may acquire large negative values for sufficiently- light plates. This corresponds to a rarefaction wave reflected from the accelerated surface of the plate. If the water cavitates, the subsequent motion of the plate will be modified.

The photos presented in Figs. 6.30 and 6.31 illustrate the interaction of a tube wall with a shock wave generated by the electro-explosion of a metal wire in water filling the tube (cylindrical steel shell). The wire is spread along the tube axis. The pictures were obtained by photographing from the end face of the tube (the radius is 1.47 cm, the thickness is 2 mm) immersed in water. Figure 6.30 shows typical photographed scans of the tube section [60].

The inner surface of the tube is seen, as the dark lines clearly visible at the first instants after the explosion. The central dark zones are formed by the metal-water vapour (explosion product). The bright zones are water. The straight lines are the shock wave trajectories. The dark zones, located within the water and near the wall of the shell, are formed by the transient cavitation.

The upper photo of Fig. 6.31 shows a typical photographed scan of the deflection of an internal point of the cylindrical shell. The shell deforms and its radius increases. As a result, the cavitation zones are formed (bottom).

Conclusions to the Sect. 6.5 During the wave breaking, the motion of a ship is practically independent of the motion of the water particles in the wave. In this case the loading (Figs. 6.22 and 6.23) may be very short, as from an explosion. Using this analogy has allowed us to understand certain important features of extreme wave action on ship hulls. Figures 6.27, 6.28, 6.29, 6.30 and 6.31 demonstrate some important peculiarities of the hull/wave interaction. In particular, we considered the explosions acting on plates and shells. There figures show the formation of the rarefaction (cavitation) zones in the water.

Fig. 6.31 The radial displacement of an internal point of the wall of a cylindrical tube under the action of an internal underwater explosion (*upper picture*) and the expansion of the explosion product (*bottom picture*) [43, 61]

Such loadings can occur from underwater explosions or from extreme surface waves. Of course, the most dangerous is the extreme breaking wave. An example of this wave is shown in Fig. 6.2.

I focus on these extreme and potentially damaging waves, whose loadings cannot be predicted very accurately in simulations. Breaking wave pressure upon a solid surface is defined by the trapped layer of air (Fig. 6.24). This result, of course, is fair for a solid body and for the hull of a ship. But in the latter case there is the new, complicating circumstance connected with deformations of the hull, which can strongly influence the impact pressure of the water. Because of the trapped atmospheric air or/and cavitation, there are large uncertainties in the determination of real loads from water waves on structural elements of ships. Thus, the uncertainties in the loadings are determined by the profile and amplitudes of the initial waves, which usually are unknown, and by the associated air bubbles and transient cavitation that is generated during the transient water-deformable surface interaction process.

The destructive effect of giant waves is well-known, and, hence, prediction of where and when they will occur is of extreme importance to all who live or work beside or upon the sea. The existence of these extreme waves makes it important to re-examine some fundamental ideas of merchant ship design, which had been developed earlier [56, 57]. In particular, experience from designing underwater naval ships (submarines), which are required to retain a high standard of operational effectiveness under explosion attack, may be used.

6.6 Experimental Studies of Hull Cavitation

The process of interaction of extreme water waves with a ship hull is very complex, since it may be associated with large strains in the structural elements, and the generation of cavitation in the water. This complex interaction process has not been studied fully, although now powerful numerical codes are available [36, 42, 43, 45, 47–52], which allow the response of a ship hull to the action of water waves to be calculated. However, there are large uncertainties in the determination of loads from extreme waves acting on the structures. Therefore, the results of the calculations need to be supported by experiment data. This is especially true for local elements of a hull such as panels and plates.

In this section we consider some important aspects of the transient interaction of deformable surfaces with water waves. This problem is examined using an experimental approach, by creating a high pressure wave in water and impacting it on plates with various properties and thicknesses. Such loadings can occur from underwater explosions or from extreme surface waves, and are important for the design of merchant and naval ships, and offshore structures. The experiments are focused on these extreme and potentially damaging waves, whose loadings cannot be predicted very accurately in simulations.

Many tests were conducted, using an apparatus whose principal features are presented in Fig. 6.32. The fundamental principle used in the apparatus is the conversion of the kinetic energy of a rapidly moving mass (a flat piston) into water pressure energy. The piston was accelerated by a compressed gas in the upper part of the cylinder.

The experimental apparatus consists of a vertical cylindrical metal tube (1) having an inner diameter of 112 mm. The tube wall thickness is 20 mm. The acoustical sound speed calculated for water in the tube was 1,265 m/s. The plane piston (2) impacts the water surface and causes the compression wave (3). The tube bottom is closed by the circular plate which is clamped around its edge. This plate can be replaced during the experiments in order to examine an influence of its properties (thickness h and stiffness) on the resulting pressures. The compression wave deforms the plate as shown in sketch. The plate is shown in its undeformed state (4) (Fig. 6.32a) and deformed state (Fig. 6.32b). An air layer ahead of the piston was compressed during its acceleration. This diminished the wave amplitude in the water to about half compared to what it would have been in the absence of an air layer. The pressures were measured at different points in the water using piezoelectric pressure sensors and recorded on an oscilloscope screen. The piezoelectric pressure sensors and displays had previously been calibrated statically and dynamically, and both calibrations gave identical results.

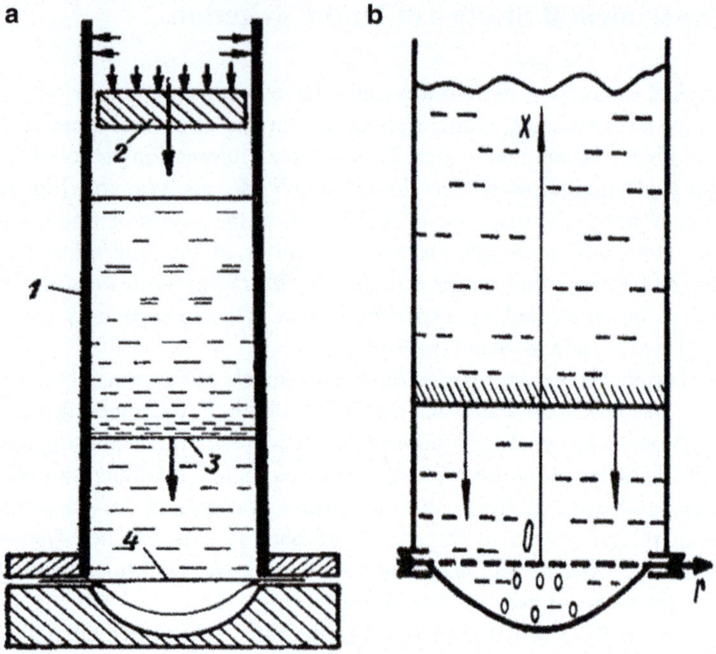

Fig. 6.32 Sketches of the experimental apparatus (**a**) and a scheme of the plate-shock wave interaction (**b**) [43, 60]

6.6.1 Elastic Plate/Underwater Wave Interaction

A series of experiments was made when the mass of the piston was 13 kg, and the height of the water column was 500 mm. The impact velocity of the piston onto the water was kept constant at 1.8–2.0 m/s. The wave amplitude was approximately 1.25 MPa in the experiments with a wavelength of 2.5–3 m.

An example oscillogram of the pressure measured on the surface of a steel plate with thickness $h=20$ mm is shown in Fig. 6.33. One can see that the pressure increases rapidly, and then decays according to an exponential law. We shall consider this pressure as the incident wave, which is the same for all tests with elastic plates.

Results from aluminium alloy plates with thicknesses $h=8$, 4, and 2 mm are shown in Fig. 6.34a–c, respectively. The pressure oscillograms measured on the tube wall at points 255 mm and 55 mm from the plate surface correspond to curves *I*, and *II* respectively. The left- hand graphs have a 2 ms time scale, and the right-hand graphs have a 0.5 ms timescale. It can be seen that a reduction in the plate thickness results in a radical change in the pressure curves with an increase in the interaction time.

We will consider primarily the curves obtained by sensor *II* close to the plate. These curves consist of the first peak, a pressure drop section, and a second peak

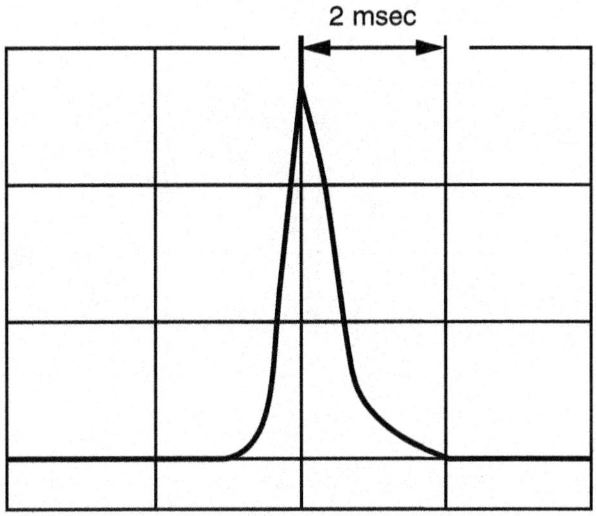

Fig. 6.33 An oscillogram of pressure in the incident wave [36]

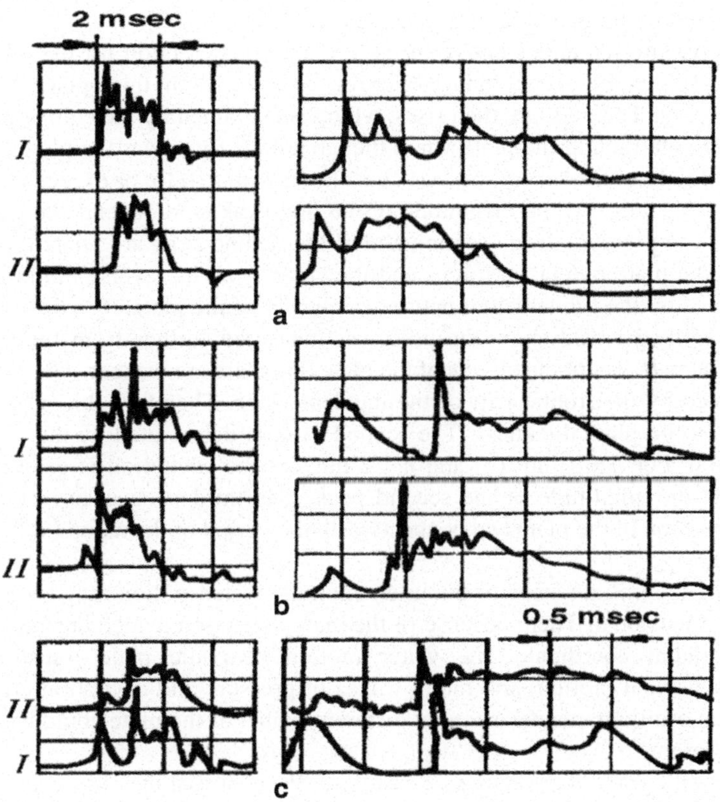

Fig. 6.34 Effect of plate thickness on the water pressure [36, 43, 62]

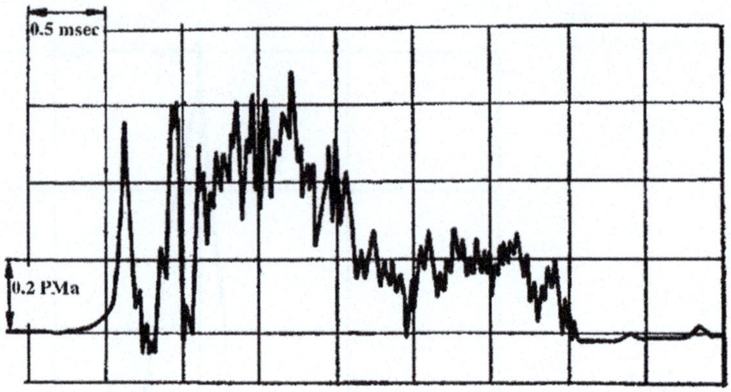

Fig. 6.35 The pressure in the plate centre (aluminium alloy, $h=2$ mm) [36, 43, 63]

which is somewhat longer. The first peak results from the pressure occurring at the very beginning of the interaction when the plate just starts to move. The interaction is evident here in that the peak amplitude depends on the plate thickness and is less than the maximum pressure of the incident wave (Fig. 6.33).

The pressure remains positive in the pressure drop section for the 8 mm thick plate (*a*). Cavitation occurs in the water for the 4 and 2 mm thick plates (cf. curves Fig. 6.34b, c). The pressure then rises, which is explained by the collision between the water column and the plate when the cavitation zone near the plate collapses. The amplitude of the second peak is close ($h=8$ mm (*a*)) to, or exceeds ($h=4$ mm (*b*) and $h=2$ mm (*c*)) the magnitude of the first peak, and depends less on h. The length of the pressure decay section and of the second peak are related to deceleration of the plate by elastic forces and the velocity of its reverse motion, i.e., are determined by the fundamental natural frequency of the plate.

The oscillograms *I* show the pressures measured further from the plate than oscillograms *II*. There the effects of the plate thickness are weaker. The first peak is determined by the frontal part of the incident wave. Therefore this peak depends weakly on the plate thickness. The second peak is determined by the water-plate interaction. For $h=4$ mm (*b*) and $h=2$ mm (*c*) the cavitation is determined the form and the amplitude of the second peak. The third peak is formed from the reverse motion of the plate and connects with the period of the natural oscillation of the plate.

During the initial period of the interaction the water motion may be considered as one-dimensional. Then, because of the plate deflection, which depends strongly on the radial coordinate, the water motion near the plate transforms into two-dimensional motion, and radial waves of pressure and cavitation are formed. Therefore oscillograms measured at different points of the plate could be expected to be different.

The oscillogram of pressure measured near the centre of the plate with $h=2$ mm is presented in Fig. 6.35. The maximum values of the pressure was approximately

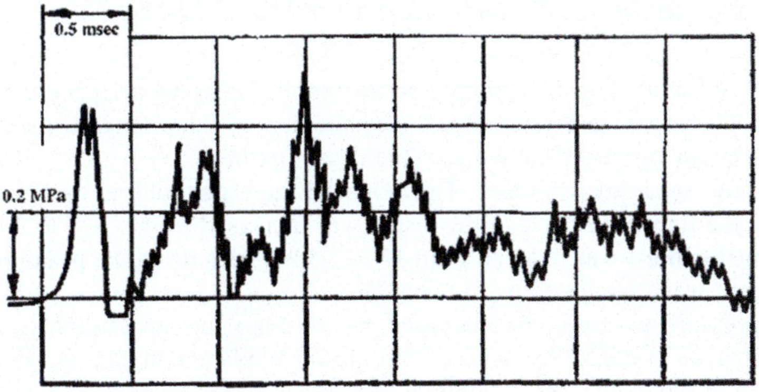

Fig. 6.36 The pressure in the plate centre (bronze alloy, $h = 1$ mm) [36, 43, 63]

0.7 MPa. One can see that this oscillogram is quite different from ones presented in Fig. 6.34c taken at the tube surface, although the oscillograms were obtained from similar plates.

Various plate thicknesses and materials were chosen for the experiments so as to ensure that there was sufficient deflection to cause the appearance of cavitation effects in the zone of elastic behaviour of the plate material. The pressure oscillogram for a plate of bronze (thickness $h = 1$ mm) is presented in Fig 6.36. The oscillogram differs from those presented in Figs. 6.34 and 6.35, as it has many peaks. These peaks are connected with strong radial motion of water along the plate during the interaction. The lifetimes of the cavitation zones in Figs. 6.35 and 6.36 are determined by the intervals between the peaks on the pressure curves. The maximum lifetime at the centre of the plate is thus about 0.2 ms.

Figures 6.34, 6.35 and 6.36 show that the cavitation measured near the plates strongly depends on mechanical and geometrical properties of the plates. In particular, if the plate is thick enough, cavitation effects are either absent or small. However, if the ratio of the radius R of the plate to its thickness h is large, the effect of cavitation on the resulting pressure may be very important. It is evident that the initial wave is damped by a thin plate. Therefore for thin plates the first peak of the pressure may be much smaller than the amplitude of the initial wave (1.25 MPa). The subsequent two-dimensional waves generated by cavitation then move along the plate surface. These waves form the peaks measured in the plate centre. Hence the lifetime of the cavitation cavities strongly depend on the plate properties.

At the same time cavitation may be generated far from the plate due to the resulting superposition of the incident (Fig. 6.33) and the reflected wave. This so-called bulk cavitation is determined by curves I in Fig. 6.34.

6.6.2 Elastoplastic Plate/Underwater Wave Interaction

The hull cavitation depends strongly on the amplitude of the incident wave and its duration. Therefore we have studied this phenomenon, examining the interaction of a circular plate with extreme waves which can rupture it.

The same apparatus was used. The height of the water column was increased to 650 mm, the impact velocity of the piston was increased from 1.8–2 to 30–33 m/s, and the piston mass was reduced from 13 to 2.3 kg. As a result the peak pressure in the incident wave increased to 15–18 MPa.

The pressure oscillograms, recorded on the tube surface at points distant by 255 mm (curve 1) and 55 mm (curve 2) from a metal plate with thickness $h = 20$ mm are shown in Fig. 6.37. One division in curves 1 and 2 correspond to 8 and 11 MPa respectively. Curve 1 has two large peaks, explained by the passage of the incident and reflected waves; curve 2's first peak is large and is caused by an approximate doubling of the wave amplitude at the plate, and the second peak which is not far from the first one. It is impossible to explain its appearance in terms of the wave reflected from the piston. The origin of the second peak is apparently related to the radial motion of the water at the plate surface, caused by a certain deformability of the plate and the tube wall, due to the very high pressure generated near the plate. There is a third insignificant peak on both the upper and lower curves.

Further experiments were carried out, using plates where the loadings resulted in elastoplastic strains.

Oscillograms of the pressure measured at the edges (curves 1) and at the centre (curves 2) of plates are presented in Fig. 6.38. The curves of Fig. 6.38a correspond to $h = 4$ mm, and the curves of Fig. 6.38b correspond to $h = 2$ mm. The pressure was measured at 15 mm from the plate surfaces. The rapid changes in the pressure curves are associated with the origination of hull cavitation in the water, and the associated radial water motion caused by the plate deformation. Since the incident wave is very large in magnitude and short in duration, the plate accelerates very quickly at the beginning of the interaction. As a result, the pressure drops to zero and the plate separates from the water. The subsequent deceleration of the plate by the strain-resistance forces, and the action of the earth's gravitation, result in the disappearance of cavitation, and the plate makes contact with the water by impact causing a new pressure peak. There is then the possibility of a second acceleration of the plate causing a separation of it from the water, and repetition of the process. Of course, this sequence is complicated by the radial motion of the pressure and the cavitation waves along the plate surface. The experiments showed that the peak pressure amplitudes depend on the thickness h of the plate, and the amplitude grows as h increases.

A comparison between the pressure curves is given in Fig. 6.39 for plates from steel (a) and aluminium alloy (b). The plate thickness is $h = 4$ mm. The pressure was measured on the wall at the distance of 55 mm from the plates. The curves are characterized by a rapid peak-like change. In the case of a steel plate, the cavitation

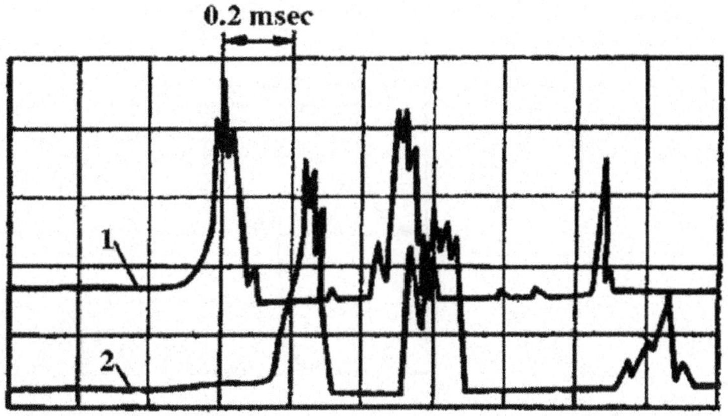

Fig. 6.37 Pressure waves measured near (2) and far (1) from the thick plate [36, 43, 63]

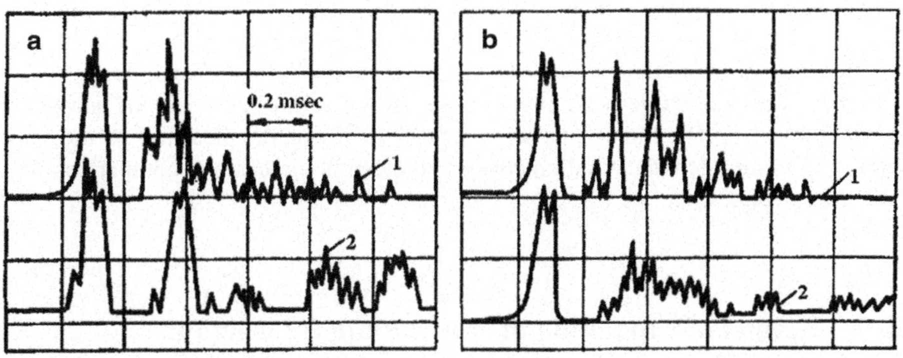

Fig. 6.38 The pressure wave measured at the edges (*curve 1*) and at the centre (*curve 2*) of the elastoplatic (aluminium alloy) plates [36, 43, 63]

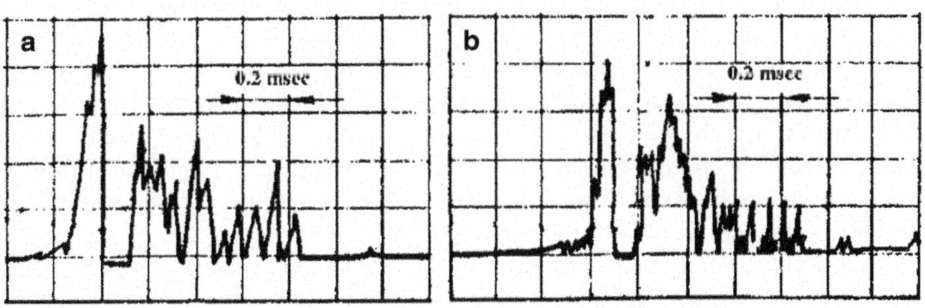

Fig. 6.39 Pressure waves on plates from steel (a) and aluminium alloy (b) [36, 43, 63]

Fig. 6.40 The hull cavitation on a thin plate [36, 43, 63]

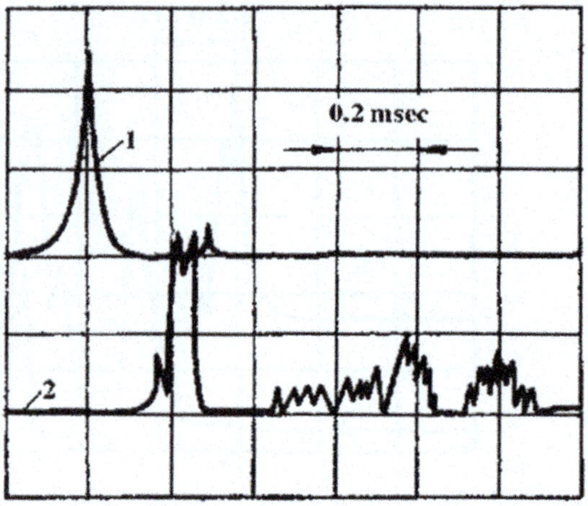

zones are more definite, and they exist for longer time than for the aluminium alloy plate.

Results for a steel plate with $h = 1$ mm is shown in Fig. 6.40. Near the plate edge (curve 1) only one large peak is generated. At the centre (curve 2) the secondary peaks are small. The high plate deformability dampens the pressure. At the same time the lifetime of the cavitation cavity at the centre increases in comparison with the thicker plates.

6.7 Results of Modelling of Wave/Plate Interaction

We showed that transient cavitation can influence very strongly on the non-stationary interaction of liquid with plates. Of course, the reasons for an appearance of an impulse of pressure in liquid can be different. In particular, pressure on the ship hull can arise both from action of an overturning wave, and from action of underwater explosion. We found above that a surface wave action, in the most dangerous cases, can resemble the action of the underwater explosion. These actions may be well described by exponential functions. These functions well describe both the pressure wave shown on Figs. 6.33 and the pressure from the action of breaking waves (Figs. 6.22 and 6.23). Examples of such description are given in Figs. 6.41 and 6.42.

In particular, the recorded pressure time histories can be modelled by the sum of exponential functions. For example, the pressure curve shown on the right in Fig. 6.23 may be modelled by the curve shown in Fig. 6.42 which can be seen to be a good model of this curve.

Thus, the actions of underwater explosions and breaking waves on solid surface may be described by similar mathematical expressions. However, the solid wall model does not take into account any hull deformation. These deformations may be

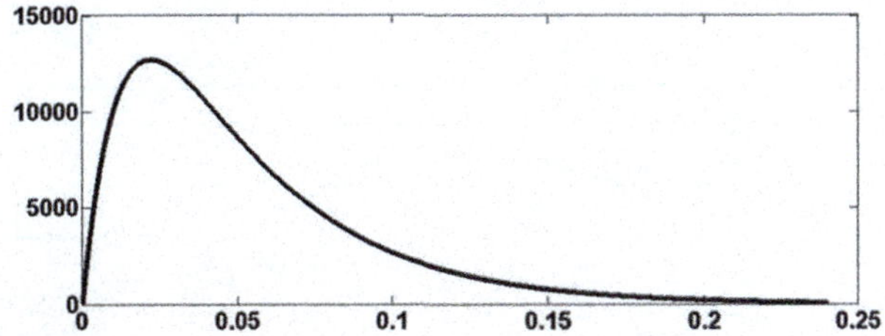

Fig. 6.41 Pressure in lb/ft^2 modelling experimental curve in Fig. 6.22

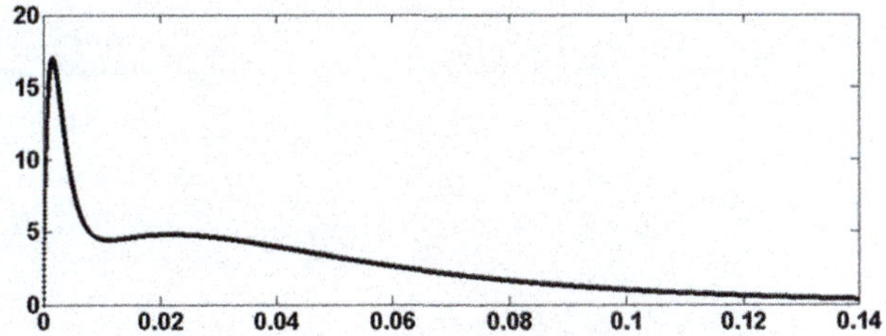

Fig. 6.42 Pressure in kPa modelling experimental curve in Fig. 6.23

described by the theory of plates and shells [36, 43] which takes into account the plasticity and the cavitation [36, 43]. We recall that the fast deformation of a hull element under the action of a very high water pressure can completely change the pressure time history. In particular, during the pressure loading process, a hull element can move faster than the water. As a result, in moving away from the water, the hull can separate from the water. This hull cavitation can increase the pressure on the deforming surface and the effects of the hull/wave interaction. The average pressure is increased because the cavitation limits the reduction of the negative pressure on the hull surface.

Using some results from [21, 26, 36, 43] we studied the damaged area shown in Fig. 6.43. This area was simulated as a circular plate [21, 26]. The results of this analysis are presented below. In particular, the effects of thickness h of a steel circular plate are shown in Fig. 6.44.

6.7.1 *Effects of Deformability*

In Fig. 6.44 the curves labelled 0.2, 0.4, 0.8 and 1.2 correspond to plate thicknesses of 0.2, 0.4, 0.8 and 1.2 cm, respectively. The crosses show the pressure P corresponding to an absolutely rigid surface.

As can be seen, the influence of the plate pliability is manifest essentially immediately. The first pressure peak and the following form of the pressure curves depend strongly on the plate thickness. In particular, for $h \geq 0.8$ cm the pressure does not drop below zero, while for $h \leq 0.4$ cm a disruption (cavitation) of the fluid is possible if it is assumed that cavitation occurs at pressures in the range from -0.2 to -0.35 MPa. The rate of displacement diminishes as time increases. As the plate

Fig. 6.43 Photograph showing a damaged area as a result of local loading of the hull by a steep wave [64]

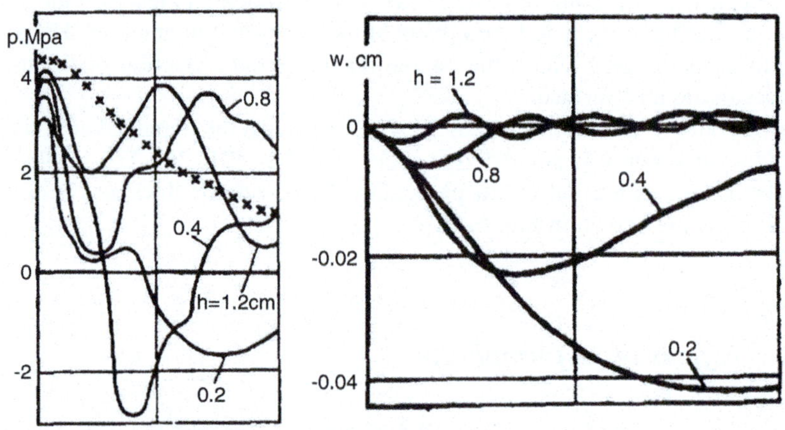

Fig. 6.44 Pressure (*left*) and displacement (*right*) curves at the centre of plates of thicknesses $h = 0.2$, 0.4, 0.8, and 1.2 cm as a function of time *t* from 0 to 17 ms for the pressure, and from 0 to 33 ms for the displacement, from a no-cavitation model [21, 26]

velocity approaches zero the plate becomes effectively more and more rigid to the water wave, and the pressure starts to rise. The water is compressed to higher positive pressures during the reverse motion of the plate and the pressure starts to exceed the pressure that would occur for a rigid surface (crosses). The absolute value and the shape of the second series of pressure peaks in Fig. 6.44 are dependent on the natural frequency of vibrations of the plate. An increase in the frequency (plate thickness) results in a diminution in the spacing between the pressure peaks. Thus, the effect of cavitation is expected to be important if the duration of the loading by the water pressure is less than the fundamental period of the plate vibration.

6.7.2 *Effects of Cavitation*

The effects of cavitation on the pressure and displacement of plates were studied in [21, 26, 36, 43]. In particular, these effects are determined by the plate properties. The study examined the influence of the plate thickness using aluminium alloy plates of different thicknesses.

Calculation results are presented in Fig. 6.45 (solid lines – cavitation model, dashed lines – no cavitation). Curves of the pressure at the plate centre, marked with the numbers 1, 2, 3 (no cavitation) and 4, 5, 6 (cavitation) are shown for plates with thicknesses $h = 0.8$, 0.4 and 0.2 cm respectively. Curves 5 and 6, which are calculated according to the cavitation model, have sections which are coincident with the line $p = 0$. For $h = 0.8$ cm (lines 1 and 4) the pressure at the centre reaches negative values but does not drop below the critical value, i.e., cavitation does not occur. Cavitation appears for $h = 0.4$ cm (lines 2 and 5). The lifetime of the cavitation zone is 8.3 ms. For $h = 0.2$ cm (lines 3 and 6) the cavitation zone exists for approximately 13.3 ms.

The second study considered the evolution of the cavitation zones on a plate. The curves in Fig. 6.46 characterize the changes in pressure and displacement at the surface of the $h = 0.2$ cm plate. The dashed curves 1, 2, 3 (no cavitation) and solid curves 4, 5, 6 (with cavitation modelled) correspond to the times 8.3, 16.6, 58.3 ms and 8.3, 16.6, 27 ms respectively. The cavitation model showed that the displacement reached its maximum value earlier (curve 6) at $t = 27$ ms, compared to the non-cavitation model which gave the maximum value later (curve 3) at $t = 58.3$ ms. The maximum displacement found with cavitation (curve 6) was almost twice the displacement calculated according to the non-cavitation model (curve 3). It is evident from Fig. 6.46 that the central part of the plate moves as a rigid body at the initial instant that the shock pressure is applied, when the displacement and pressure change rapidly with radial position only near the edge. Therefore a low pressure zone is generated near the edge (curves 1 and 4). At $t \approx 8.3$ ms the cavitation forms a ring-like low pressure zone (curve 4). For $t \approx 16.6$ ms the cavitation zone covers the central part of the plate. The velocity of the plate motion later decreases as it reaches its maximum displacement (curve 6) and the cavitation zone diminishes. There is no cavitation at the beginning of the plate's return motion when $t \approx 27$ ms.

The calculations show that the effect of cavitation on the impact deformation of hull elements may be very important. In particular, the cavitation tends to increase

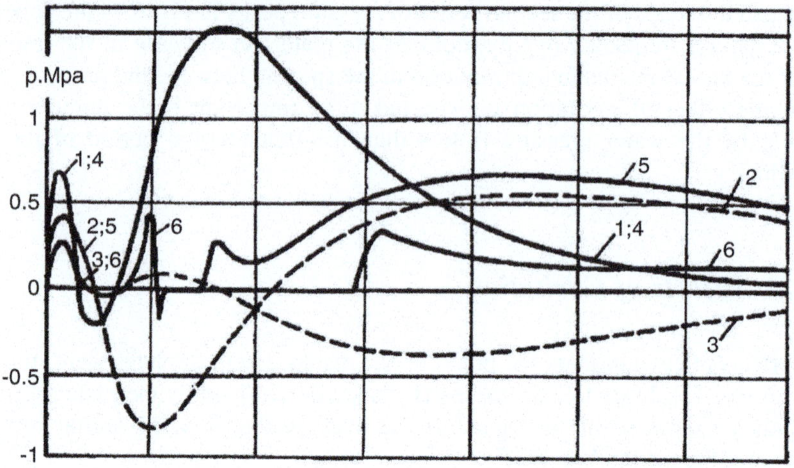

Fig. 6.45 Curves characterizing the change in pressure on the wet surface of plates as a function of time *t* from 0 to 0.0583 s (*solid lines* cavitation model, *dashed lines* no cavitation) [21, 26]

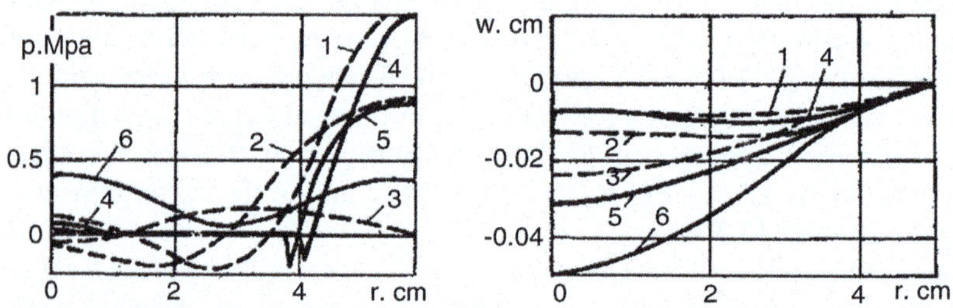

Fig. 6.46 Change in pressure (*left*) and displacement (*right*) on the surface of the $h = 0.2$ cm thick aluminium alloy plate as a function of radial position *r* (*solid lines* – cavitation model, *dashed lines* – no cavitation) [21, 26]

the breaking wave pressure on the ship hull. On the whole this calculations agree with the results of publications about the effects of underwater explosions impacting various kinds of vessels.

6.7.3 Effects of Plasticity

Here we present a comparison between results of calculations and experimental data. These last are obtained with the help of the apparatus shown in Fig. 6.32. The mass of the piston was 13 kg, and the height of the liquid column 50 cm. The impact velocity on the water was 2 m/s. In Fig. 6.47 results for a bronze alloy plate ($h = 1$ mm) are presented. The dashed curves are calculated according to the elastic model of the plate material, and the solid curves are calculated by means of the

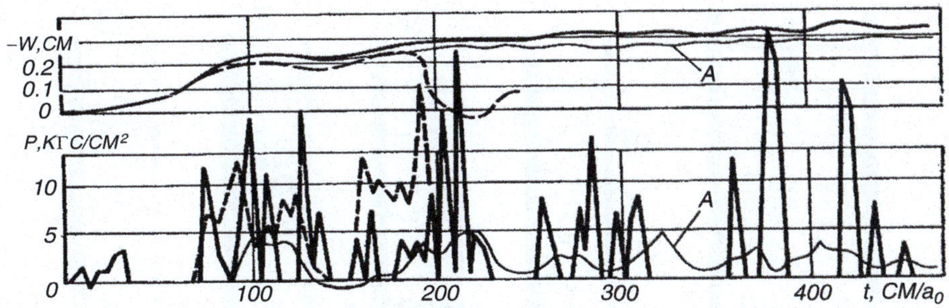

Fig. 6.47 A comparison of measured and calculated data for a thin metal plate [33, 36, 43]

Fig. 6.48 Pressure and displacement curves calculated in the centre of a plate. They show the influence of the boundary deformability on the generation and the evolution of the surface transient cavitation [33, 36, 43]

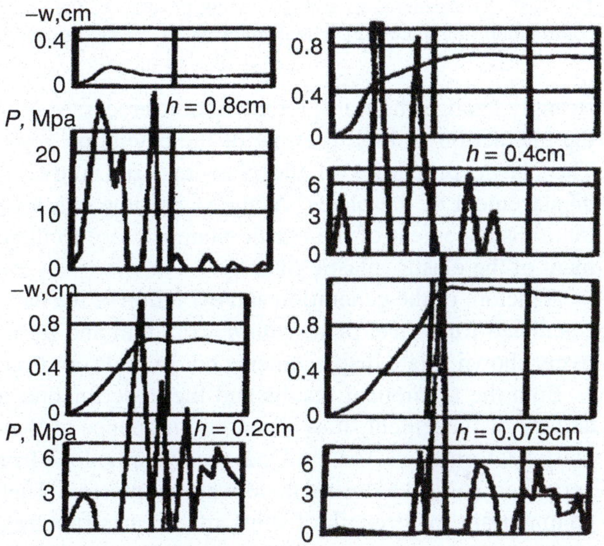

elastic–plastic model. The thick solid line for the pressure p is calculated at the plate centre. The curve $w(A)$ represents the displacement measured near the plate centre. The thin solid curve $p(A)$ is the pressure measured in the water at a distance of about 1.5 cm from the plate centre.

One can see that there is a good qualitative agreement between the experimental data and the calculations done according to the elastic–plastic model. The lifetime of the cavitation zones is determined by the intervals between the peaks in the pressure curves. It is seen that the locations of the peaks and the cavitation zones on the pressure curves practically coincide. The elastic model may be used only to predict the first moments of the interaction. The maximum displacement of the elastic plate, however, is close to the residual displacement of the elastic–plastic plate.

In Fig. 6.48 the influence of the plate thickness h on the underwater wave/plate interaction is shown. The plates (aluminium alloy) having $h = 0.8$, 0.4, 0.2 and 0.075 cm are studied. It is seen that the deformability of the water boundary can

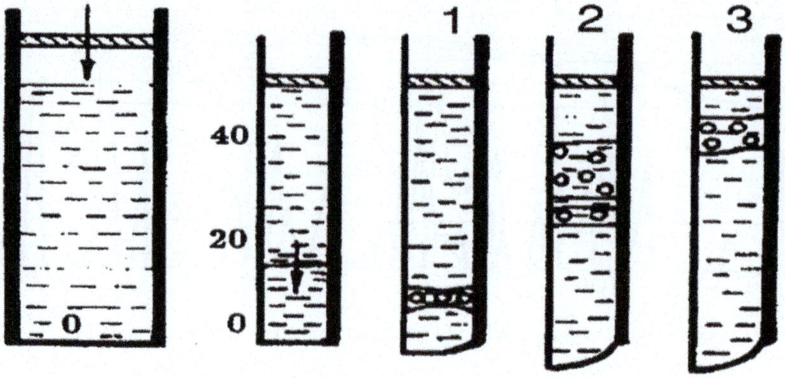

Fig. 6.49 The generation and the evolution of a cavitation zone during the transient plastic deformation of a plate [43]. The piston mass is 2.3 kg, the impact velocity is 32 m/s

strongly change the transient cavitation processes on this boundary. The effect of the interaction and the lifetime of the cavitation zones increases when h reduces. The plastic properties of the plate material allow one to calculate the residual displacement of the plates. At the same time, we found that the calculated maximum displacement of the elastic plate is close to the calculated residual displacement of the elastic–plastic plate.

A picture of the generation and evolution of the cavitation zones within the tube, calculated for a steel plate with $h = 0.2$ mm and three instants of time, is qualitatively shown (plate displacements not in scale) in Fig. 6.49.

Both the theoretical results and the experimental oscillograms of the pressure and the displacement show the strong influence of the cavitation interaction on the plastic deformation. I remind that the dynamic load from the breaking wave may be very short [64] like during underwater explosion. Therefore presented results may be important for strength of ships and ocean structures. Most complex engineering systems are constructed often from different plates, so that I focused on these elements. Figure 6.50 shows the final plastic displacement of the middle section of metal plates (width 78 = mm, length >>78 mm, thickness = 0.6 mm) loaded by underwater shock wave having different energy contents E. The numerical calculations were made using the 2D cavitation model proposed by Galiev [36, 40] (curve 1), Shauer's model [65] (curve 2), and the model of an ideally elastic liquid (curve 3). The experimental data are represented by the dots. It can be seen that cavitation can increase the final displacement of the plate by a factor of 2÷3. Curve 1 describes the experimental data very well. Shauer's model (curve 2) is valid only for small plastic deformation.

Thus, we can conclude that transient cavitation is an important phenomenon which occurs during extreme wave/hull interaction.

A wide spectrum of two- and three-dimensional problems of structure/underwater shock wave interaction has already been studied [36–52, 58–65]. In these works, the water is considered as a ruptured liquid, or a gas/liquid mixture, or as a bubbly liquid.

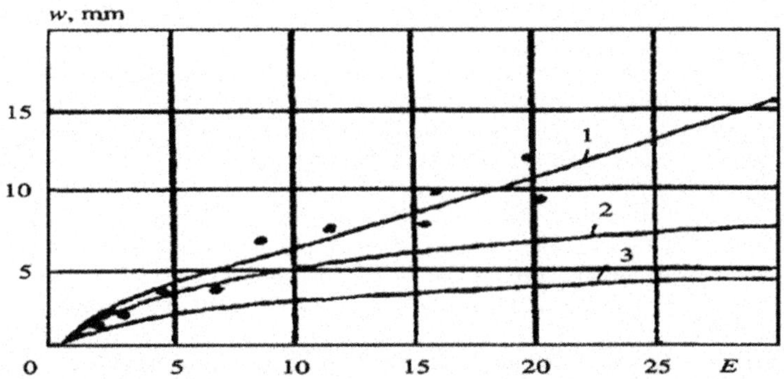

Fig. 6.50 Comparison of the maximum displacement of the plate centre. Black points (*dots*) are experimental points, the lines are calculated by means of Shauer's model [64], Galiev's model [36, 40], and the elastic liquid model (*3*)

Conclusions According to the materials of this section:

1. In the majority of cases the deformation of a ship hull by a short duration water wave pressure pulse is accompanied by hull cavitation. The effect of cavitation may be important if the duration of a loading is less than the period of the fundamental frequency of the hull element oscillations;
2. The cavitation zones can cover practically the all wet surface and completely change the water loading onto the hull element, compared to the pressures that would be calculated if the cavitation were ignored;
3. The interaction is not a 1-D process. The circular plate deformation generates radial pressure and cavitation waves, which create and collapse the cavitation cavities.

References

1. Darwin C (1890) Journal of Researches into the Natural History and Geology of the Countries Visited during the Voyage of H.M.S. Beagle Around the World, under the Command of Captain FitzRoy, R N T Nelson and Sons, London
2. Kurkin AA, Pelinovsky E (2004) Волны-убийцы: Факты, Теория и Моделирование (Freak waves: facts, theory and modelling). NNSTU, Nizhny Novgorod
3. Liu PC, MacHutchon KR (2006) Are there different kinds of rogue waves? OMAE2006-92619
4. Liu PC, MacHutchon KR, Wu CH (2006) Exploring rogue waves from observations in South Indian Ocean. http://www.ifremer.fr/web-com/stw2008/rw/fullpapers/liu.pdf
5. Fochesato C, Grilli S, Dias F (2007) Numerical modelling of extreme rogue waves generated by directional energy focusing. Wave Motion 44(5):395–416
6. Dysthe K, Krogstad HE, Muller P (2008) Oceanic rogue waves. Annu Rev Fluid Mech 40:287–310
7. Olagnon M, Prevosto M (eds) (2009) Rogue waves 2008. Ifremer http://www.ifremer.fr/web-com/stw2008/rw/
8. Kharif C, Pelinovsky E, Slunyaev A (2009) Rogue waves in the ocean. Springer, Berlin
9. Denchfield SS, Murphy AJ, Temarel P (2009) A deep-water beach laboratory generation of abnormal waves. In: Olagnon M, Prevosto M (eds) Rogue waves. Ifremer. http://www.ifremer.fr/web-com/stw2008/rw/full/SSDenchfield.pdf

10. Akhmediev N, Ankiewicz A, Taki M (2009) Waves that appear from nowhere and disappear without a trace. Phys Lett A 373:675–678
11. Akhmediev N, Soto-Crespo JM, Ankiewicz A (2009) Extreme waves that appear from nowhere: on the nature of rogue waves. Phys Lett A 373:2137–2145
12. Zakharov VE, Dyachenko AI, Shamin RV (2010) How probability for freak wave formation can be found. Eur Phys J-Spec Top 185(1):113–124
13. Kibler B, Fatome J, Finot C, Millot G, Dias F, Genty G, Akhmediev N, Dudley M (2010) The Peregrine soliton in nonlinear fibre optics. Nat Phys 6:790–795
14. Vennell R (2010) Resonance and trapping of topographic transient ocean waves generated by a moving atmospheric disturbance. J Fluid Mech 650:427–442
15. Nikolkina I, Didenkulova I (2011) Rogue waves in 2006–2010. Nat Hazards Earth Syst Sci 11:2913–2924
16. Latifah AL, van Groesen E (2012) Coherence and predictability of extreme events in irregular waves. Nonlinear Process Geophys 19:199–213
17. Chabchoub A, Hoffmann N, Onorato M, Akhmediev N (2012) Super rogue waves: observation of a higher-order breather in water waves. Phys Rev X 2:011015
18. Galiev ShU, Mace BR (2014) Lagrangian description of extreme ocean waves. Известия Уфимского Научного Центра Российской Академии Наук (Herald of Ufa Scientific Center, Russian Academy of Sciences (RAS)) 4:6–13
19. Clauss G, Schmittner C, Henning J (2003) Simulation of rogue waves and their impact on marine structures. Proceedings of MAXWAVE final meeting, October 8–10, Geneva
20. Kjeldsen SP (2004) Measurements of freak waves in Norway and related ship accidents. Internet
21. Galiev SU, Flay RG (2009) The transient interaction of plates with extreme water waves: effects of deformability and hull cavitation. In: Brebbia C (ed) Fluid structure interaction V. MIT Press, Cambridge
22. Denchfield SS, Hudson DA, Temarel P, Bateman W, Hirdaris SE (2009) Evaluation of rogue wave induced loads using 2D hydroelasticity analysis. In: Proceedings of the 5th international conference on hydroelasticity in marine technology. University of Southampton, Southampton, pp 347–360
23. Clauss G (2010) Freak waves and their interaction with ships and offshore structures. In: Ma Q (ed) Advances in numerical simulation of nonlinear water waves. World Scientific, Hackensack
24. Bennett SS, Hudson DA, Temarel P (2013) The influence of forward speed on ship motion in abnormal waves: experimental measurements and numerical predictions. J Fluids Struct 39:154–172
25. Bennett SS, Hudson DA, Temarel P (2012) A comparison of abnormal wave generation techniques for experimental modelling of abnormal wave-vessel interactions. Ocean Eng 51:34–48
26. Galiev SU, Flay RG (2014) Interaction of breaking waves with plates: the effect of hull cavitation. Ocean Eng 88:27–33
27. Lavrenov IV (2003) Wind-waves in ocean. Springer, New York
28. Galiev SU (2011) Геофизические Сообщения Чарльза Дарвина как Модели Теории Катастрофических Волн (Charles Darwin's geophysical reports as models of the theory of catastrophic waves). Centre of Modern Education, Moscow (in Russian)
29. Holliday NP, Yelland MJ, Pascal R, Swail VR, Taylor PK, Griffiths CR, Kent E (2006) Were extreme waves in the Rockall Trough the largest ever recorded? Geophys Res Lett 33(5), L05613
30. Lamb H (1932) Hydrodynamics, 6th edn. Dover Publications, New York
31. Galiev SU, Galiyev TS (2001) Nonlinear transresonant waves, vortices and patterns: from microresonators to the early Universe. Chaos 11:686–704
32. Smith RA (1998) An operator expansion formalism for nonlinear surface waves over variable depth. J Fluid Mech 363:333–347

33. Galiev SU (2008) Strongly-nonlinear wave phenomena in restricted liquids and their mathematical description. In: Perlidze T (ed) New nonlinear phenomena research. Nova Science Publishers, New York
34. Galiev ShU (2009) Modelling of Charles Darwin's earthquake reports as catastrophic wave phenomena. http://researchspace.auckland.ac.nz/handle/2292/4474
35. Luo J-J (2011) Climate science: ocean dynamics not required? Nature 477:544–546
36. Galiev ShU (1977) Динамика Взаимодействия Элементов Конструкций с Волной Давления в Жидкости (Dynamics of structure element interaction with a pressure wave in a fluid). Naukova Dumka, Kiev (in Russian). The same, American Edition, Department of the Navy, Office of Naval Research, Arlington (1980)
37. Cole RH (1948) Underwater explosions. Princeton University Press, Princeton
38. Zamislaev BV, Iakovlev YS (1967) Динамические Нагрузки при Подводных Взрывах (Dynamical loads during underwater explosions). Sudostroenie, Leningrad
39. Bleich HH, Sandler IS (1970) Interaction between structures and bilinear fluid. Int J Solid Struct 6:617–639
40. Galiev SU (1975) Напряженное Состояние Периодически Подкрепленного Цилиндра при Действии Подводной Волны (Stress–strain state of a periodically reinforced cylinder subjected to an underwater shock). Naukova Dumka, Kiev (in Russian)
41. Galiev SU (1980) Models of a cavitating liquid in nonsteady hydroelastic plasticity. Strength Mat 10:1206–1214
42. Newton RE (1981) Effects of cavitation on underwater shock loading -Plane problem. Final report. NPS-69-81-001. Naval Postgraduate School, Monterey
43. Galiev SU (1981) Динамика Гидроупругопластических Систем (Dynamics of hydroelastoplastic systems). Naukova Dumka, Kiev (in Russian)
44. Pertsev AK, Platonov EG (1987) Динамика Пластин и Оболочек (Dynamics of plates and shells). Sudostroenie, Leningrad
45. Galiev SU (1997) Influence of cavitation upon anomalous behavior of a plate/liquid/underwater explosion system. Int J Impact Eng 19(4):345–359
46. Johnson W (1998) Comments on 'Influence of cavitation upon anomalous behavior of a plate/underwater explosion systems'. Int J Impact Eng 21(1):113–115
47. Galiev SU, Romashchenko VA (1999) A method of solving nonstationary three-dimensional problems of hydroelasticity with allowance for fluid failure. Int J Impact Eng 22:469–483
48. Felippa CA, DeRuntz JA (1984) Finite element analysis of shock-induced hull cavitation. Comput Methods Appl Mech Eng 44:297–337
49. Sandberg G (1995) A new finite – element formulation of shock – induced hull cavitation. Comput Methods Appl Mech Eng 120(1–2):33–44
50. Felippa CA, Park KC, Farhat C (1999) Partitional analysis of coupled mechanical systems. Report CU-CAS-99-06. Center for Aerospace Structures, College of Engineering, University of Colorado. http://www.colorado.edu/engineering/CAS/Felippa.d/FelippaHome.d/Publications.d/Report.CU-CAS-99-06.pdf
51. Shin YS (2004) Ship shock modelling and simulation for far-field underwater explosion. Comput Struct 82:2211–2219
52. Zhang Z-H, Wang Y, Zhang L-J, Yuan J-H, Zhao H-F (2011) Similarity research of anomalous dynamic response of ship girder subjected to near field underwater explosion. Appl Math Mech Engl Ed 32(12):1491–1504
53. Bascom W (1980) Waves and beaches. Anchor, New York
54. Russell RCH, Macmillan DH (1970) Waves and tides. Greenwood Press, Westport
55. Peregrine DH (2003) Water-wave impact on walls. Annu Rev Fluid Mech 35:23–43
56. 15th international ship and offshore structures congress 2003, vol 1, San Diego, 11–15 Aug 2003. Dynamic Response, Elsevier Science, pp 193–264 (September 15, 2005)
57. 16th international ship and offshore structures congress 2006, vol 2, 20–25 Aug 2003. Naval Ship Design. University of Southampton, School of Engineering Sciences, Southampton, pp 213–263 (June 2006)

58. Galiev SU (1988) Нелинейные Волны в Ограниченных Сплошных Средах (Nonlinear waves in bounded continua). Naukova Dumka, Kiev (in Russian)
59. Galiev SU, Borisevich VK, Potapenko AN, Plisko-Vinjgradskii AF (1984) A method of calculating the load from nonlinear waves on a tank cover. Strength Mat 5:663–669
60. Galiev ShU (1994) Influence of cavitation on transient deformation plates and shells by the liquid. Proceedings of IUTAM symposium on impact dynamics, Peking University Press, Beijing
61. Galiev SU, Demina VM, Shelom VK (1979) Impulse expansion of cylindrical shells by a liquid. Strength Mat 12:1379–1383
62. Ilgamov MA, Pavlov AA (1974) Исследование ударной волны в жидкости (Research on a shock wave in liquid). Proceedings of seminar of the theory of shells, vol 4. Kazanskii Physical-Techical Institute, pp 181–195
63. Galiev SU, Pavlov AA (1977) Experimental investigation of the cavitation interaction of a compression wave with a plate in liquid. Strength Mat 8:988–993
64. Buchner B, Voogt A (2004) Extreme waves can have a very steep front. http://www.ifremer.fr/web-com/stw2004/rogue/papers/buchner.pdf
65. Schauer HM (1951) The after flow theory of the reloading of air-backed plates at under-water explosions. Proceedings of first U.S. National Congress of Applied Mechanical, Illinois Institute of Technology, Chicago, pp 887–892

Chapter 7
Modelling of Extreme Waves in Natural Resonators: From Gravity Waves to the Origin of the Universe

We can allow satellites, planets, suns, universe, nay whole systems of universe, to be governed by laws,... ***(Darwin.*** *Notebooks)*

Many objects in Nature may be considered as natural resonators. When an object is stimulated at its resonant frequency, the amplitude of the vibrations is greatly increased. We connected above the appearance of extreme waves with this phenomenon. In particular, there natural resonators exist including sediment layers, ocean, hills, volcanoes, the ocean shelf and so on. Their resonant effects can cause rapid and very large amplitude growth of seismic waves. For example, our model of a natural resonator explained particularities of the tsunami evolution described by Darwin (cf. Chap. 5) and the behaviour of Tarzana hill during an earthquake (cf. Sect. 4.1.4).

In this Chapter we will give further remarks about experimental observations of this catastrophic amplification, and our theoretical research about it. On the other hand, it is our purpose to emphasize the common character of some results presented in the book. We considered above extreme seismic and ocean waves. We have shown that there are similarities of highly-nonlinear wave phenomena in many areas of wave dynamics. However, up to now highly-nonlinear wave phenomena have been poorly understood. Is there a universal, generic theory that can describe all these wave phenomena? We believe that such a theory has been created for 1-dimensional waves which are sufficiently long. We think that many extreme waves are the result of resonant nonlinear amplification of initially small-amplitude waves. We also believe that this theory, when applied to the nonlinear Klein-Gordon equation might shed light on the emergence and initial evolution of the Universe.

For example, Darwin described the local amplification of the seismic waves which ruptured the small island Quiriquina. I suggested that this island could be regarded as a natural resonator [1]. Darwin wrote about *...the parallel and steep sides of the narrow island...* [2, p. 370]. Very long seismic waves excited the shelf and vibrated the base of the island. Qualitative scheme of waves which might be excited on the island surface are shown in Fig. 7.1. Generally speaking, shock-like waves might have been excited in the island material during of the 20 February

Sh.U. Galiev, *Darwin, Geodynamics and Extreme Waves*,
DOI 10.1007/978-3-319-16994-1_7

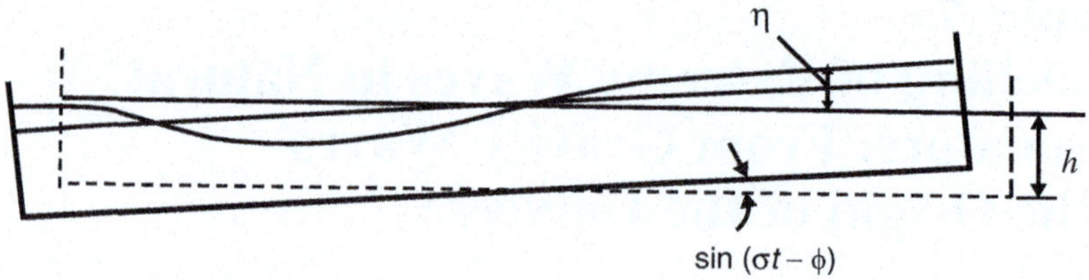

Fig. 7.1 Scheme of earthquake-induced oscillations and waves in certain natural resonators

1835 Chilean earthquake [1, 3]. Analogously, the earthquake-induced motion of ground surface can force surface waves in sediment valleys (basins) and lakes (Fig. 7.1).

In this Chapter we outline briefly an approximate, but universal, generic theory of extreme waves, which can appear in different processes from earthquakes till the origin of the Universe [1, 3–15]. We use well-documented experiments and observations so to check the theory.

7.1 Governing Relations Describing Extreme Seismic Waves in Certain Natural Resonators

In this book we have often compared results of observations and the experimental data with the calculations, to demonstrate the resonant nonlinear nature of the appearance of extreme waves. We did not linger on the mathematical description of the resonant extreme waves, since the theory of them was presented in many publications [1, 3–15]. However, we consider that it will be useful to outline some elements of the theory, here at the end of the book.

7.1.1 One-Dimensional Highly Nonlinear Wave Equation and the Nonlinear Generalization of d'Alembert's Solution

Many long waves may be described by the highly nonlinear equation [16],

$$u_{tt} = c_0^2 u_{aa}(1 + u_a)^{-\gamma-1}. \tag{7.1}$$

Here u is a displacement, a is a Lagrangian coordinate, t is time, $u_a = \partial u/\partial a$, $u_t = \partial u/\partial t$ and c_0 is the wave speed. Taking into account quadratic and cubic nonlinear terms, we can write the Eq. (7.1) in the form

$$u_{tt} - c_0^2 u_{aa} = \beta u_a u_{aa} + \beta_1 u_a^2 u_{aa} + \mu u_{aat} + k u_{aatt} + \bar{s} u_t. \tag{7.2}$$

Here $\beta = -(\gamma + 1)c_0^2$ and $\beta_1 = 0.5(\gamma + 1)(\gamma + 2)c_0^2$. We introduced in (7.2) small phenomenological terms which take into account the effects of viscosity μu_{aat}, dispersion $k u_{aatt}$ and of bottom friction $\bar{s} u_t$. The last term may be very important for waves propagating in thin layers.

Approximate solutions of (7.2) were presented in [3, 4, 11, 13, 17]. These solutions describe 1-dimensional waves in solid bodies, gas, liquids and gassy materials.

Here we will present a scheme of the solution of the Eq. (7.2). The equation is being rewritten using new variables:

$$r = t - a c_0^{-1} + \beta u c_0^{-3}/4, \quad s = t + a c_0^{-1} - \beta u c_0^{-3}/4. \tag{7.3}$$

Then we find new expressions for the terms of (7.3). For example,

$$\begin{aligned} u_t &= u_r r_t + u_s s_t = u_r\left(1 + \beta u_t/4C_0^3\right) + u_s\left(1 - \beta u_t/4C_0^3\right) \\ &\approx (u_r + u_s)\left[1 + A_*(u_r - u_s) + A_*^2(u_r - u_s)^2\right], \end{aligned} \tag{7.4}$$

where $A_* = \beta c_0^{-3}/4$ and the subscripts r and s refer to partial derivatives with respect to r and s. Then, we can approximately rewrite (7.2) in the form

$$\begin{aligned} u_{rs} &= A_*(u_s u_{ss} - u_r u_{rr}) + 0.25 c_0^{-2}[\mu(u_{rrr} - u_{rrs} - u_{ssr} + u_{sss}) \\ &\quad + k(u_{rrrr} - 2u_{rrss} + u_{ssss})] + a_* u_r u_s (u_{rr} + u_{ss}) \\ &\quad + b_*\left[u_{rr}(u_s)^2 + u_{ss}(u_r)^2\right] + c_*\left[u_{rr}(u_r)^2 + u_{ss}(u_s)^2\right] + 0.25\bar{s}(u_r + u_s). \end{aligned} \tag{7.5}$$

Here $a_* = 5A_*^2 - 2D_*, b_* = -2A_*^2 + D_*, c_* = -4A_*^2 + D_*, D_* = \beta_1/4c_0^4$. A solution of (7.5) is sought as a sum:

$$u - u_1 + u_2 + u_3 + \ldots, \tag{7.6}$$

where $u_1 >> u_2 >> u_3$. Thus, the solution of (7.5) is being sought by the perturbation-type method. This method has a broad application in physics. At the same time there is a serious difficulty with the application of this method. This difficulty is the so-called 'secular terms' which are generated by nonlinearity and grow up infinitely when $t \to \infty$. We shall meet the secular terms later. Substituting (7.6) into (7.5) and equating terms of the same order we obtain a system of linear differential equations:

$$u_{1rs} = 0, \tag{7.7}$$

$$\begin{aligned} u_{2rs} &= A_*(u_{1s}u_{1ss} - u_{1r}u_{1rr}) + 0.25a_0^{-2}\left[\mu(u_{1rrr} - u_{1rrs} - u_{1ssr} + u_{1sss})\right. \\ &\quad \left. + k(u_{1rrrr} - 2u_{1rrss} + u_{1ssss})\right] + 0.25\overline{s}(u_{1r} + u_{1s}), \end{aligned} \tag{7.8}$$

$$\begin{aligned} u_{3rs} &= A_*(u_{1s}u_{2ss} + u_{2s}u_{1ss} - u_{1r}u_{2rr} - u_{2r}u_{1rr}) + a_*u_{1r}u_{1s}(u_{1rr} + u_{1ss}) \\ &\quad + b_*\left[u_{1rr}(u_{1s})^2 + u_{1ss}(u_{1r})^2\right] + c_*\left[u_{1rr}(u_{1r})^2 + u_{1ss}(u_{1s})\right]^2. \end{aligned} \tag{7.9}$$

We took into account only the nonlinear effects in (7.9). The d'Alembert's solution of the wave equation (7.7) is

$$u_1 = J(r) + j(s). \tag{7.10}$$

Further, a solution of (7.8) is given by

$$\begin{aligned} u_2 &= J_2^*(r) + j_2^*(s) - 0.5A_*\left(sJ'^2 - rj'^2\right) \\ &\quad + 0.25C_0^{-2}\left[\mu\left(sJ'' + rj''\right) + k\left(sJ''' + rj'''\right)\right] + 0.25\overline{s}(sJ + rj). \end{aligned} \tag{7.11}$$

Here $J = J(r)$, $j = j(s)$ and $J_2^*(r), j_2^*(s)$ are functions of integration. The prime denotes the differentiation with respect to the appropriate variable: r or s (7.3). The solution (7.11) depends on t and coordinate a. The secular terms are eliminated if

$$J_2^*(r) = J_2(r) + 0.5rA_*J'^2 - rC_0^{-2}\left(\mu J'' + kJ'''\right)/4 - 0.25r\overline{s}J, \tag{7.12}$$

$$j_2^*(s) = j_2(s) - 0.5sA_*j'^2 - sC_0^{-2}\left(\mu j'' + kj'''\right)/4 - 0.25s\overline{s}j. \tag{7.13}$$

Using (7.10), (7.11), (7.12) and (7.13) we find u_3 from (7.9)

$$\begin{aligned} u_3 &= J_3^*(r) + j_3^*(s) - A_*sJ'J_2' + A_*rj'j_2' + a_*\left(jJ'^2 + Jj'^2\right)/2 \\ &\quad + (b_* - A_*^2/2)\left(J'\int j'^2 ds + j'\int J'^2 dr\right) + (c_*/3 - A_*^2/2)\left(rj'^3 + sJ'^3\right) \\ &\quad + A_*^2(s^2/2 - rs)J''J'^2 + A_*^2(r^2/2 - rs)j''j'^2 \\ &\quad + A_*\iint\left(j'j_2'' + j''j_2' - J'J_2'' - J''J_2'\right)drds. \end{aligned} \tag{7.14}$$

Here $J_2 = J_2(r)$, $j_2 = j_2(s)$ and $J_3^*(r), j_3^*(s)$ are functions of integration. The secular terms are eliminated if

$$J_3^*(r) = J_3(r) + rA_*J'J_2' - r(c_*/3 - A_*^2/2)J'^3 + r^2A_*^2J''J'^2/2, \tag{7.15}$$

$$j_3^*(s) = j_3(s) - sA_*j'j_2' - s(c_*/3 - A_*^2/2)j'^3 + s^2A_*^2j''j'^2/2. \tag{7.16}$$

The approximate solution of (7.2) is

$$\begin{aligned} u = {} & J + j + J_2 + j_2 + J_3 + j_3 + 0.5A_*(r-s)\left(J'^2 + j'^2\right) + c_0^{-2}(s-r)\left[\mu(J'' - j'')\right. \\ & \left. + k(J''' - j''')\right]/4 + 0.25\bar{s}(s-r)(J-j) + A_*(r-s)(J'J_2' + j'j_2') \\ & + 0.5a_*\left(jJ'^2 + Jj'^2\right) + (b_* - 0.5A_*^2)\left(J'\int j'^2 ds + j'\int J'^2 dr\right) \\ & + (c_* - 1.5A_*^2)(r-s)\left(j'^3 - J'^3\right)/3 + 0.5A_*^2(r-s)^2\left(J''J'^2 + j''j'^2\right) \\ & + A_*\iint (j'j_2' + j''j_2 - J'J_2' - J''J_2)\,dr\,ds, \end{aligned} \tag{7.17}$$

where $J_3 = J_3(r), j_3 = j_3(s)$ and $J_1, j_1, J_2, j_2, J_3, j_3$ are unknown functions defined by boundary and initial conditions. We consider (7.17) as the approximate d'Alembert's type solution of the highly nonlinear wave equation (7.2).

Let $J_2 = J_3 = j_2 = j_3 = 0$, then

$$\begin{aligned} u = {} & J + j + \frac{1}{2}A_*(r-s)\left(J'^2 + j'^2\right) + \frac{1}{4}c_0^{-2}(s-r)\left[-\mu\left(J'' - j''\right)\right. \\ & \left. + k\left(J''' - j'''\right)\right] + \frac{1}{4}\hat{s}\,(s-r)(J-j) + \frac{1}{2}a_*\left(jJ'^2 + Jj'^2\right) \\ & + \left(b_* - \frac{1}{2}A_*^2\right)\left(J'\int j'^2 ds + j'\int J'^2 dr\right) \\ & + \frac{1}{3}(c_* - 1.5A_*^2)(r-s)\left(j'^3 - J'^3\right) + \frac{1}{2}A_*^2(r-s)^2\left(J''J'^2 + j''j'^2\right), \end{aligned} \tag{7.18}$$

Let us consider an unidirectional travelling wave of displacement u. In this case $j = 0$ and we have

$$\begin{aligned} u = {} & J + 0.5(r-s)A_*J'^2 + 2ac_0^{-4}\left(0.5J''^2 + J'J'''\right) + (s-r)C_0^{-2}(\mu J'' + kJ''')/4 \\ & + 0.25(s-r)\bar{s}J + 2ac_0^{-1}(c_*/3 - 0.5A_*^2)J'^3 + 2a^2c_0^{-2}A_*^2J''J'^2. \end{aligned} \tag{7.19}$$

Here $r - s = -2ac_0^{-1} + \beta uc_0^{-3}/2$. The expression (7.19) explicitly demonstrates the nonlinear effect. Indeed, according to the linear theory $u = J$ and the wave propagates with an constant profile. Because of the nonlinear terms in (7.19), the profile deforms as the coordinate a varies. For example, if we have at a given point, say $a = 0$, a sinusoidal wave with frequency ω, then, according to (7.19), harmonics with frequencies 2ω, 3ω, 4ω, and 5ω appear in $a > 0$.

7.1.2 The Equations of Continuity and State for Different Waves and Materials

Surface gravity waves In this case the wave elevation η is connected with u by the equation

$$\frac{\eta + h}{h} = (1 + u_a)^{-1}. \tag{7.20}$$

Here h is the layer thickness. For gravity waves, $c_0^2 = gh$ and $\gamma = 2$ in (7.2).

Body waves in one-phase materials In this case the density ρ is connected with u by the continuity equation

$$u_a = \rho_0 \rho^{-1} - 1. \tag{7.21}$$

The pressure p is determined by the adiabatic relation:

$$\frac{p + B}{p_0 + B} = (1 + u_a)^{-\gamma}. \tag{7.22}$$

Here subscript 0 refers to undisturbed values of p and ρ. One can see that the state equation (7.22) resembles qualitatively the Eq. (7.20) for surface waves. The body wave speed is determined as

$$c^2 = \gamma (p + B) \rho^{-1}. \tag{7.23}$$

If p and ρ are small enough, the expression (7.23) is simplified to the linear form:

$$c_0^2 = \gamma (p_0 + B) \rho_0^{-1}. \tag{7.24}$$

For many gases $\gamma = 1.4$, for water $\gamma = 7.15$, for steel γ is approximately 5. We stress that c_0 is determined as (7.24) in (7.1) and (7.2), (7.3).

The modelling of waves in gassy materials Thus, the values γ and c, c_0 may be different for different media. Therefore properties of multiphase media depend on properties of the phases and its interaction. For example, for weakly-cohesive, water-saturated and gas-saturated materials, the transformation of one phase into another and the exchange of momentum and energy between components may be important. However, explicit expressions describing these transformations and exchanges are usually not known. Therefore, the behaviour of these materials is often described within the framework of well-known classical models [18]. Namely, the space-averaged values are introduced for strains, stresses, displacements, pressure, and density. The averaging takes place over a volume element of the mixture, which can contain many particles (for example, gas bubbles), but has a small size with respect to the characteristic length of the propagating wave. The number of the particles per unit volume does not change. The particles do not move with respect to each other. In this case, the weakly-cohesive materials can be approximately considered as a continuum, and described by the wave equation (7.1). The mechanical properties of the material are described by the state equation.

This equation should take into account the most important peculiarity of weakly-cohesive materials. The properties should resemble properties of gas within the rarefaction zones. However, within the compression zones the properties should resemble the properties of solid materials. In [11, 13, 19] two different state equations were used so that to describe such behaviour of the weakly-cohesive gassy materials. The first equation is determined by the adiabatic relation (7.22), where γ and B depend on properties of the phases. The second version of the state equation is more complex than (7.22). On the whole, it was found that the two different state equations do correspond to each other. The variations of the sound speed calculated according to these equations are shown in Fig. 7.2.

The curves in Fig. 7.2 demonstrate the dependence of the sound speed in the mixture on the gas concentration nV_0. The gas concentration was changed within the interval $0 \leq nV_0 \leq 1$. If $nV_0 = 0$, then the material contains only condensed phases, and if $nV_0 = 1$, then the material transforms into gas.

We note that $c = 1{,}450$ m/s at $nV_0 = 0$ for all of the curves of Fig. 7.2. At the same time, these curves reach different values of c at $nV_0 = 1$, because there c depends strongly on p and p_0.

According to Fig. 7.2 the influence of the gas on the wave propagation may be very important. That agrees with well-known results of many investigations. The curves of Fig. 7.3 illustrate the influence of gas on properties of water-based mud located on depths up to 5 km [20] (left) and of bubbly liquid [21] (right).

The sound velocity in mud can reduce strongly, if the gas phase fraction is slightly larger than 0 or slightly smaller than 1. It emphasized that the curves of Fig. 7.3 correspond qualitatively to the curves of Fig. 7.2.

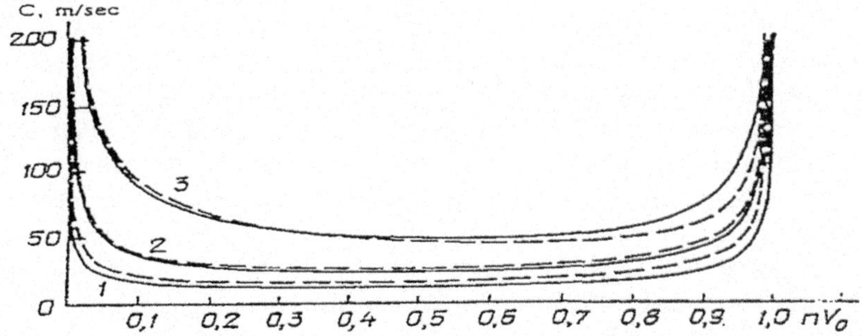

Fig. 7.2 Curves showing the change of the sound speed of bubbly water c with increasing gas component nV_0. The curves are calculated according to two different state equations (*dashed lines* and *solid lines*) for different values of p and p_0: $p = 0.06$ MPa, $p_0 = 0.1$ MPa (curves 1); $p = 0.16$ MPa, $p_0 = 0.2$ MPa (curves 2); $p = 0.3$ MPa, $p_0 = 0.5$ MPa (curves 3) [11, 13, 19]. On the whole, it was found that the two different state equations do correspond to each other

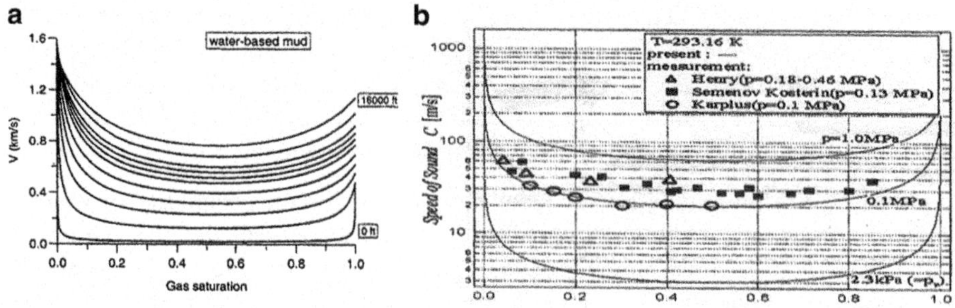

Fig. 7.3 Sound velocity of water-based drilling mud versus gas saturation. The depth varies, the frequency is 25 Hz [20] (*left*). Sound speed of bubbly liquid under the isothermal condition [21] (*right*)

Thus, we have considered the government equations and the state equations. In particular, it is shown that the simple Eqs. (7.22) and (7.23) can describe complex properties in different environments. The solution (7.18) describes nonlinear waves excited in different media and different situations. Below we will study cases of extreme amplification of waves, using (7.18).

7.1.3 Experimental Modelling of Seismic Waves in Natural Resonators

Resonance can occur when a forced frequency matches one of the natural frequencies of the natural resonator. However, earthquake- induced frequencies may be quite different from the natural frequencies.

Therefore, it is difficult to find a frequency of the resonant amplification of a seismic shock. It is difficult to predict the amplitude and form of local earthquake-induced extreme waves in the resonators. It is difficult to check the results of calculations, because of absence of well-documented data of observations.

Hence, to simplify the problem, we shall limit ourselves to natural resonators which may be approximated by simple 1-dimensional geometries (Figs. 7.1 and 7.4). Waves in such systems have been studied in many fundamental researches (see, for example, [22–35]). Their results can be valid for elongated natural resonators excited by earthquakes. We will use certain fundamental results to discuss the waves in natural resonators.

We noted above that vertical earthquake-induced oscillations can excite surface waves in seas, lakes, long deposit layers and volcano conduits. The scheme of such excitation is shown in Fig. 7.1. These waves will be additionally studied in the Sect. 7.2. However, there the vertical base oscillations can also force body waves. These waves may be modelled as nonlinear resonant perturbations in finite tubes

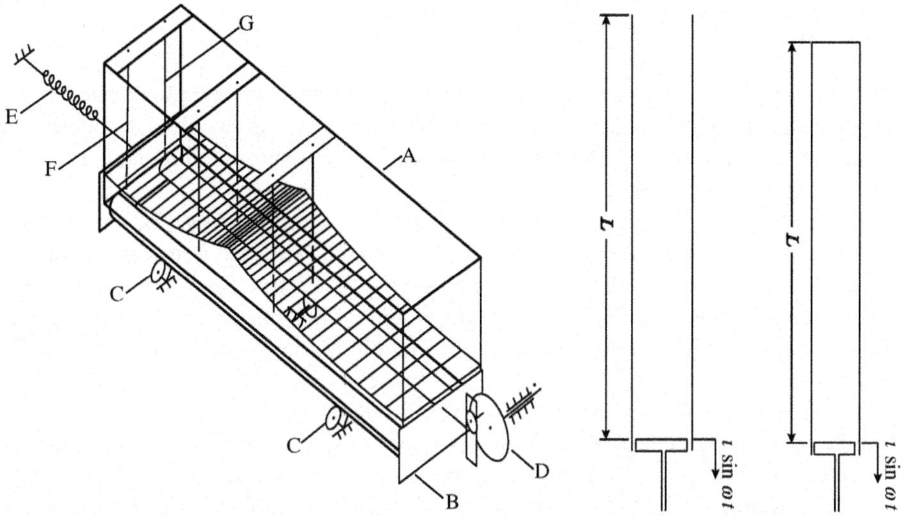

Fig. 7.4 One-dimensional schemes modelling some natural resonators. The basin (tank) modelling the earthquake-induced horizontal oscillations of sediment valleys and lakes (*left*) [25]. The open and closed vertical tubes with the oscillating piston as a model of layers vertically excited by solid beds (*right*)

containing gas, liquid or granular material (see Figs. 5.13 and 7.4, right). Some resonant waves excited in gas-filled closed tubes are considered in the Sect. 7.3. We suggest that similar resonant waves may be excited in gassy magma-filled conduits [36]. The resonant amplification of body waves in closed and semi-open resonators can explain many seismic- induced phenomena. For example, we explained by this amplification the anomalous accelerogram of the Iwate-Miyagi 2008 earthquake in Japan (see Sect. 4.1.3). Indeed, volcano conduits, ocean water and surface sediment layers can be considered as semi-open resonators where trapped vertical seismic waves may be strongly amplified. This model has been discussed in this book a few times. In the Sect. 7.4 we consider it again, using new calculations, data of Natanzon's experiments and of the Iwate Miyagi 2008 earthquake in Japan.

In the Sect. 7.5 the coastal amplification of ocean waves will be briefly considered. The Sect. 7.6 is devoted to the experiments of Sir Geoffrey Taylor.

Thus, in Sects. 7.2, 7.3, 7.4, 7.5 and 7.6 a few results of our theory and experimental data are presented, which illustrate additionally the geophysical results presented in the book. We will consider the appearance of extreme waves, using our theory and data of experimental studies of strongly-nonlinear resonant waves.

We emphasize that most of the calculations of extreme waves presented in the book were based on the Eq. (7.2) and the solution (7.18). Therefore, we decided to check the precision of (7.2) and (7.18) additionally, with the help of data from well-documented experiments.

7.2 Resonant-Surface Gravity Waves

During 10–20 s, a deposit layer can be transformed into liquefied state by an earthquake (see Sect. 4.1.2). After that, surface waves in the layer can be modelled as surface water waves. Therefore the water wave experimental data are very interesting for seismology.

Below we will compare our theoretical results determined by (7.2) and (7.18) with the experimental data of Chester and Bones [25]. They excited water waves in a rectangular tank, with length was 60 cm and width 15 cm. The forcing amplitude and the depth were varied during the experiments. The amplitude defined a contribution of the nonlinearity on the wave formation. Variation of depth changes the dispersive effect.

The theory predicts that in the case of small depth and strong excitation, waves should occur with steep front resembling shock waves in gas. On the other hand, periodic solitons should be excited if the depth is large enough. An unexpected result was found in the case of strong excitation with depth not very small – extreme waves with a height about equal to the water depth were observed.

We sought a periodical solution of Eq. (7.2) satisfying the following boundary conditions,

$$u = 0, \quad \text{at} \quad a = 0 \tag{7.25}$$

$$u = l\cos\omega t, \quad \text{at} \quad a = L, \tag{7.26}$$

Here l is the amplitude of exciting oscillations and $L >> l$. The most of the experimental data presented here were obtained for the first resonant frequency, $\pi c_0/L$. Near this frequency the solution (7.18) describes the appearance of a variety of nonlinear waves. Their forms are determined by the water layer thickness and the forced amplitude l.

Long wave Expressions describing the experimental data follow from (7.18) and (7.20), if quadratic nonlinearity and viscous effects are taken into account. For thin layers these expressions are

$$\eta = -hc_0^{-1}(u_s - u_r) \tag{7.27}$$

and

$$u_r = \alpha_*^{-0.5}\varepsilon^{0.5}\left\{2R\pi^{-1} + \tanh\left[2\omega^{-1}\chi^{-1}\sqrt{\alpha_*\varepsilon}\sin\left(\frac{1}{2}\omega r + \phi\right)\cos\phi\right]\cos\frac{1}{2}\omega r\right\}. \tag{7.28}$$

Here R is the transresonant parameter, $\phi = \arcsin R$, the constant χ takes into account the viscous effects. The other constants in (7.28) are defined in [6, 11, 13]. The expression for u_s is similar (7.28).

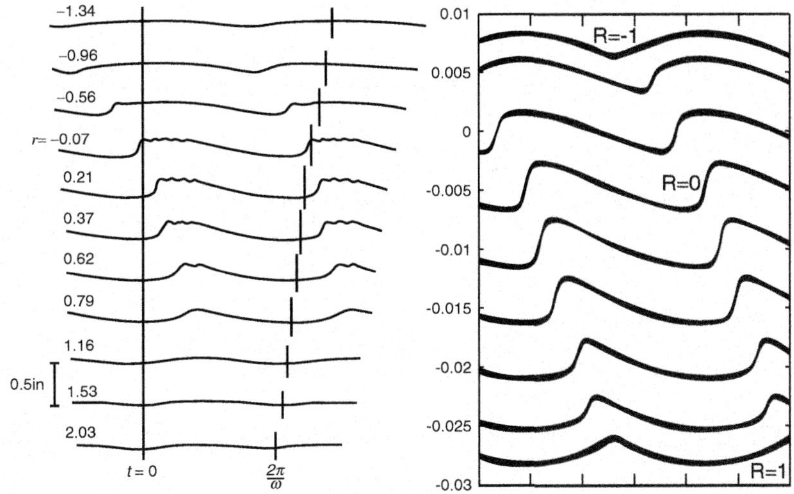

Fig. 7.5 Experimental recordings of the evolution of the surface waves within the resonance band (*left*) [25] and theoretical results (*right*). The thickness of the water layer $h \approx 1.25\,\text{cm}$, the forcing amplitude is 0.31 cm

The expression (7.28) is valid only within the resonant band, where $-1 \leq R \leq 1$. If χ is very small, then (7.28) describes a very steep wave front.

The measured profiles of water waves are given in Fig. 7.5 (left). It is easy to see that they correspond to the calculations using (7.27) and (7.28) (Fig. 7.5, right). Thus, if the dispersion is very small (very thin layer), surface waves can resemble shock waves in gas (see Fig. 7.9).

Effect of weak dispersion We found [7, 13] that weak internal viscosity and the very weak dispersion cannot strongly change the steep front and the wave profile formed by the quadratic nonlinearity. However, the influence of the dispersion increases when the depth increases. Data of experiments and calculations are presented in Fig. 7.6 for this case. The transresonant evolution of water waves differed strongly the evolution shown in Fig. 7.5.

The left and central curves are experimental, but the right curves are theoretical. Like the left curves in Fig. 7.5 the left profiles in Fig. 7.6 were taken from [25]. The thickness of the water layer was increased (5 cm), however the forcing amplitude was reduced (0.0775 cm) from the data of Fig. 7.5. Some periodical solitons appear instead of the shock-like waves in the tank, when the water depth was multiplied by 4. The appearance of similar waves was also demonstrated in [26] (Fig. 7.6, central). The right curves were calculated according to (7.27) and

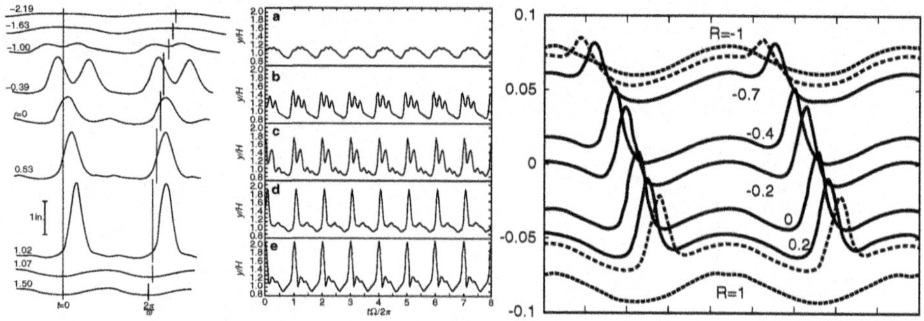

Fig. 7.6 The effect of the dispersion. Evolution and catastrophic amplification of the surface waves crossing the resonant band. The *left curves* are taken from [25], the *central curves* are given in [33]. The *right curves* were calculated using (7.27) and (7.29).

$$u_r = \alpha_*^{-0.5}\varepsilon^{0.5}\left\{2R\pi^{-1} + 1.4\left[3\,\mathrm{sech}^2(\gamma\sin(N\omega r - \phi)) - 1\right]\cos^2\frac{1}{2}\omega r\right\}. \quad (7.29)$$

Here $\phi = \arcsin R$. The other constants in (7.29) are defined in [7, 11, 13]. The expression (7.29) is valid only within the resonant band, where $-1 \le R \le 1$.

Spatial resonant oscillations of the water waves are shown in Fig. 7.7.

One solitary wave is excited near the first fundamental resonance of the container (left). Near the second fundamental resonance two solitary waves are excited (right). These waves are quite different from the waves excited in very shallow water (Fig. 7.5). Because of the dispersion, a sharp crest is formed in the wave top. It is known that the dispersion can strongly increase the wave amplitude relative to the pure nonlinear wave. Thus, the waveforms and the crest angle can strongly depend on the dispersive effect, according to the calculations and the experiments.

Effect of the cubic nonlinearity The cubic nonlinear term in (7.2) is responsible for the formation of breaking water waves. Results of the solution of the Eq (7.2) are presented in Fig. 7.8 (right). On the left the data of experiments [32] are submitted. We emphasize that a closed tank is considered.

Remark Figures 7.5, 7.6, 7.7 and 7.8 show that strongly-nonlinear waves can be excited in water tanks. Generally speaking, a few oscillations with resonant frequencies might be enough to form similar strongly-nonlinear waves in tanks. After these oscillations the strongly-nonlinear waves can proceed for some time, since dissipative losses of energy in water are usually small. This result is quite applicable for resonant oscillations in sedimentary layers. For example, in the 1985 Mexico City earthquake the source was near the coast some 400 km west of Mexico City. The quake caused little damage near the coast. As the seismic waves propagated inland, they started to gradually attenuate. Nevertheless, on reaching the city the waves suddenly amplified, inducing severe shaking at some areas which continued to shake for several minutes after the seismic waves had passed. There are a few similar well-documented effects during earthquakes.

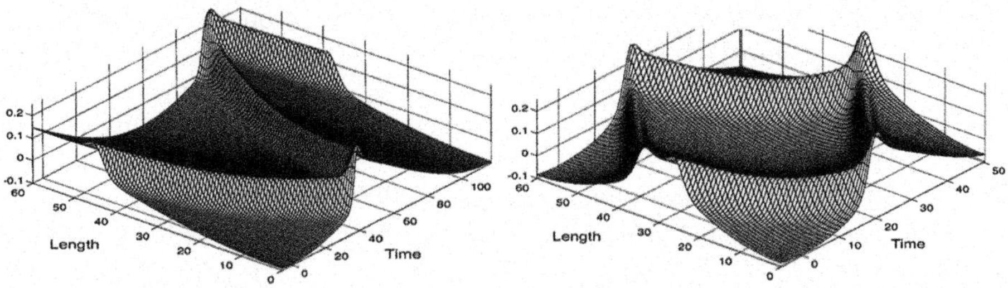

Fig. 7.7 Solitary waves travel back and forth in the tank, being repeatedly reflected from the boundaries [11]

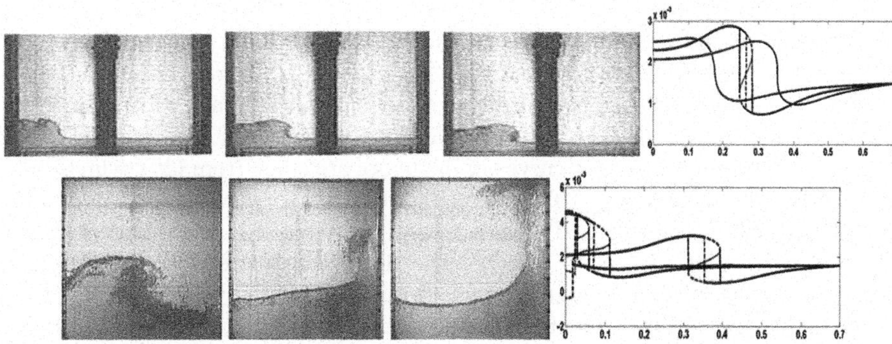

Fig. 7.8 Two examples of the formation and the evolution of breaking waves in a tank. The experimental curves are presented at the *left* [32]. The theoretical curves are presented at the *right*

The dispersive property of the layers may be very important. The solitary wave amplitude can greatly exceed the amplitude of the resonant waves excited in thinner layers. In particular, near a resonance the amplitude of the solitary waves can exceed ten and more times the amplitude of nonresonant waves (see Fig. 7.6, left). For example, if the height of a harmonic ocean wave is 7 m, then, according to our research, the height of the resonant waves can be about 70 m.

We considered the surface resonant waves in finite resonators, and strong amplification of the waves within the resonant bands is shown. The appearance of extreme waves is determined by resonant effects.

7.3 Resonant Waves in Closed, Gas-Filled Tubes as a Model of Vertical Earthquake-Induced Body Waves

Physical reasons for generation and transformation of resonant waves can differ dramatically. Nevertheless, the equations and analytical solutions describing these processes are often similar (cf. Sect. 7.1). That was not surprising for us. Many years ago, Lamb [16] stressed the similarity between the one-dimensional wave equations for shallow water and gas. Therefore, the resonant waves excited in gas-filled tubes can resemble nonlinear ocean waves. On the other hand, waves in gassy magma of volcano conduits can be modelled as resonant waves in closed or semi-open gas-filled tubes.

We will consider in this section the waves produced in a closed, gas-filled tube by oscillations of a piston at one end (Fig. 7.4 right). In this case the expression (7.22) yields

$$p - p_0 = -\gamma c_0^{-1}(p_0 + B)(u_s - u_r), \tag{7.30}$$

where u_r and u_s are similar (7.28).

The linear resonance Forced waves excited in the vicinity of the first linear resonance of the closed tube are shown in Fig. 7.9.

Figure 7.9 shows the transresonant evolution of a harmonic wave into shock waves. Waves were excited by a piston in a pipe in the cases of $R = -2$; -1, -0.97, -0.6, 0, 0.5 [35]. Three upper curves of the figure are taken from [29], which show completely the evolution of the waves in the case of change of the forcing frequency near a resonance. Those waves differ from the waves measured for $R = -2$; -1, -0.97, -0.6, 0, 0.5, because those waves are influenced also by some average current existing in the pipe [29].

We will simulate the experimental data shown in Fig. 7.9 using the solution (7.30). Results of the calculations are presented in Fig. 7.10.

The weakly-nonlinear harmonic waves begin to evolve into cnoidal waves (Fig. 7.10, $R = 1$), when the transresonant parameter R is reduced to $R = 1$. Then shock waves can appear travelling to and fro in the tube, being repeatedly reflected from the piston and from the closed end.

Effects of the nonlinearity and viscosity on the transresonant evolution of the pressure waves are shown in Fig. 7.10. The discontinuous lines are calculated for very small viscosity. The smooth lines are calculated for two different viscosities. The discontinuous lines transform into the harmonic curves, if the viscosity of the gas is large enough.

The spatial nonlinear oscillations of the wave in the tube calculated for the fundamental resonance are shown in Fig. 7.11. Waves are calculated for $R = 0$, 0.866 and 1. The transresonant evolution of the wave profile is demonstrated. It is seen that the shock amplitude is increased when R is reduced to zero.

Near the second resonant frequency, two shock waves appear in the tube (Fig. 7.12).

The collision of shock waves explains the extreme amplification of the pressure at the point $L/2$ (Fig. 7.12).

The nonlinear resonance It is found [35] that shock waves may be also generated in the tube at the vicinity of nonlinear resonances, when $\omega L/c_0 = 0.5N\pi$ and $N = 1$, 3, 5, ... (Fig. 7.13).

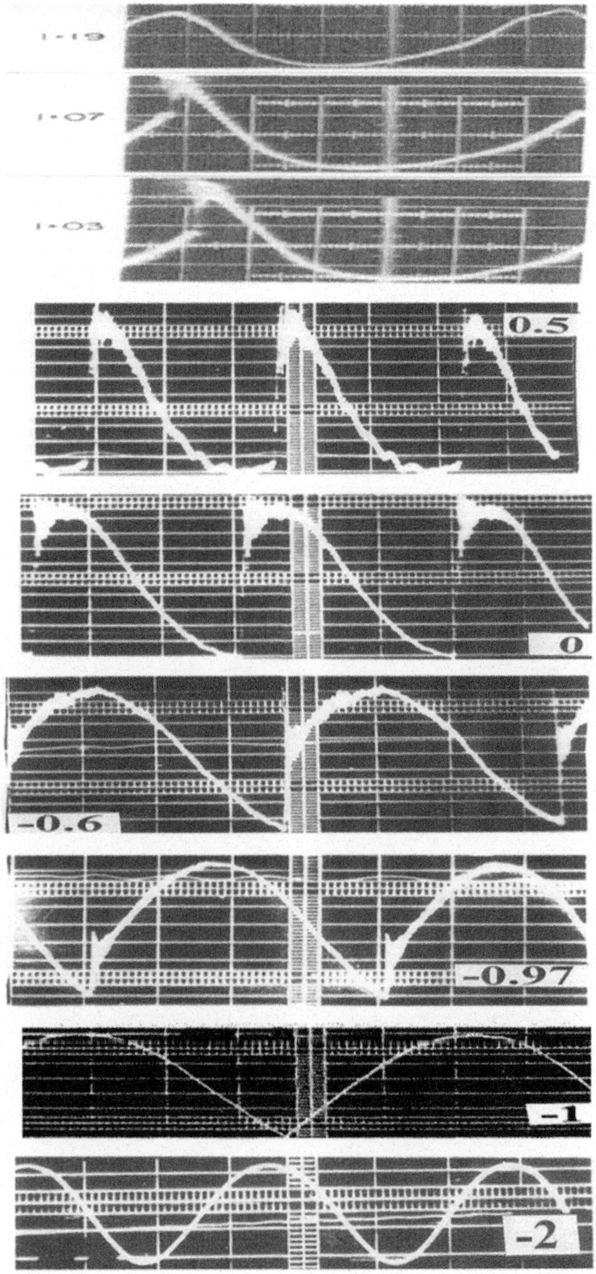

Fig. 7.9 The experimental pressure profiles at the closed end, measured for different values of the transresonant parameter R ($R = -2$; -1, -0.97, -0.6, 0, 0.5) [11, 13, 35]. The curves 1.9, 0.7 and 0.3 are taken from [29]

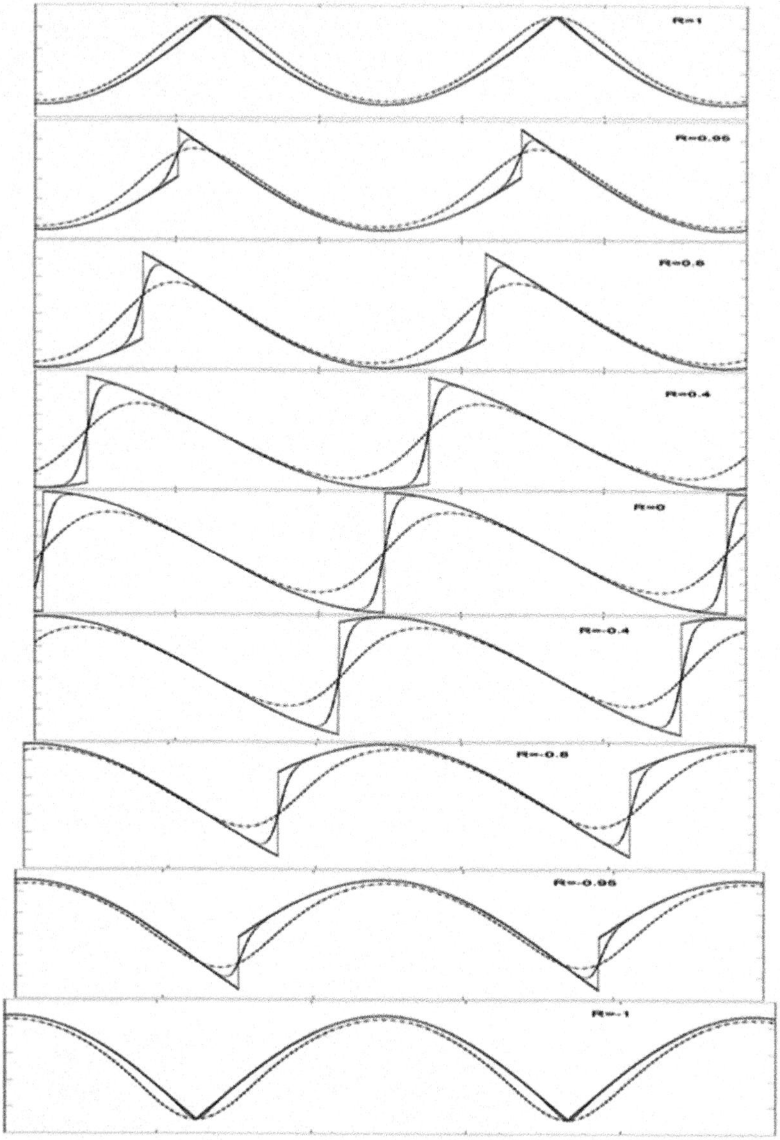

Fig. 7.10 Resonant waves in a closed tube, calculated for different transresonant parameter R and the viscous effect [13]

The curves in Fig. 7.13 are symmetrical around $R = 0$. The principal difference from Fig. 7.9 is that there are discontinuities with smaller jump, located between the strong discontinuities in each piston period. Moreover, the entire wave structures are less intense.

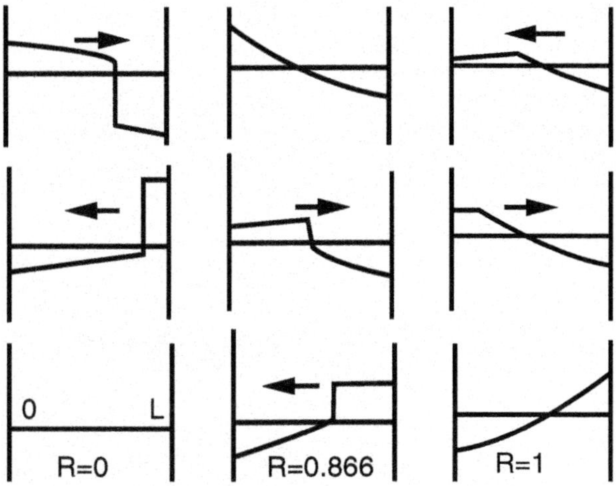

Fig. 7.11 Transresonant evolution of the wave excited in the tube when $R = 0$, 0.866 and $R = 1$. R is the transresonant parameter. The shock wave travels in the tube, being repeatedly reflected from the piston and the closed end, if $R < 1$. There is an amplification of the pressure when the shock wave reflects from the boundaries [3, 13]

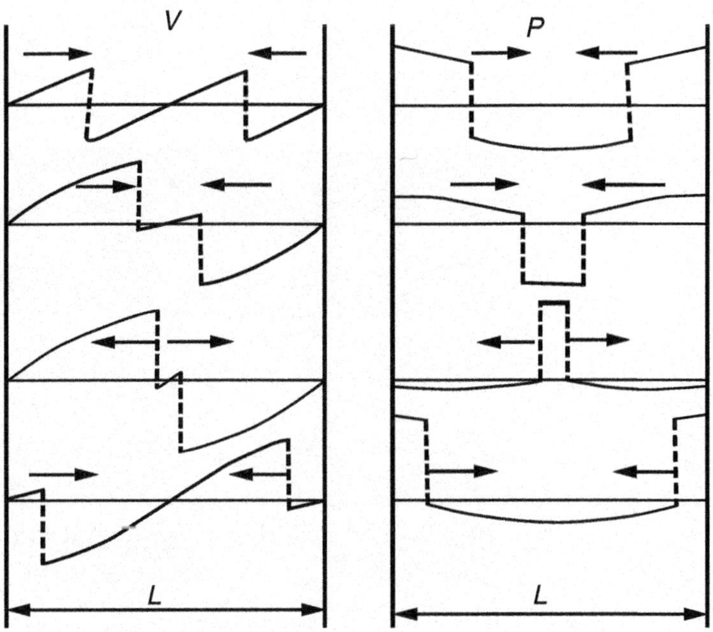

Fig. 7.12 The resonant shock waves of the velocity (*left*) and the pressure (*right*) excited at the second fundamental resonance of the gas column [4]

Results of calculations are in Fig. 7.14. The very thin lines describe the linear harmonic waves. The thicker lines are defined by the nonlinear component of the solution. The thick lines correspond to the full solution of Eq. (7.2). The transresonant evolution of waves is more complex in this case, than the evolution of waves in Fig. 7.10.

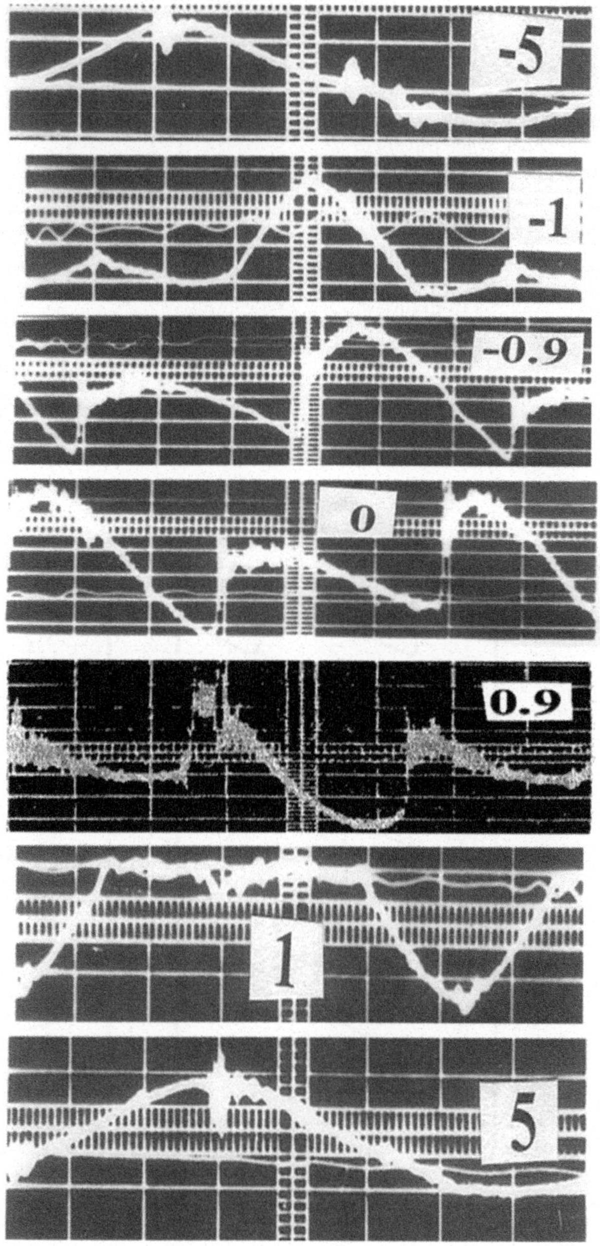

Fig. 7.13 The piston operating at nearly the first nonlinear resonance of the closed tube. The experimental pressure profiles at the closed end were measured for different transresonant parameters $R = -5; -1, -0.9, 0, 0.9, 1, 5$ [4, 13, 35]

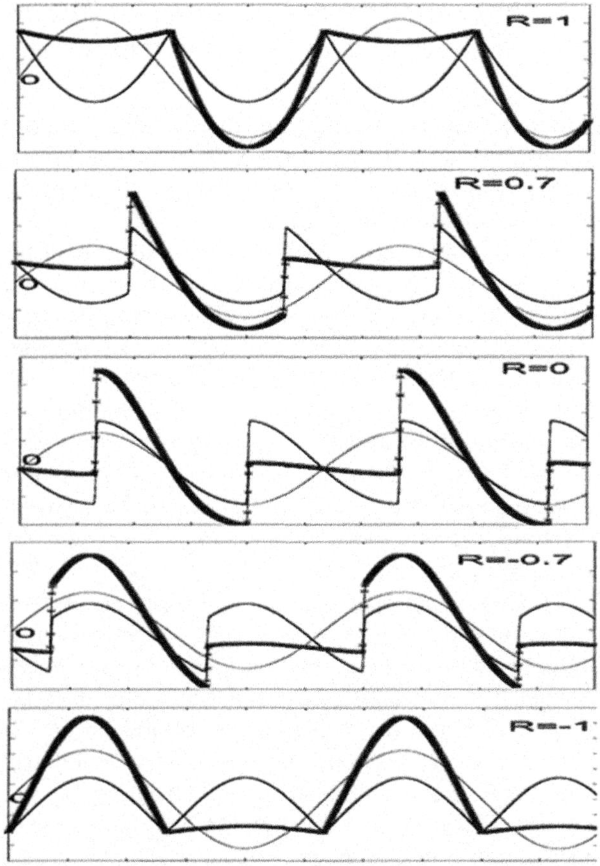

Fig. 7.14 Results of modelling the transresonant evolution of waves near a quadratic resonance [13, 35]

We considered the resonant waves in the closed pipe. The experiments and the theory show the strong amplification of the waves within the resonant bands. In the case of quadratic resonance the amplification is two times; in case of the linear resonance the amplification is three and more times. Therefore, we can claim that extreme waves have appeared in the pipe within the resonant bands.

We think that the experiments with gas-filled tubes can qualitatively describe resonant oscillations of gassy magma in volcano conduits. If a frequency of the earthquake-induced oscillations of a volcano base coincides with the fundamental resonance of the magma column, then a strong amplification of oscillations both the magma and the volcano can occur [36].

Remark It is interesting that on a water surface some 'table-top' waves (Fig. 5.18) can be excited, which are similar to waves observed in gas (Fig. 7.14, $R = 1$).

7.4 Extreme Waves in Semi-open Resonators: Ocean Sediment Layers and Volcano Conduits

Different cases of loosening and disruption of surface layers were considered in the Chaps. 4 and 5. It was emphasized that these processes can take place during earthquakes. Vertical surface waves can be trapped by the layers, and they are amplified due to resonance.

In this section, semi-open resonators filled with liquid or with weakly-cohesive gassy materials are considered. Generally speaking, these resonators can simulate behaviour of water layers and sediment valleys during seaquakes and earthquake. On the other hand, volcano conduits can be considered sometimes as semi-open resonators. This resonators may be excited by the earthquake shocks [36]. Mechanical properties of gassy magma can resemble properties of liquids and gas, especially within the rarefaction zones. It is known that magma properties sometimes resemble properties of some liquid [37–39]. During strong vertical vibrations, gassy materials can locally demonstrate the properties of gas or liquids. They may determine the amplitudes and the forms of resonant waves. Perhaps, wave properties of gassy magma may be described by curves of Figs. 7.2 and 7.3.

Here the semi-open natural resonators are modelled as open tube filled with water, or with weakly-cohesive gassy materials. Rough schemes of earthquake-induced oscillations of these resonators are presented in Fig. 7.15 (cf. Fig. 5.13). In particular, the rarefaction (cavitation) zone can be localized near the exciter (Fig. 7.15A(b)). In this case, the behaviour of a natural resonator during an earthquake can resemble the bouncing of solid balls on a vibrating base. We think that similar motion was excited during the Iwate-Miyagi 2008 earthquake in Japan (Figs. 2.4 and 4.13) and in Natanzon's experiments (Fig. 2.3). Generally speaking, a few rarefaction zones can appear in natural resonators during earthquakes (see Figs. 7.15B and 5.13). The appearance of cavitation within the rarefaction zones is well shown in Fig. 2.5 (left).

Within the rarefaction zones the volume of the gassy material can increase greatly. As a result, the free surface of the ruptured material may be periodically uplifted, as happened at the top of the volcano Santiaguito in Guatemala (Fig. 4.36). On the other hand, the lift of the rarefaction zones can explain the volcanic eruptions during earthquakes, and surface explosions of the gas bubbles during seaquakes (see, also, Figs. 1.7 (right), 2.1 and 2.6).

For the model shown in Fig. 7.15 (left) the boundary conditions may be written as

$$u_a = 0, \quad \text{on} \quad a = 0 \tag{7.31}$$

$$u = lF(t), \quad \text{on} \quad a = L. \tag{7.32}$$

Here l is the amplitude of exciting oscillations and $L >> l$. We assumed that on the free surface of the layer the pressure is zero (7.31). The bottom surface of the layer oscillates according to a law $F(t)$.

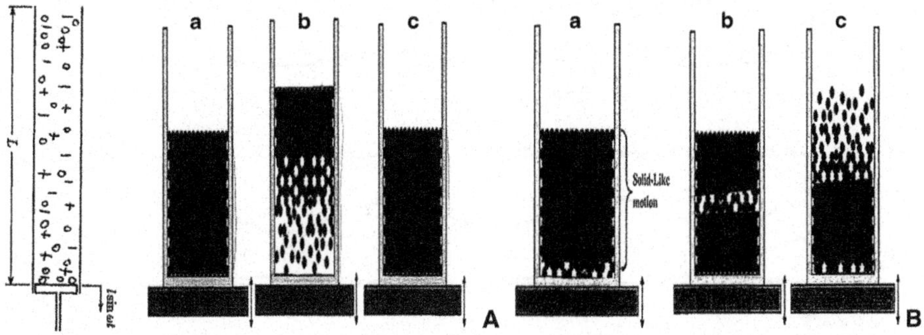

Fig. 7.15 A model simulating the oscillations of gassy material during earthquakes (*left*). Scheme of periodical formation of a rarefaction (cavitation) zone near a vibrating base (*A*). Scheme of lifting of the rarefaction (cavitation) zones in natural resonators accompanied by eruptions (*B*)

Natanzon's experiments Figure 5.11a shows a sketch of the experimental apparatus which Natanzon used. The boundary conditions (7.31), (7.32) correspond to his large-scale experiments (see Fig. 2.3 and Sect. 5.2.1). The length of the water column was varied from 4 to 7.5 m. Therefore the fundamental frequency of the water columns approximately changes from 300 to 160 Hz. During these experiments the formation of the rarefaction (cavitation) zones was well-documented.

Data of Natanzon's experiments were simulated in [4, 11] using the Eq. (7.2). Different models of bubbly liquid are used. Some results of those investigations are shown in Figs. 7.16 and 7.17.

In Fig. 7.16 examples of pressure curves are presented which are calculated near the piston. The excited frequency changes from 5.5 to 22 Hz. The curves are calculated for gas content equal to 0.5 % of the water volume. Three types of harmonic oscillations may be generated in the tube according to the theory, if the forced frequency $\nu < 9\,\mathrm{Hz}$. Two of them are shown in Fig. 7.16 for frequencies 5.5, 7.5, 8.5 Hz. When the frequency was being increased from 5.5 Hz the amplitudes of the curves were being increased too. However, there is the critical frequency 9 Hz, when the two solutions (harmonic curves) begin to contact and form the single solution. As a result, shock – like waves (water hammers) are generated in the system. If the frequency increases further the amplitude reduces, and the shock-like waves transform into the harmonic waves. Thus, a resonance phenomenon occurs in the tube.

One can see that there are large variations of the pressure waves in the frequency range extending from 5.5 to 22 Hz. It is surprising that the strong wave amplification took place. Indeed, the excited frequencies were smaller than the first fundamental frequency of the water columns by about 10 times. The amplification may be explained by the periodical generation of cavitation in the water near the piston (see Fig. 7.15). The rarefaction zone (cavitation) was excited near the piston for all the above water columns when the piston acceleration exceeds g.

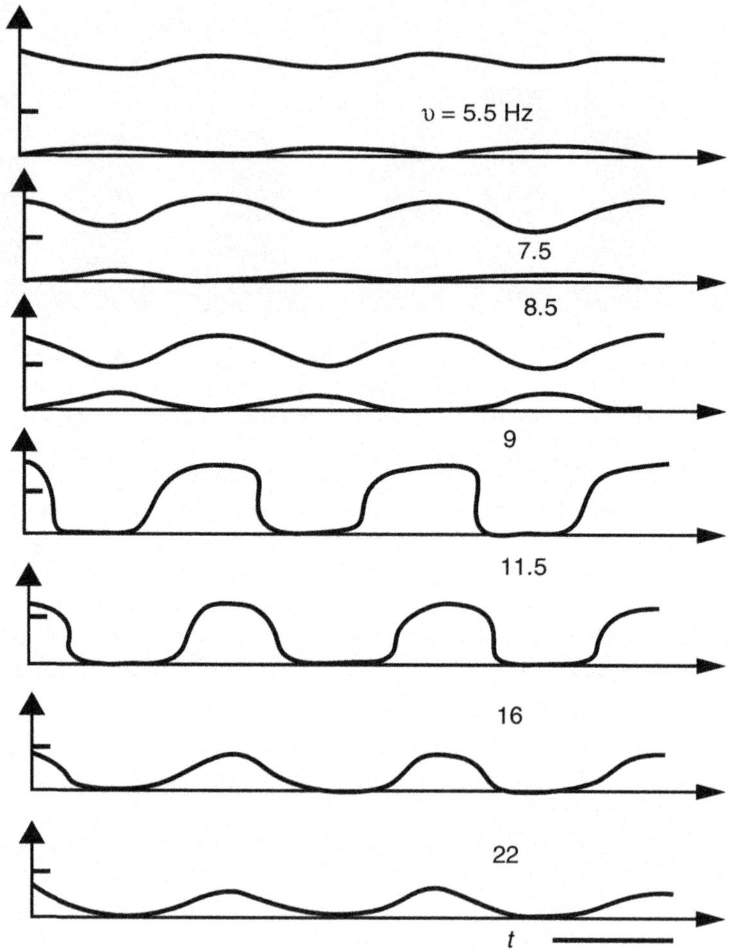

Fig. 7.16 Pressure curves calculated for different forcing frequency ν near the piston [4, 11]

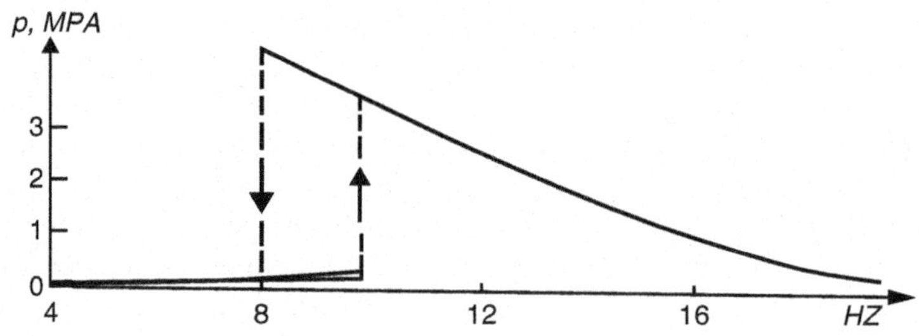

Fig. 7.17 Amplitude – frequency curves of the water column calculated for Natanzon's experiment [4, 11]

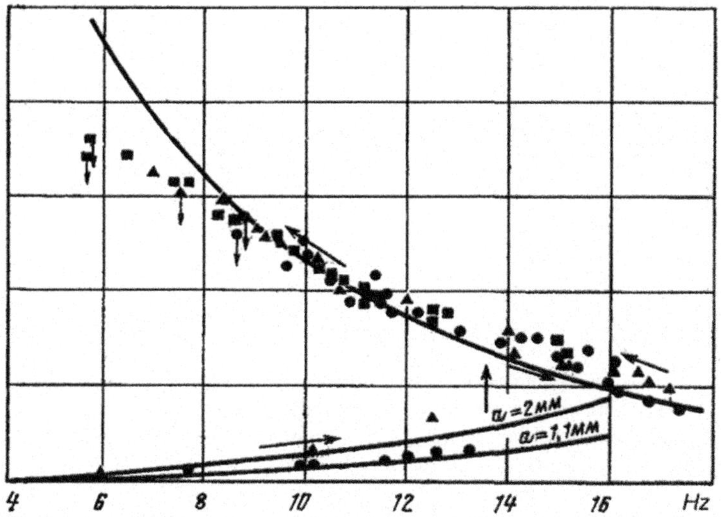

Fig. 7.18 Amplitude – frequency curves of the water column measured for open tube [11, 26]. Here a is the forcing amplitude

Figure 7.17 summarizes the results of analytic investigations presented in [4, 11]. These curves qualitatively agree with experimental data shown in Fig. 7.18 [11, 26]. The amplitude – frequency curves measured for the tube length $L = 3.7$ m and $l = 0.0011, 0.002$ m (Fig. 7.18) correspond to the calculated amplitude – frequency curves (Fig. 7.17).

We also simulated the experimental data using the solution (7.18). The cubic nonlinear effects were taken into account. We assumed $\nu = 14$ Hz as the resonant frequency. The results are presented in Figs. 7.19, 7.20 and 7.21. In particular, Figs. 7.19 and 7.20 show the transresonant evolution of waves u_a near the piston. First we studied the case when the excited frequency and the transresonant parameter R were increased (Fig. 7.19). The dashed curves in Fig. 7.19 were calculated according to the acoustic solution. The other curves were calculated according to the analytic solution (7.18). It is seen that the amplitude of the dashed curves increases very strongly when $R \to 0$. In particular, if $R \approx -0.00002$ the amplitude of the acoustical solution practically coincides with the amplitude of the nonlinear solution. We assume that this strong increase corresponds to the jumps up in Figs. 7.17 and 7.18. If R increases further, the amplitude slowly reduces (see curves $R = 0.1, 0.9$ in Fig. 7.19).

Then the case has been considered in which the frequency was reduced (Fig. 7.20). In this case the amplitude increases slowly in an interval $-0.99 < R < 0.9$. When $R \approx -1$ the amplitude suddenly reduces. We assume that this process corresponds to the jumps down in Figs. 7.17 and 7.18. If the frequency reduces further, the saw-like waves and then the smooth waves are excited. If $R = -1.5$ then the acoustical oscillations are generated. One can see that Figs. 7.19 and 7.20 qualitatively correspond to the experimental data. They

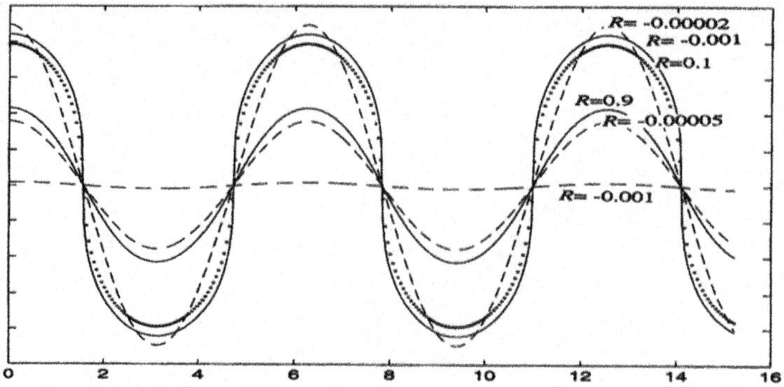

Fig. 7.19 Transresonant evolution of waves u_a near oscillating piston when the excited frequency increases. The *dashed curves* were calculated according to the acoustic solution, and the *other curves* were calculated according to the analytic solution (7.18). The amplitude of the *dashed curves* increases very strongly when R increases from −0.001 to −0.00002 [1, 11]

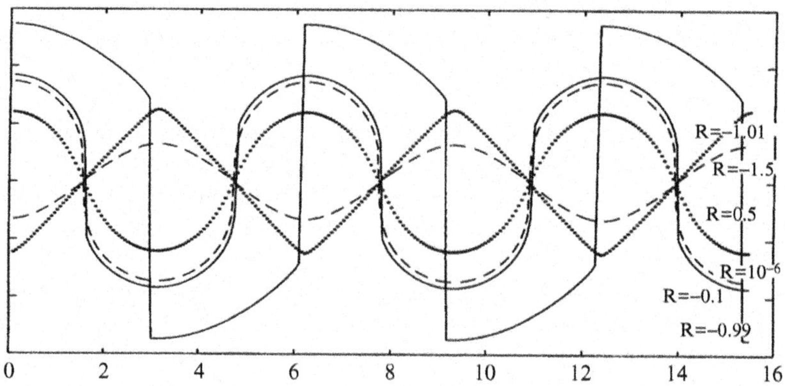

Fig. 7.20 Transresonant evolution of waves u_a near oscillating piston when the excited frequency decreases [1, 11]

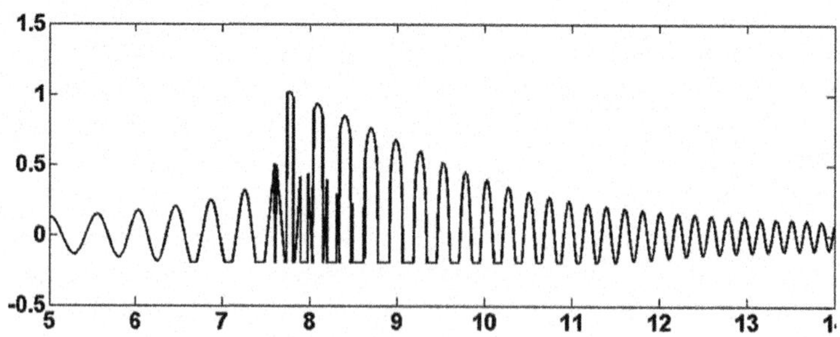

Fig. 7.21 Pressure curve simulating the data of Fig. 2.3

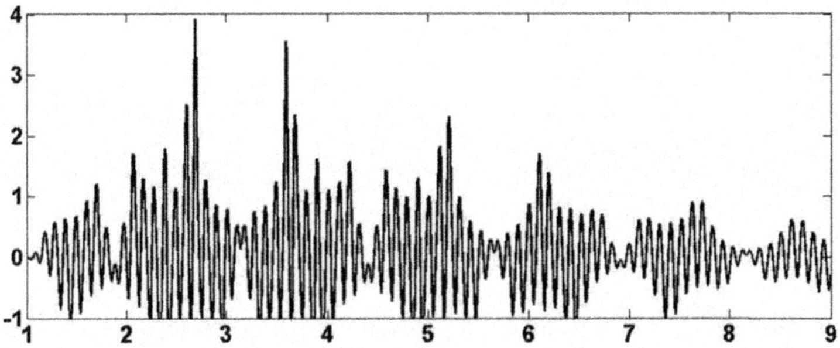

Fig. 7.22 Vertical component acceleration calculated for the Iwate-Miyagi 2008 earthquake in Japan

describe strong amplification of the wave amplitude near the resonance ($R = 0$), the generation of discontinuous waves, when $R < 0$, and the hysteretic phenomenon.

The pressure waves were also simulated, with cavitation taken into account. Study of transient cavitation is connected with major problems of mathematical modelling of the physical phenomenon. For instance, the site and the time of the gas bubbles increase (or decrease) in the bulk of liquid are unknown. Consequently, all approaches taking into account the initiation of cavitation are based on more or less simplifying statements [4, 11, 13, 40, 41].

Here we use the strongly-nonlinear state equation of the cavitation liquid, presented in [4, 11, 13, 40, 41]. A corresponding curve of the pressure calculated near the piston is drawn in Fig. 7.21. On the whole, it describes the data of Fig. 2.3.

Modelling of accelerogram of free ground surface of the Iwate-Miyagi 2008 earthquake in Japan. The technique developed for the description of water oscillations in the long vertical pipes allows us to calculate earthquake-induced waves in the top layers of the ground. Really, in both cases excited oscillations can resemble the oscillations shown in Fig. 7.15A. The separation of the top layers from the basis is well-documented from nuclear explosions (see Sect. 4.1.5). We think that similar separations can repeat several times in cases of strong earthquakes, when the vertical acceleration in some points of the upper material exceeds some critical value.

The result of calculations is presented in Fig. 7.22. On the whole, it qualitatively describes the data of measurements presented in Fig. 4.13 (left).

7.5 Resonant Coastal Waves

Now unidirectional waves are considered which cross the resonant band. We studied tsunamis using this approach earlier.

The evolution of coastal waves is perhaps the most striking and beautiful observable phenomenon in Nature. This evolution is a result of bottom-wave interaction. The waves coming into a coast can have varying speeds. The speed

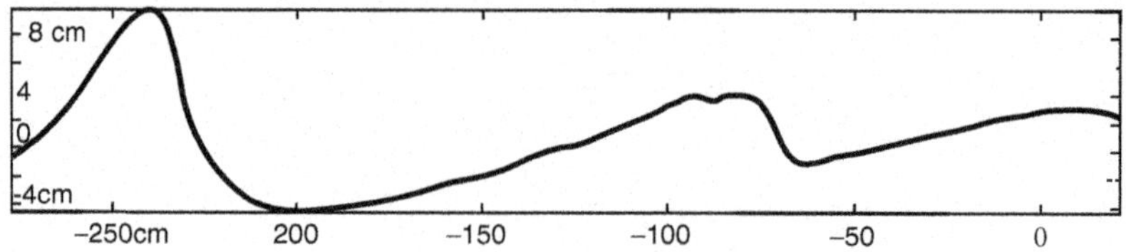

Fig. 7.23 A typical profile of waves excited in experiments [42]

increases when the wave moves in shallow water. When the speed coincides with the shallow wave speed ($\sqrt{gh}$), then a resonance occurs.

We will consider long waves propagating over water, where the undisturbed depth exhibits a gradual spatial variation at the scale of the wavelength. This variation involves essentially no reflected waves. There are only slight deviations of the properties of these waves from the properties of waves propagating above a constant depth at this scale. Nevertheless, the distorting effects of such small deflections are accumulating over long distances up to significant values.

We have used the following equation, which takes into account a gentle variation of the liquid thickness [13],

$$u_{tt} - ghu_{aa} - 2gh_a u_a - gh_{aa} u = \beta u_a u_{aa} + \beta_1 {u_a}^2 u_{aa} + k u_{aatt} + \bar{s} u_t. \tag{7.33}$$

Here the thickness h is a function of the coordinate a, and the terms $2gh_a u_a$ and $gh_{aa}u$ take into account the bottom slope. A solution of (7.33) was looked for as a sum of unidirectional waves,

$$u = A(a)J(r) + \bar{J}(r). \tag{7.34}$$

Here $J(r)$ is the primary (initial) wave, defined by a disturbance of the ocean far from the resonant band. The coefficient $A(a)$ is determined by the Green-type law which takes into account the bottom friction [12, 13]. The nonlinear component of the wave (7.34) determines the evolution of the primary wave crossing the resonant coastal band. This component is approximately described according to (7.19).

Results of this approximation were compared with experiments. A detailed description of the experiments can be found in [42]. The typical profile of waves excited in the experiments is shown in Fig. 7.23.

Figure 7.24 compares the measured waves (the left column [42]) and calculated waves (the right column).

In Fig. 7.24 the solid lines are calculated according to the nonlinear model (7.33); the dashed lines are calculated according to the linear model (the Green-type law with bottom friction). It is seen that the nonlinear resonant model (7.33) describes the experimental data.

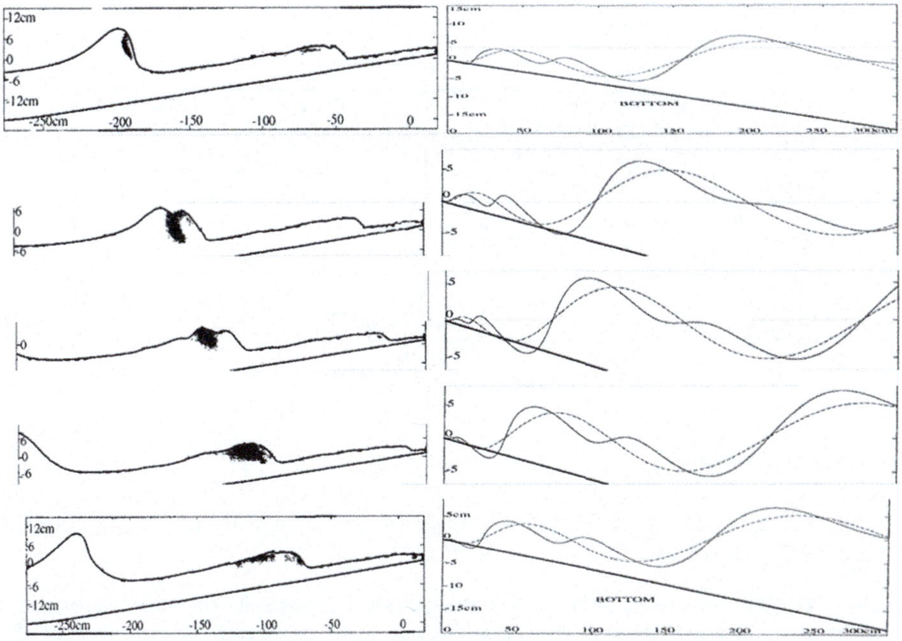

Fig. 7.24 The transresonant evolution of coastal waves: measured (*left*) [42] and calculated waves (*right*) [12, 13]

7.6 The Experiments of Sir Geoffrey Taylor

In Taylor's experiment waves were produced in a tank by wave makers operating with small amplitude and at frequencies where great amplification occurred, owing to resonance.

The apparatus used is shown in Fig. 7.25. The detailed description of it is in [43], where many interesting experimental results are presented for different forced amplitudes and frequencies.

We are interested in the cases when the waves became extreme. Some of them are shown in Fig. 7.26 (left). Figure 7.26 (21) shows the beginning of the violent phase of motion. This photograph shows the central conical crest rising in the middle of the tank. The water drop forms above the wave crest. Figure 7.26 (22–26) shows various stages of this motion. In the whole the noted waves are similar to some waves which were studied in the Chap. 5.

The results of the modelling are presented in Fig. 7.26 (right). We assumed the solution (7.18). In the whole the theoretical results correspond to the experiments. We will not go into detailed comparison of the experimental and theoretical results, since the aim of this section is to demonstrate the power of the most simple variant of our theory presented in the Sect. 7.1. However we underline the appearance of a drop (particle) above the top of the wave (0.2425, right). The appearance of

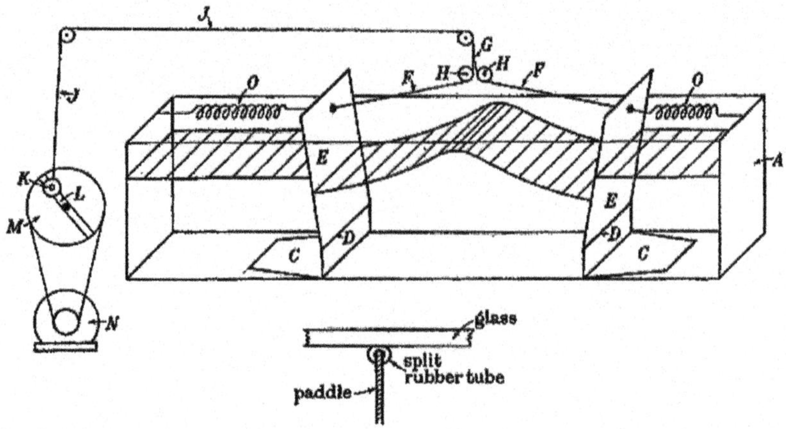

Fig. 7.25 Wave tank [43]

particles during a wave evolution will be discussed as an important element of the origin of the Universe.

Final remark We have presented and discussed the results of well-documented experiments where extreme nonlinear waves were observed. We think that the appearance of similar waves in local natural resonators can explain some catastrophic results of shallow earthquakes. The theory presented in [1, 3–15, 17, 19, 35] enables us to predict the local catastrophic amplification of seismic waves.

7.7 The Introduction and Versions of the Nonlinear Klein-Gordon Equation (NKGE)

Darwin's life was devoted to solving two great mysteries. The first was the age of Earth. At the time of Darwin, most people believed that this age was about 6,000 years. During the *Beagle* voyage many evidences had been found which show the Earth being much older than that. This started the conflict between Darwin and the Bible. Thus, Darwin solved together with Charles Lyell and other brilliant naturalists the great mystery of the age of the Earth [44–47]. However, most of his life Darwin devoted to solving the second great mystery – the mystery of the appearance of species. Natural selection not only clashed with religious beliefs based on the Bible, but it also conflicted with predictions of the classical physics.

Darwin was interested in other great puzzles of Nature and the Universe all his life. He tried to find keys to puzzle of '*that mystery of mysteries*' .... – *the first appearance of new beings on this earth* .... Darwin wrote to Joseph Hooker on March 29, 1863 ... *one might as well think of origin of matter*. ... However, Darwin did not have any well-documented facts about the evolution of life or matter so that to try to develop his interests. Such facts have been found during recent decades.

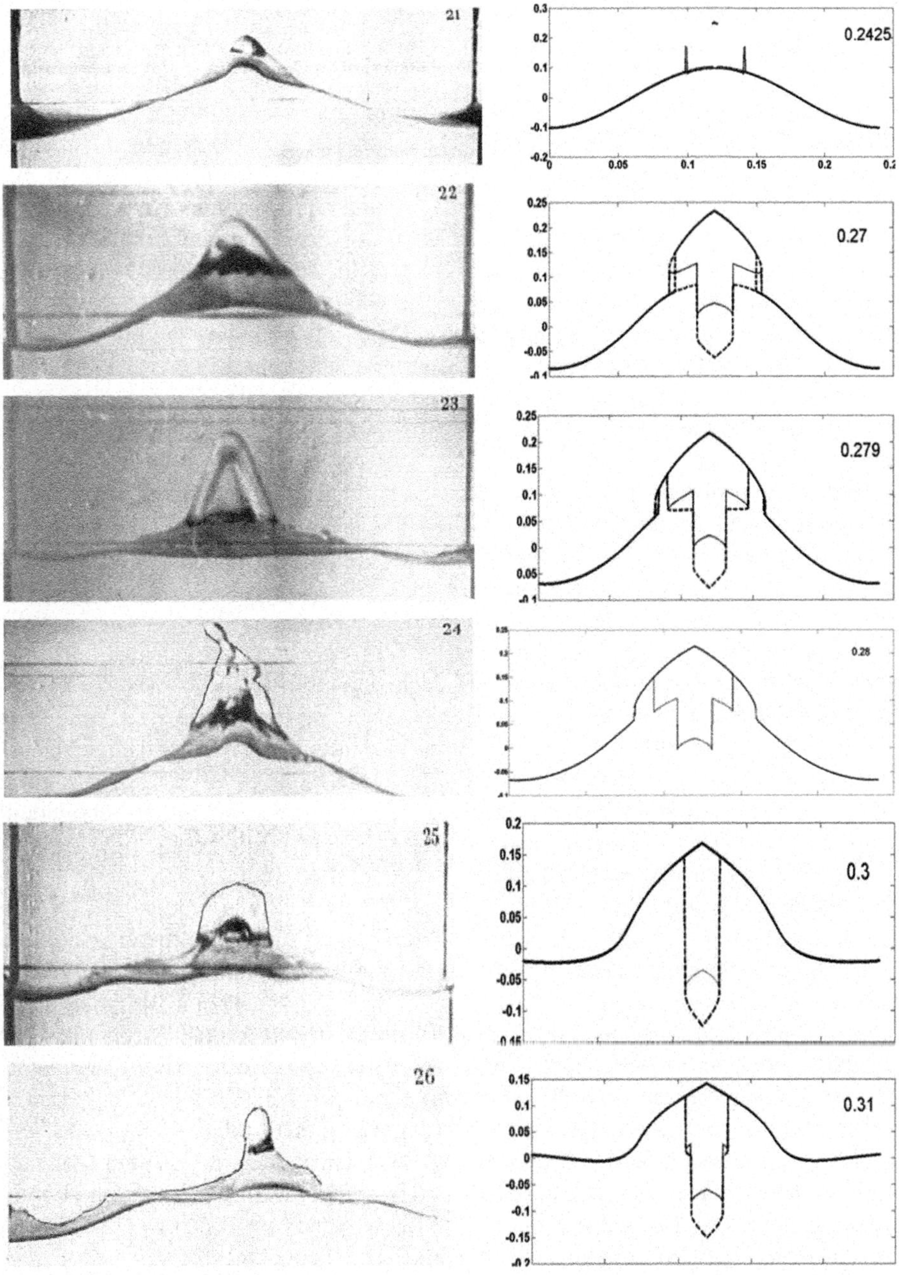

Fig. 7.26 Extreme surface waves: experimental data (*left*) [43], theoretical data (*right*)

They unravelled some secrets of Nature and the Universe. The researchers even try to unravel the mystery of the origin of the Universe. How did the Universe come to be? It is perhaps the greatest Great Mystery, and the root of all of the others.

Many tales, myths, theories were made up in order to explain the origin of the Universe. Up to this moment the origin of the Universe continues to be the greatest puzzle. However, over the last 70–60 years many scientific models of the origin were developed. The most popular model is the standard hot Big Bang model [48–59]. According to this model the Universe emerged from a point of infinite energy and density, a singularity where the laws of physics break down. Therefore, we cannot study the origin of the Universe using the modern physics equations. We cannot have any information about this point and the pre-explosion history. The standard Big Bang model was then supplemented by the 'inflation' theory – an extremely rapid expansion of space [54, 55, 60]. Thus, the Big Bang model replaces the puzzle of the origin by puzzles of the singularity and the inflation.

But the question continues to be open – what was there before 'the Big Bang'. According to another group of theories the Universe appeared from nothing instantaneously in a form of a clot of energy or matter [61]. This suggestion allows us to use physical law to study the beginning of the Universe's evolution. The singularity was excluded in these models.

There are theories which declare that the Universe was born not just once, but multiple times in an endless cycle of fiery death and rebirth. Cyclic-universe models were popular in the 1920–1930 years. But they were replaced by the idea of the Big Bang. Now they appear again [62, 63]. Some researchers say that our Universe may have begun as a Big Bounce rather than just the Big Bang [64–66]. They suggest that the Universe didn't necessarily begin from a singularity, but rather 'rebounded' from the collapse of a previous universe.

Of course, what we described above are the most popular theories. Several new models were developed during the last years [67–75]. Some models use an idea about a metaverse [67–69]. The Universe can appear inside the metaverse like a gas bubble in boiling water [70]. The birth might be caused by quantum effects [60, 61, 71–73].

It was impossible to check the facts of those theories since the astronomers did not have much information about very early Universe up to the very recent time. However, over the last decade more and more precise studies of temperature anisotropies of the cosmic microwave background (CMB) radiation allowed scientists a glimpse into how the Universe was born. Of course, new cosmic portraits contain new strange anomalies. For example, a huge swathe of sky at a slightly cooler temperature than the rest (the Axis of Evil) which was observed in the 2013 Planck map. There is no theory which explains the appearance of some anomalies (see, for example, Fig. 7.45). Other anomalies might exist and an upcoming generation of telescopes could detect them. On March 17, 2014 the Harvard-Smithsonian Center for Astrophysics announced the indirect detection of the 'primordial gravity waves' and vortex-like structures on CMB (Fig. 7.71) [56–59, 76].

Many of the models do not predict gravitational waves. If the 17 March 2014 results are confirmed, they will rule out a number of cosmic evolution models. On the other hand, these results support the models which describe the ripples of the spacetime. We can suggest that cosmology of the universe's origin was in many respects the physics of extreme oscillations and waves [77].

Thus, there are a few well-documented facts which any theory should explain. For example, all theories should describe and explain the fact of the high speed of expansion of the Universe. Therefore, each theory should have a time of very fast expansion. We underline that this time might be accompanied by generation, propagation and interaction of the greatest extreme waves. However, it is clear that the origin of the Universe will continue to puzzle the cosmologists for a very long time [78–90].

A model of the origin presented in [14, 84] is considered below. A scalar field described by a nonlinear Klein-Gordon equation is considered as a fundamental field in this model. The origin of the Universe is being studied as an evolution of an initial scalar field into other fields (particles), energies, waves and vortices. Using approximate solutions of the nonlinear Klein-Gordon equation we study the models for the scalar fields and the scalar potentials. The evolution of the initial field is determined by the quantum perturbations, bifurcations and resonances. We believe that this evolution gives a coherent model for the emergence and initial evolution of the Universe.

In particular, some solutions may be considered as a description a weakly oscillating pre-Universe. At any moment this multidimensional structure gives the origin to the billions of 'seed' of rapidly evolving universes. One of them accidentally formed our Universe. A scenario is developed, when the Universe begins in a state that differs greatly from that of theories of the Big Bang and the inflation. The Universe appeared having a finite volume.

Cosmologists can describe the history of the entire universe –from the present day all the way back to a fraction of a fraction of a second after the Universe beginning –using only a few equations and parameters (the Lambda Cold Dark Matter model). However, they cannot solve the main puzzle – the origin of the Universe. We decided to consider this puzzle using only a concept of energy.

Of course, this is here only a rough model. Modelling the birth of the Universe is a bit like playing whack-a-mole. All time the new results knock out the old theories.

Versions of nonlinear Klein-Gordon equation (NKGE) The classical NKGE has the following form

$$\hat{\Phi}_{tt} - c_*^2 \sum_{i=1}^{I} \hat{\Phi}_{ii} = -\partial V(\hat{\Phi})/\partial \hat{\Phi} . \tag{7.35}$$

Here c_* denotes a constant, I is an integer value ($I \geq 1$) and V is the scalar field potential. The indexes denote the differentiation; $\hat{\Phi}_{ii} = \partial^2 \hat{\Phi} / \partial x_i^2$, here x_i are the coordinates. Different expressions for the function V can be found in many books [51, 52, 86–90].

We assumed that

$$V(\hat{\Phi}) = V_0 + \alpha(\hat{\Phi}^2 - 1)^2 + \beta\hat{\Phi}^4 + C^*\hat{\Phi}, \tag{7.36}$$

Here V_0, C^*, α and β are constants. If the scalar field and the potential are known, one can calculate the pressure p and the energy density ρ according to expressions [90],

$$p = \frac{1}{2}\hat{\Phi}_t^2 - \frac{1}{6}(\nabla\hat{\Phi})^2 - V(\hat{\Phi}), \tag{7.37}$$

$$\rho = \frac{1}{2}\hat{\Phi}_t^2 + \frac{1}{2}(\nabla\hat{\Phi})^2 + V(\hat{\Phi}). \tag{7.38}$$

Expressions 7.37 and 7.37 show that the density and the pressure depend very strongly on the gradient of a scalar field and $\hat{\Phi}_t^2$. In points where the field changes rapidly the pressure and the density can be very large. There they can create new fields corresponding to the birth of high-energy particles.

A solution of (7.35) is represented as a sum,

$$\hat{\Phi} = \overline{\Phi} + \Phi. \tag{7.39}$$

Here $\overline{\Phi}$ is a stationary component and Φ is a dynamic component of the scalar field. Let $C^* = 0$ in (7.36).

Using (7.39) we can obtain from (7.35) two equations for $\overline{\Phi}$ and Φ:

$$\overline{\Phi}_{tt} - c_*^2\sum_{i=1}^{I}\overline{\Phi}_{ii} = -m^2\overline{\Phi} + \lambda\overline{\Phi}^3, \tag{7.40}$$

$$\Phi_{tt} - c_*^2\sum_{i=1}^{I}\Phi_{ii} = -m^2\Phi + \lambda\left(3\overline{\Phi}^2\Phi + 3\overline{\Phi}\Phi^2 + \Phi^3\right). \tag{7.41}$$

The following notations were introduced:

$$m^2 = -4\alpha \quad \text{and} \quad \lambda = -4(\alpha + \beta). \tag{7.42}$$

Here m and λ are constants.

New variables and a simplified model of NKGE Equation (7.35) is not integrable in the general case. Therefore we sought approximate solutions. The following new variables were used to describe different scenarios of the evolution of the field:

$$\xi = B\sin^2\omega t - K\sum_i^I (a_i + \sin\vartheta x_i)^2,\ \eta = -\overline{K}\sum_i^I (a_i + \sin\vartheta x_i)^2. \tag{7.43}$$

Using (7.43) we rewrite Eqs. (7.40) and (7.41) in the form

$$-2c_*^2\vartheta^2\overline{K}^2\left(\sum_{i=1}^I a_i^2 + \frac{1}{4}I\right)\overline{\Phi}_{\eta\eta} + m^2\overline{\Phi} - \lambda\overline{\Phi}^3 = 0, \tag{7.44}$$

$$\left[\frac{1}{2}\omega^2B^2 - 2c_*^2\vartheta^2K^2\left(\sum_{i=1}^I a_i^2 + \frac{1}{4}I\right)\right]\Phi_{\xi\xi}$$
$$= -m^2\Phi + \lambda\left(3\overline{\Phi}^2\Phi + 3\overline{\Phi}\Phi^2 + \Phi^3\right). \tag{7.45}$$

Here B, ω, K, a_i, ϑ and $\overline{K}$ are constants. In the Eqs. (7.44) and (7.45) we ignored terms explicitly dependent on harmonics containing t and x_i. We hope that these equations allow us to study qualitatively the change of the scalar field inside the multidimensional spacetime.

Below the interaction of $\overline{\Phi}$ and Φ will be ignored in (7.36) and (7.45).

Remark 1 The linear Klein-Gordon equation was named after the physicists Oskar Klein and Walter Gordon who in 1926 proposed that it describes relativistic electrons. This equation is considered as relativistic version of the Schrödinger equation. The nonlinear Klein-Gordon equation (7.35) is used in nonlinear optics, plasma physics, fluid mechanics and cosmology.

Remark 2 It is important that the Klein-Gordon equation explicitly contains the d'Alembertian operator. This operator can equal zero or a very small value (the 'small' divider or the 'resonance' term). In particular, in (7.45) the value $\frac{1}{2}\omega^2B^2 - 2c_*^2\vartheta^2K^2\left(\sum_{i=1}^I a_i^2 + \frac{1}{4}I\right)$ may be the small divider. At the same time this value determines the width of the resonant band.

7.8 A Landscape of the Scalar Potential

The localised solutions of (7.44) and (7.45) are sought in the form

$$\overline{\Phi} = \overline{A}\operatorname{sech}\eta,\ \Phi = A\operatorname{sech}\xi. \tag{7.46}$$

Using (7.46) we approximately found from (7.44) and (7.45) that

$$\overline{A} = A = 0, \overline{A}_{\pm} = A_{\pm} = \pm\sqrt{2}\lambda^{-0.5}m, \tag{7.47}$$

$$\vartheta^2\overline{K}^2 = 2m^2 c_*^{-2} I^{-1},\ \omega^2 B^2 = -2m^2 K^{-2} + c_*^2 \vartheta^2 K^2 \left(4\sum_{i=1}^{I} {a_i}^2 + I\right). \tag{7.48}$$

The three amplitudes (7.47) are determined by the values m and λ. We cannot really comprehend what scalar fields are represented by these formulas. It is however possible that these three fields are certain analogues of true and false vacuums. Values $\vartheta, \overline{K}, B, \omega$ and K are not completely determined by (7.48). We will consider these values as arbitrary.

The expressions (7.46), (7.47) and (7.48) allow us to consider the scalar potential. Namely, the two-dimensional landscape of the scalar potential will be studied. Substituting (7.39) into (7.36) we found approximately that

$$V(\hat{\Phi}) = V(\overline{\Phi}) + V^*(\Phi), \tag{7.49}$$

where

$$\begin{aligned} V(\overline{\Phi}) &= V_0 + \alpha\left(\overline{\Phi}^2 - 1\right)^2 + \beta\overline{\Phi}^4 \text{ and} \\ V^*(\Phi) &= -\alpha + \alpha\left(\Phi^2 - 1\right)^2 + \beta\Phi^4. \end{aligned} \tag{7.50}$$

We recall that the interaction of $\overline{\Phi}$ and Φ is ignored in this theory.

In Figs. 7.27 and 7.28 the results of the model calculations are presented. The scalar potentials $V(\overline{\Phi})$ (7.49) and $V^*(\Phi)$ (7.50) are determined by the constants: $V_0 = 7.5$, $\alpha = -18$ and $\beta = -10$. For the stationary and dynamical parts of the scalar field we assume the following values of constants (7.47) and (7.48): $\vartheta = 0.15$ and $\overline{A} = A = 1, \overline{K} = 2, \omega = 1, B = 30, K = 100$. We also assume that $a_i = 0$.

The stationary part of the scalar field potential describes a landscape which consists of the hills and the valleys (Fig. 7.27 left). The highest energy density is reached at the top of the hill. The lowest energy density is reached in the valleys. At the hill tops there are craters.

The dynamic part of the potential corresponds to a multidimensional sphere (bubble, clot) which has a very thin wall. The scalar field changes only near the sphere wall. Within the sphere the field is practically constant. The dynamic part was calculated at three dimensionless moments of time: 0.1, 1.57 and 2.4 (Fig. 7.27 left).

The actual landscape is determined by the sum of the stationary and the dynamic components of the potential. The dynamic part of the potential oscillates inside of an energy barrier, which is formed by the crater wall (Fig. 7.28).

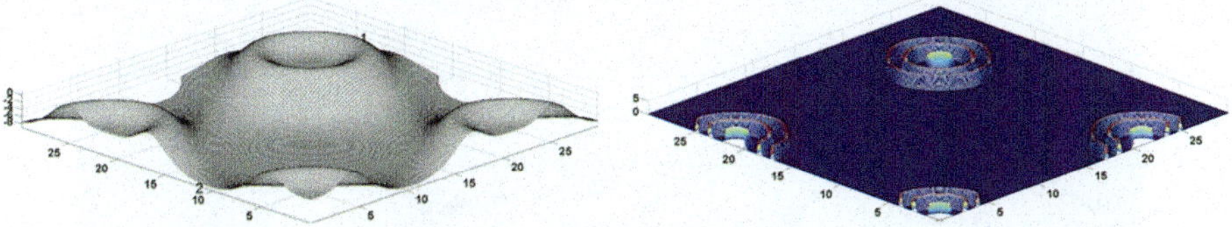

Fig. 7.27 The two-dimensional landscape of the stationary part of the potential. Number 1 corresponds to the hill tops, and the number 2 – the valleys of the potential (*left*). The oscillations of the dynamic part (*right*) [14, 15]

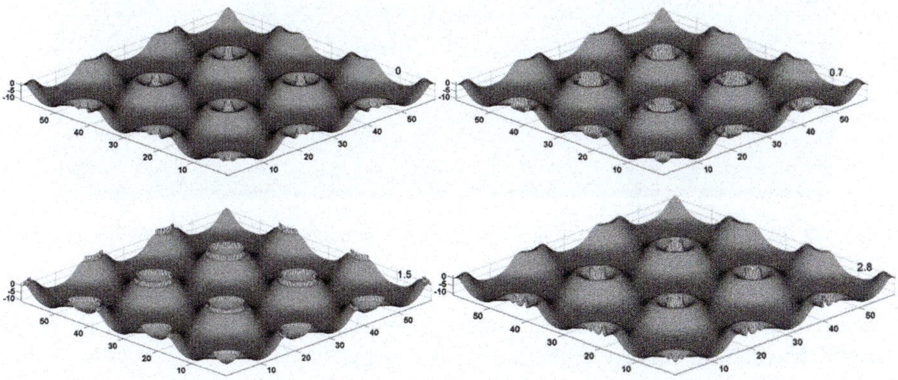

Fig. 7.28 The two-dimensional maps of the combined landscape calculated for different times. The calculations were made at four dimensionless moments of time: 0.1, 0.7, 1.5 and 2.8 [14, 15]

We can tell that the energy clot (sphere) oscillates inside of the energy well. This clot cannot cross the barrier unless it is given a large enough energy influx.

Thus, we have determined the shape of the energy landscape which contains the oscillating clots of energy.

We approximately described the version of the scalar field dynamics. We believe that similar dynamics may take place for different fields having different sizes and exist in different spacetimes.

Remarks We will consider the scalar field, which is described by the Klein-Gordon equations, as a fundamental field. According to the quantum field theory, absolutely everything is made of a field or a combination of them. What we call 'particles' are tiny vibrations in these fields.

Every particle is, deep down, a vibrating wave in a particular field. The photons that carry electromagnetism are waves in the electromagnetic field that stretches through space. Gravitons are waves in the gravitational fields, gluons are waves in the gluon fields, and so on. Just as sound waves propagate through the air, waves propagate through quantum fields, and we observe them as particles.

Each field exerts a tiny influence on the others. Two fields or a few fields can resonate together, as a result a new field and a particle may be formed in spacetime.

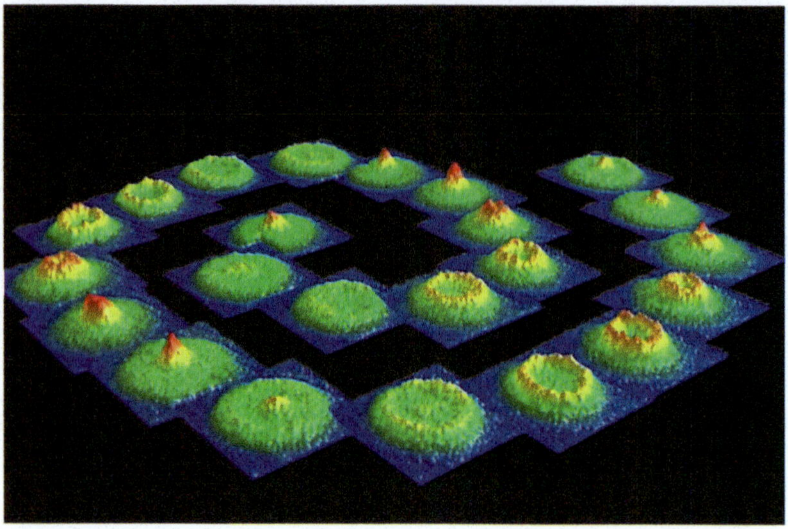

Fig. 7.29 Oscillations of a rotating Bose-Einstein condensate [82], http://www.lkb.ens.fr/recherche/atfroids/anglais/vortex_an.html

Scalar field waves can have a huge energy. The energy depends on the wave length. The shorter the wave, the larger its energy.

Mass is just a form of energy. Therefore very short waves can carry a lot of energy and a lot of mass. In particular, the dynamic component of the field (Fig. 7.27 right) corresponds to a huge energy and mass.

We assume that NKGE describes all quantum wave phenomena. Quantum waves explain the appearance of mass and different energies. We assume that solutions of NKGE describe the situation before the origin of our Universe, the process of its birth and the beginning of evolution of our Universe.

There is a certain similarity between the oscillations of the dynamic part of the potential (Fig. 7.28) and the data from the experiments studying the wave processes in the Bose-Einstein condensate [91] and on a granular layer, shown in Figs. 7.29 and 5.43.

7.9 The Tunnelling of the Energy Bubble Through the Potential Wall

The scalar potential landscape described above is subject to quantum fluctuations. Let us consider an impact of a quantum fluctuation on the energy bubble.

7.9.1 Instant Quantum Action

First we consider Eq. (7.45) subject to an instant quantum fluctuation. In this case we have

$$\left[\frac{1}{2}\omega^2B^2 - 2c_*^2\vartheta^2K^2\left(\sum_{i=1}^{I}a_i{}^2 + \frac{1}{4}I\right)\right]\Phi_{\xi\xi} + \partial V(\Phi)/\partial\Phi = f(\Phi)C\delta(\overline{\xi}). \tag{7.51}$$

Here $\overline{\xi}$ is a point subject to the quantum action, $\delta(\overline{\xi})$ is the Dirac delta function (the impulse function), C is the amplitude of the quantum fluctuation and $f(\Phi)$ is an arbitrary function. Let

$$\Phi = A\mathrm{sech}\, k\xi \quad \text{and} \quad f(\Phi) = \mathrm{sech}^2 k\overline{\xi}\, \sinh k\overline{\xi}. \tag{7.52}$$

Let us consider the field in the vicinity of $\overline{\xi}$. We assume that there the value of the field can change discontinuously as a result of a quantum kick (fluctuation). This discontinuous change in the field is computed by integration (7.51) from $\xi_j = \overline{\xi} - \varepsilon$ to $\xi_{j+1} = \overline{\xi} + \varepsilon$:

$$\begin{aligned}&\left[\frac{1}{2}\omega^2B^2 - 2c_*^2\vartheta^2K^2\left(\sum_{i=1}^{I}a_i{}^2 + \frac{1}{4}I\right)\right]\left[\Phi_\xi(\overline{\xi}+\varepsilon) - \Phi_\xi(\overline{\xi}-\varepsilon)\right]\\ &+\int_{\overline{\xi}-\varepsilon}^{\overline{\xi}+\varepsilon}(m^2A\mathrm{sech}\, k\xi - \lambda A^3\mathrm{sech}^3 k\xi)d\xi = C\int_{\overline{\xi}-\varepsilon}^{\overline{\xi}+\varepsilon} f(\Phi)\delta(\overline{\xi})d\xi.\end{aligned} \tag{7.53}$$

Let $\varepsilon \to 0$ and $A_{\overline{\xi}+\varepsilon} >> A_{\overline{\xi}-\varepsilon}$. In this case the Eq. (7.53) approximately yields that

$$\begin{aligned}&\left(A_{\overline{\xi}+\varepsilon}\right)^3 + 2\lambda^{-1}k^2\left[\frac{1}{2}\omega^2B^2 - 2c_*^2\vartheta^2K^2\left(\sum_{i=1}^{I}a_i{}^2 + \frac{1}{4}I\right)\right]A_{\overline{\xi}+\varepsilon} + 2C\lambda^{-1}k\\ &= 0.\end{aligned} \tag{7.54}$$

Here $2C\lambda^{-1}k$ is a constant. Its value is determined by the amplitude of the quantum fluctuation C. We will study cases when $2C\lambda^{-1}k$ is varied from 10^{-15} till 10^{-30} and

$$\frac{1}{2}\omega^2B^2 \approx 2c_*^2\vartheta^2K^2\left(\sum_{i=1}^{I}a_i{}^2 + \frac{1}{4}I\right). \tag{7.55}$$

We will consider a small fluctuation of $2\lambda^{-1}k^2\left[\frac{1}{2}\omega^2B^2 - 2c_*^2\vartheta^2K^2\left(\sum_{i=1}^{I}a_i{}^2 + \frac{1}{4}I\right)\right]$.

Let its amplitude varies from 10^{-5} to 10^{-15}. As a result the increase in the amplitude of the scalar field may be up to 10^{-2}.

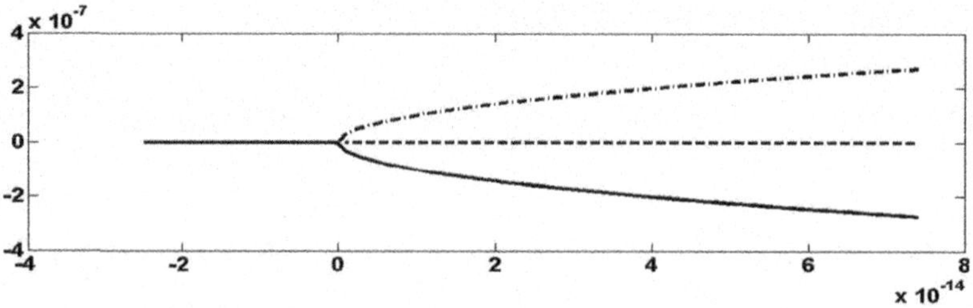

Fig. 7.30 Typical bifurcations of the amplitude of the dynamic part calculated for different values $2\lambda^{-1}K^2\left[\frac{1}{2}\omega^2B^2 - 2c_*^2\vartheta^2K^2\left(\sum_{i=1}^{I}a_i^2 + \frac{1}{4}I\right)\right]$ [14, 15]

We note again that NKGE explicitly contains the d'Alembertian operator. Therefore, the amplitude of the dynamic part can change very strongly when the value $\frac{1}{2}\omega^2B^2 - 2c_*^2\vartheta^2K^2\left(\sum_{i=1}^{I}a_i^2 + \frac{1}{4}\right)$ corresponding to this operator changes its sign. Thus, the field changes when the resonant condition (7.55) takes place. This resonant situation corresponds to a bifurcation (Fig. 7.30).

7.9.2 *Finite Time Quantum Action*

Figure 7.30 shows the possibility of a very strong change in the amplitude of the dynamic part. It becomes possible for the energy bubble to escape the potential well. However, we think that this process of tunnelling through the energy barrier is not instant. Let us assume that the quantum action is described by a function $f(\xi)$. In this case the Eq. (7.51) yields

$$\left[\frac{1}{2}\omega^2B^2 - 2c_*^2\vartheta^2K^2\left(\sum_{i=1}^{I}a_i^2 + \frac{1}{4}\right)\right]\Phi_{\xi\xi} + \partial V(\Phi)/\partial\Phi = f(\xi). \tag{7.56}$$

Changing $f(\xi)$ we will model the tunnelling processes. For the calculations we assume $\left[\frac{1}{2}\omega^2B^2 - 2c_*^2\vartheta^2K^2\left(\sum_{i=1}^{I}a_i^2 + \frac{1}{4}\right)\right]\Phi_{\xi\xi} = 0$ in (7.56). In this case we have

$$\Phi^3 + \overline{R}\Phi = \lambda^{-1}f(\xi). \tag{7.57}$$

Here $\overline{R} = -\lambda^{-1}m^2$ and a function $\lambda^{-1}f(\xi)$ models the quantum action (see, also, (7.70) in Sect. 7.10). We assumed that

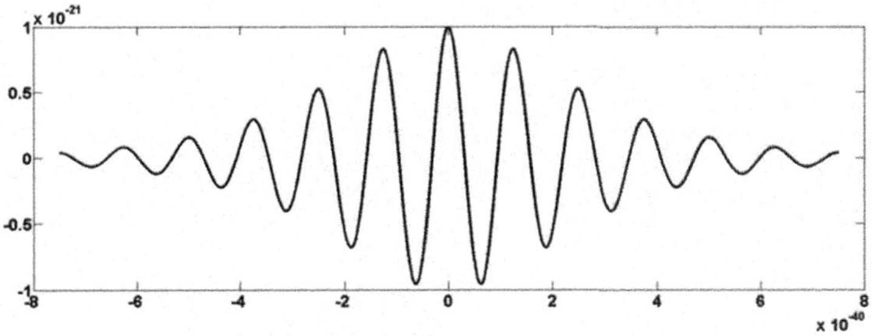

Fig. 7.31 The quantum fluctuation as a group of waves (7.58) where $l = -10^{-21}$, $\Omega = 5 \times 10^{39}$, $\overline{\Omega} = 5 \times 10^{40}$ and $J = 1$ [14, 15]

$$f(\xi) = l\lambda\left(\mathrm{sech}\Omega\xi \cos \overline{\Omega}\xi\right)^J, \tag{7.58}$$

where l, Ω, $\overline{\Omega}$ and J are constants. Thus, the quantum action is being described by the group of waves. We expect that the amplitude and the form of oscillations of the dynamic part can change strongly inside the potential well as a result of the quantum action. Equation (7.57) has the three solutions. If the transresonant parameter $\overline{R} = -\lambda^{-1}m^2$ is small (resonance), then the solutions may be discontinuous. However, using of them we can construct continuous resonant solutions having certain folds.

Below several examples of the action are presented.

At first we consider three cases of the calculations when $J = 1$ in (7.58).

1. Let the parameters of the quantum fluctuation (7.58) be: $l = -10^{-21}$, $\Omega = 5 \times 10^{39}$, $\overline{\Omega} = 5 \times 10^{40}$, $J = 1$ (Fig. 7.31).

Let $\overline{R} = -10^{-14}$. For this case the solutions of the Eq. (7.57) are shown in Fig. 7.32.

The three independent fields are described by (7.57) if the amplitude of the quantum action is small enough. However, the fields begin to interact when the quantum oscillations increases. As a result the very complex composite field is formed. Its profile resembles the mushroom-like waves.

The same behaviour is observed in the following two cases where we consider a quantum fluctuation with a different set of parameters.

2. Let the parameters of the quantum fluctuation be: $l = -3 \times 10^{-39}$, $\Omega = 5 \times 10^{39}$, $\overline{\Omega} = 5 \times 10^{40}$, $J = 1$.

In this case the resonant interaction of the fields take place if $\overline{R} = -10^{-25.75}$ (Fig. 7.33).

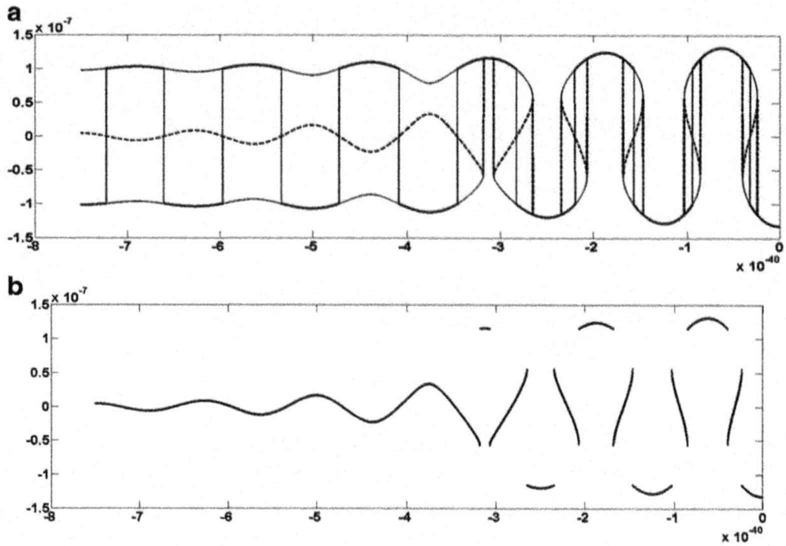

Fig. 7.32 The evolution of the scalar fields at the resonance ($\lambda^{-1}m^2 = 10^{-14}$) [14, 15, 84]. The composite (multivalued) solution (**a**), the discontinuous solution (**b**)

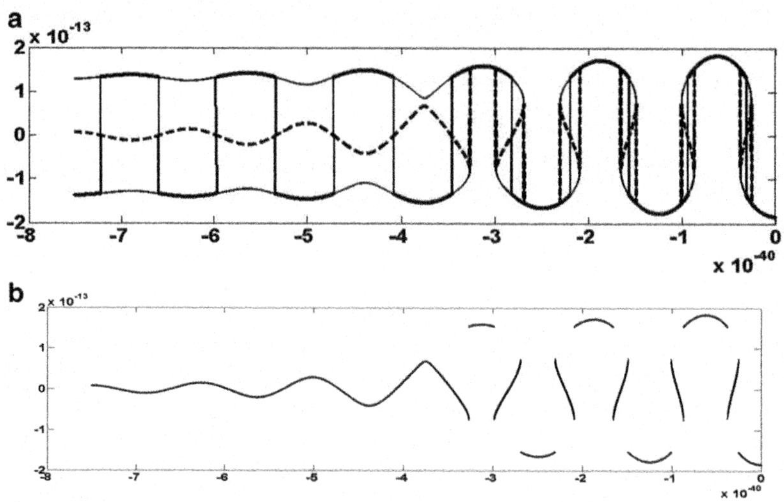

Fig. 7.33 The evolution of the scalar fields at the resonance ($\lambda^{-1}m^2 = 10^{-25.75}$) [14, 15, 84]. The composite (multivalued) solution (**a**), the discontinuous solution (**b**)

3. Let the parameters of the quantum fluctuation be: $l = -3 \times 10^{-51}$, $\Omega = 5 \times 10^{39}$, $\overline{\Omega} = 10^{41}$, $J = 1$. In this case the resonant condition is satisfied at $\overline{R} = -3 \times 10^{-34}$ and the resulting fields are illustrated by Fig. 7.34.

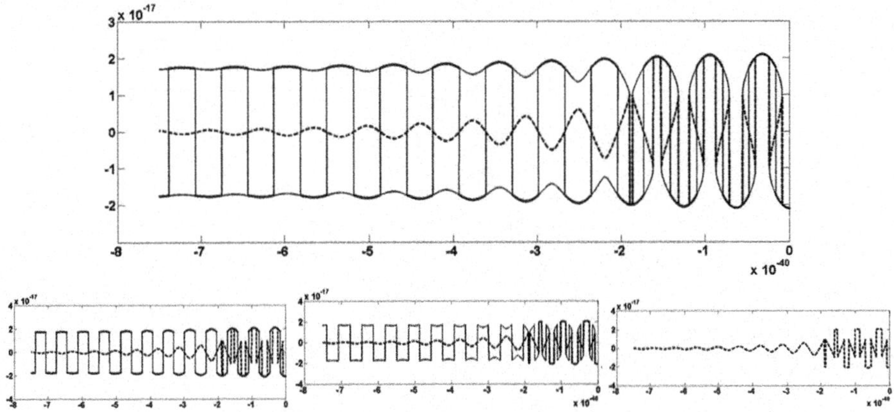

Fig. 7.34 The evolution of the scalar fields (solutions) during the tunnelling: The interaction of three fields (*upper picture*); the interaction of two fields (*low pictures, left* and *mid*); the discontinuous oscillations of one field (*low picture, right*) [15]

In the cases considered above (Figs. 7.32, 7.33 and 7.34) we found resonant parameters at which the three source independent scalar fields begin to interact and form a new composite field. The amplitude of the composite field's oscillations increases beyond that of the original fields. It may increase to a point where it becomes possible for the energy bubble to escape the potential well (the crater) (Figs. 7.27 and 7.28).

The nature of the oscillations themselves change. They become very complex and can potentially contain jump discontinuities.

In addition to the cases described above we consider two cases with a different form of the quantum action. The results confirm the above findings.

Let the parameters of the quantum fluctuation be: $J=2$, $l=-3\times10^{-39}$, $\Omega=5\times10^{39}$, $\overline{\Omega}=5\times10^{40}$. In this case the function $f(\xi)$ (7.58) describes a group of peaks (Fig. 7.35).

In this case the interaction of the fields takes place if $\overline{R}=-10^{-25.75}$ in (7.57). The evolutions of the solutions excited by the quantum action (Fig. 7.35) are shown in Fig. 7.36.

Let the parameters of the quantum fluctuation be: $J=3$, $l=-3\times10^{-39}$, $\Omega=5\times10^{39}$, $\overline{\Omega}=5\times10^{40}$ (Fig. 7.37).

In this case the resonant condition for the Eq. (7.57) is satisfied at $\overline{R}=-10^{-26.45}$. The resulting fields are presented in Fig. 7.38.

We studied the instability of a scalar field which is caused by a quantum fluctuation. We can expect that the wave shapes shown in Figs. 7.32, 7.33, 7.34, 7.36 and 7.38 may form in different unstable systems during impact actions. And indeed, there are certain similarities of the wave forms presented in Figs. 7.32, 7.33, 7.34 and 7.38 with the wave shapes generated due to the Richtmyer–Meshkov instability of incompressible liquids (see Fig. 5.15).

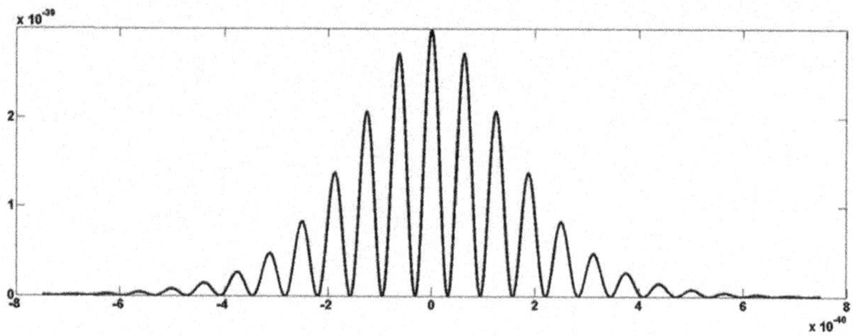

Fig. 7.35 The quantum fluctuation as a group of peaks: $l = -3 \times 10^{-39}$, $\Omega = 5 \times 10^{39}$, $\overline{\Omega} = 5 \times 10^{40}$ and $J = 2$ [14, 15]

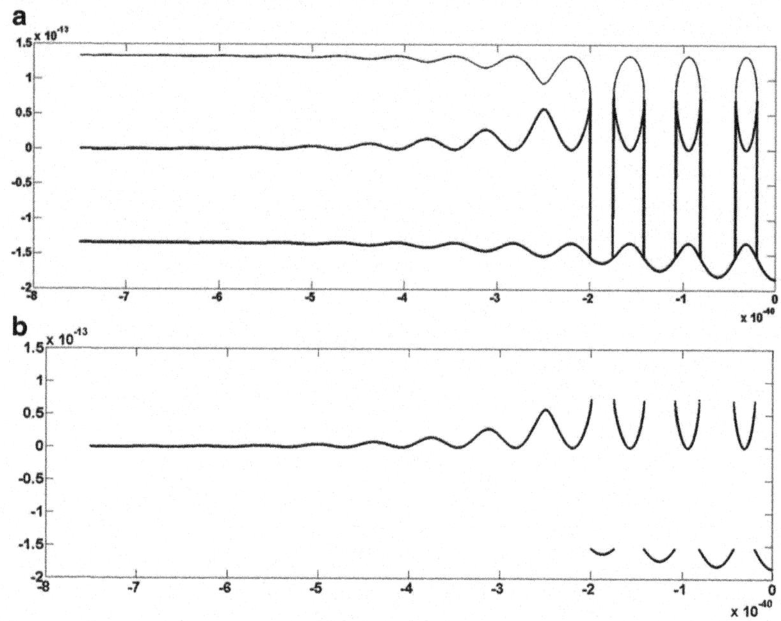

Fig. 7.36 The evolution of the scalar fields during the action of the group of peaks: $\lambda^{-1}m^2 = 10^{-25.75}$ [14, 15, 84]. The composite (multivalued) solution (**a**), the discontinuous solution (**b**)

We found that the amplification of the fields and the strongly nonlinear evolution of their forms took place when some resonant conditions for coefficients $\lambda^{-1}m^2$ and l of the Eq. (7.57) were satisfied. We considered cases when these resonant conditions take place. The influence of the change of the quantum fluctuations on the amplitude and the form of the resulting nonlinear scalar fields (waves) was studied.

An important result of the calculations is that a very small quantum action can form the composite scalar field within the resonant band. Near the resonant band the

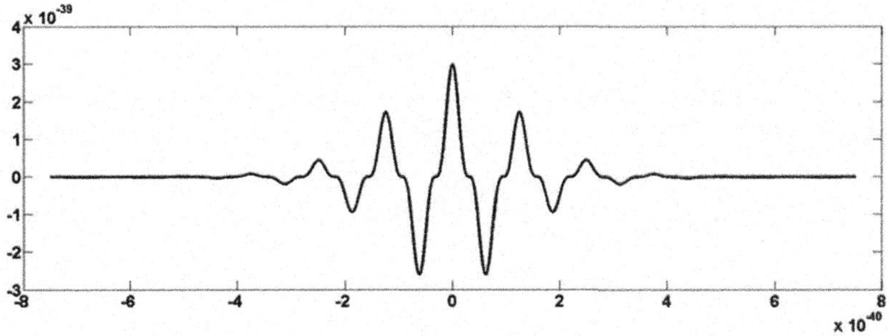

Fig. 7.37 The quantum fluctuation: $l = -3{*}10^{-39}$, $\Omega = 5{*}10^{39}$, $\overline{\Omega} = 5{*}10^{40}$ and $J = 3$ [14, 15]

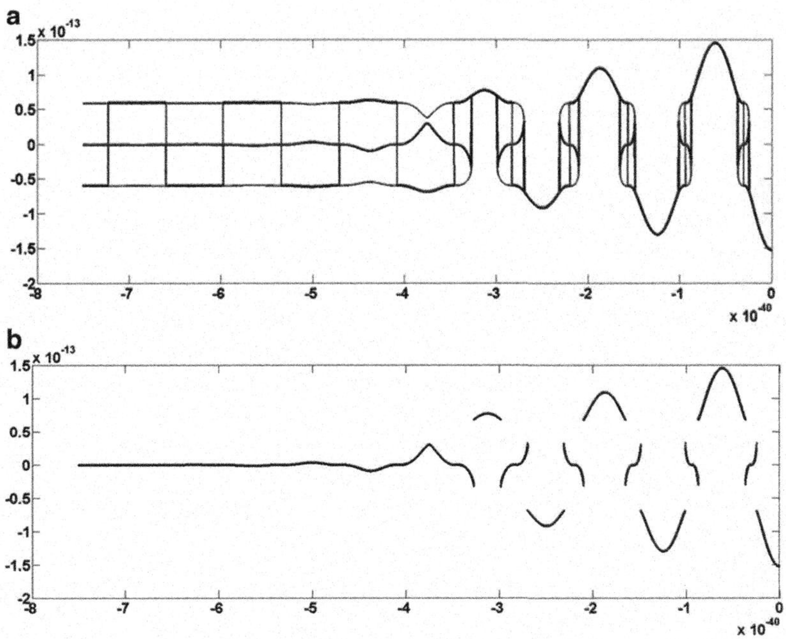

Fig. 7.38 The evolution of the scalar fields at the resonance: $\lambda^{-1}m^2 = 10^{-26.45}$ [14, 15, 84]. The composite (multivalued) solution (**a**), the discontinuous solution (**b**)

field evolution may be described by three branches of the solution. Generally speaking, the branches may interact. According to the calculations the energy can jump between the upper and lower branches (between positive and negative values of the energy).

Remark The calculated results did not qualitatively change when Ω and $\overline{\Omega}$ were reduced to $\Omega = 5{*}10^{27}$, $\overline{\Omega} = 5{*}10^{28}$.

7.9.3 Transresonant Tunnelling

Different ways of the evolution of the scalar field and the tunnelling of the energy bubble are described in the Sect. 7.9.2. There $\overline{R}$ was a constant in (7.57). Let us consider cases when $\overline{R}$ changes its sign according to linear law (Fig. 7.39):

We also assumed instead of (7.58) the following law for the quantum actions

$$f(\xi) = l\lambda \tanh \Omega\xi \cos^{J} \overline{\Omega}\xi, \tag{7.59}$$

Let us consider two cases of the action. The first when $l = -10^{-38}$, $\Omega = 5 \times 10^{39}$, $\overline{\Omega} = 0.76 \times 10^{41}$, $J = 1$ in (7.59) (Fig. 7.40, left). The next is $l = -10^{-38}$, $\Omega = 5 \times 10^{39}$, $\overline{\Omega} = 0.76 \times 10^{41}$, $J = 2$ (Fig. 7.40 right).

1. Let $J = 1$. The results of the solution of (7.57) are presented in Fig. 7.41 for two values of l. It was found that during the quantum action the scalar field amplitude increases from the order of 10^{-38} to the order of 10^{-13}.

Figure 7.41 demonstrates the typical resonant situation when the amplification weakly depend on the forced amplitude. At the same time Fig. 7.41(a left) shows the appearance of three scalar fields as a result of the transresonant evolution. These fields interact very strongly within interval from 3* 10^{-40} till 6.5* 10^{-40}. On the whole Fig. 7.41 (a left) corresponds to the processes which were shown in Figs. 7.32, 7.33, 7.34 and 7.38, but Fig. 7.41 (a right) corresponds to Fig. 7.30. The reduction of the amplitude l yields the simplification of the transresonant process according to Fig. 7.41.

There are certain similarities of the wave form presented in Fig. 7.41 (a left) with the wave shapes shown in Figs. 7.42 and 5.15.

Thus, the effects of the quantum action can remind the Richtmyer–Meshkov instability of an interface of two-liquid system.

2. Similar results are for $J = 2$ and $\Omega = 5 \times 10^{39}$, $\overline{\Omega} = 0.76 \times 10^{41}$ (Fig. 7.43).

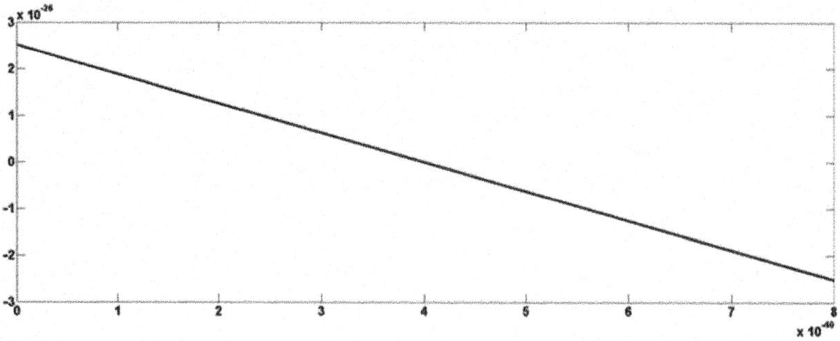

Fig. 7.39 Variation of $-\overline{R}$ in (7.57) during the tunnelling

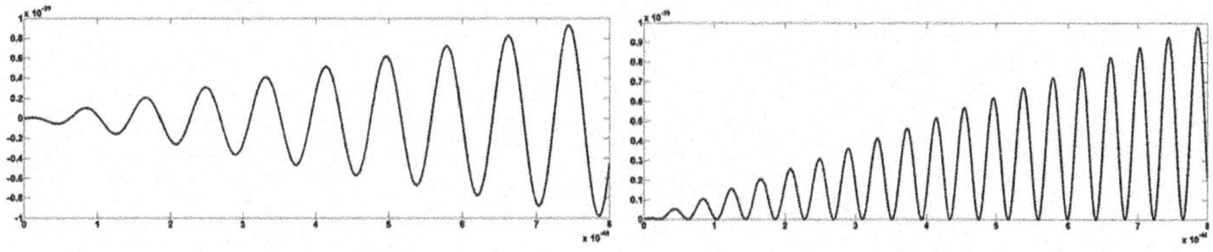

Fig. 7.40 The quantum fluctuations as groups of waves

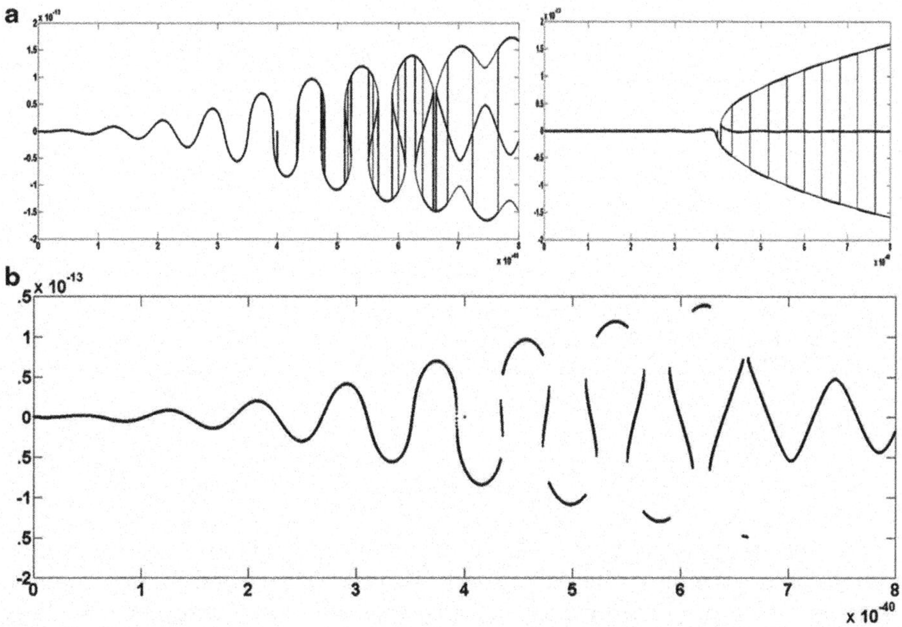

Fig. 7.41 Result of the quantum action, when $J = 1$. Examples of the amplification of the scalar field from the order of 10^{-38} to the order of 10^{-13} if $l = -10^{-38}$ (**a** *left* and **b**) and $l = -10^{-40}$ (**a** *right*) [14, 15, 84]. The composite (multivalued) solutions (**a**), the discontinuous solution (**b**)

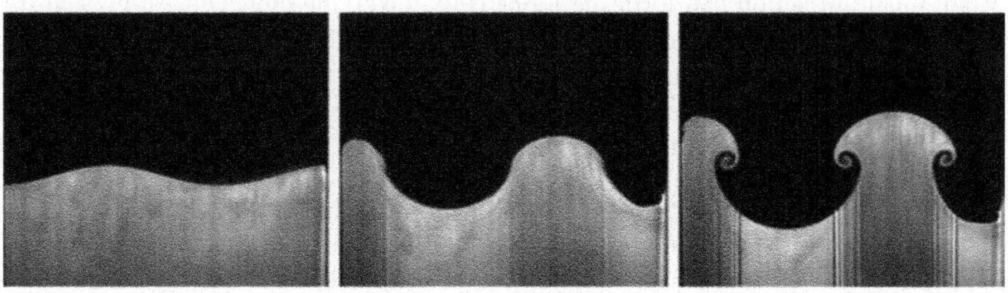

Fig. 7.42 A sequence of images of waves generated due to the Richtmyer–Meshkov instability of incompressible liquids [92] (see, also, Figs. 5.15 and 5.16)

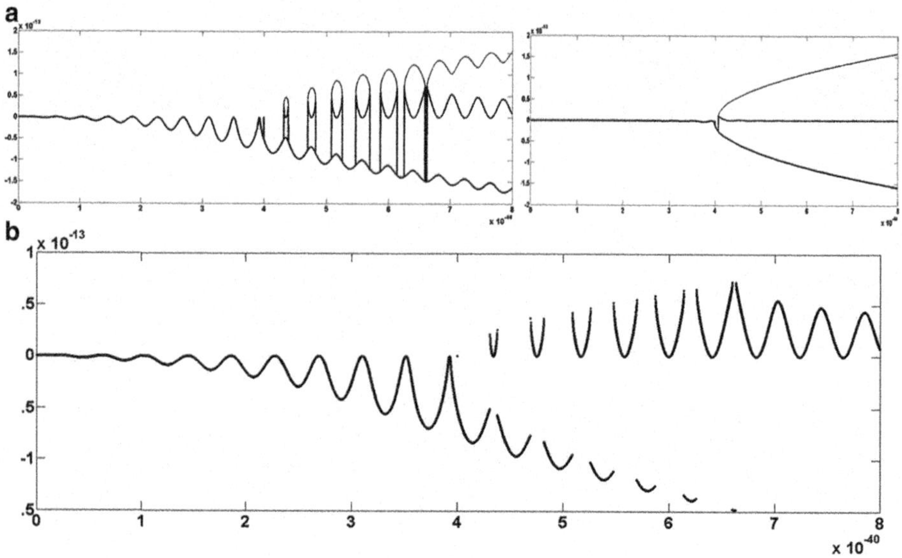

Fig. 7.43 Result of the quantum action when $J=2$. Examples of the amplification of the scalar field from the order of 10^{-38} till the order of 10^{-13} if $l=-10^{-38}$ (**a** *left* and **b**) and $l=-10^{-40}$ (**a** *right*) [14, 15, 84]. The composite (multivalued) solutions (**a**), the discontinuous solution (**b**)

On the whole Fig. 7.43 (a left) resembles Fig. 7.36. Figure 7.43 (a right) corresponds Fig. 7.30.

We considered the different cases of the resonant action of the quantum perturbation on the scalar field. The energy of the bubble may be increased very strongly. Therefore the bubble can leave the potential wall. We tried to illustrate this escape by calculations presented in Fig. 7.44. It shows a picture of the crossing of the energy barrier by the energy bubble (clot).

It is seen from Fig. 7.44 that the ways of tunnelling through the potential wall may be quite different. Another important result of the calculations is that a very small quantum action can form the loops (Fig. 7.41), the clots (Fig. 7.43) and the finite elements of the scalar field. On the other hand there are complex nonlinear interaction among elements of the scalar field within the resonant band. We will assume as a rough hypothesis that the very energetic scalar field begins to tunnel (to radiate) from the potential wall.

Remark According to (7.41) the interaction of the components $\overline{\Phi}$ and Φ may be important for the tunnelling and a form of the potential landscape. Thus, the landscape of the static part of the scalar field influences the tunnelling process. The shape of the energy barrier could be marked in the cosmic microwave background (CMB) radiation emitted in the very early stages of the formation of the Universe. Perhaps, the influence of the interactions noted above is reflected in a

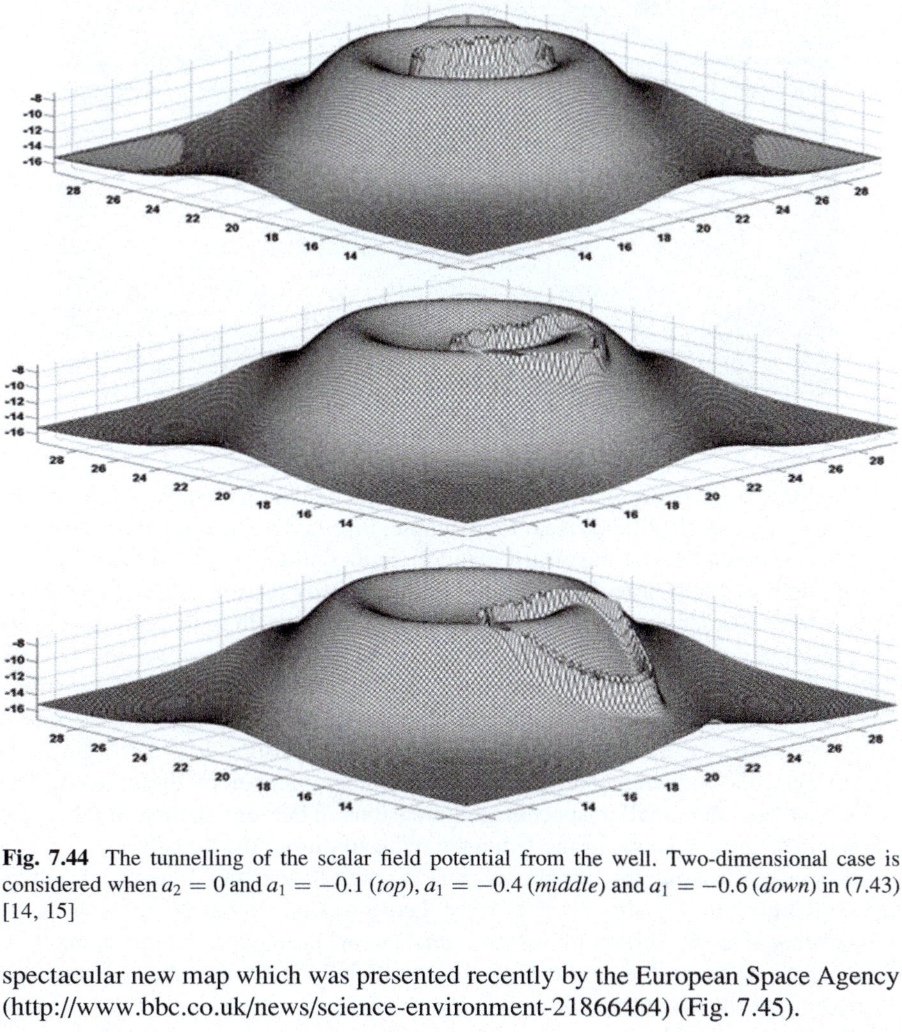

Fig. 7.44 The tunnelling of the scalar field potential from the well. Two-dimensional case is considered when $a_2 = 0$ and $a_1 = -0.1$ (*top*), $a_1 = -0.4$ (*middle*) and $a_1 = -0.6$ (*down*) in (7.43) [14, 15]

spectacular new map which was presented recently by the European Space Agency (http://www.bbc.co.uk/news/science-environment-21866464) (Fig. 7.45).

7.9.4 The Fragmentation of Multidimensional Spacetime During the Tunnelling

We considered many examples of the effect of the quantum fluctuations on the scalar field. According to the calculations tiny quantum perturbations can increase the energy of the field very strongly. We found resonant parameters when the three independent scalar fields begin to interact and form a new composite field oscillating with a great amplitude. The nature of the oscillations themselves changes. They become very complex and can potentially contain jumps of discontinuity.

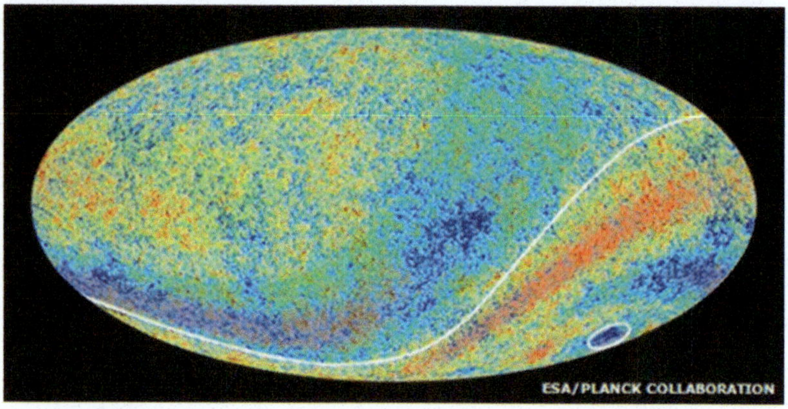

Fig. 7.45 The north/south differences and a 'cold spot' in CMB

Thus, the resulting oscillations of the composite field include the elements of the previously independent fields (see, for example, Fig. 7.34). The transition of the one element to another is not necessarily smooth and continuous. Discontinuous transitions are possible. These discontinuities could destroy the spacetime. The boundaries between dimensions were less stable during the tunnelling than they are now. As a result the multidimensional spacetime can be split into many two-dimensional (space + time) elements. The spacetime fabric was transformed and fragmented when the field energy have been increased strongly. Generally speaking, it agrees with the 'vanishing dimensions' theory [93–96] (see, also, [79]). According this theory systems with higher energy have a smaller number of dimensions. The higher energy, the smaller spacetime dimensions. Thus our theory implies that the number of dimensions in the Universe reduced during the tunnelling.

Generally speaking, the initial energy field could have many dimensions, which are curled up into tiny shapes. The string theory proclaims that the number of the dimensions may be different, for example, 5, or 11 or 26. As the number of dimensions reduces, the volume of the initial bubble (clot, sphere, drop) can increase after the fragmentation very, very strongly.

The process could be visualized by imagining a three-dimensional drop of oil impacting the surface of water. As a result of the impact the oil drop is separated into many elements (particles) which spread over the two-dimensional surface of water. These elements of oil occupy in the two-dimensional space much bigger volume than they had in the initial moment. It is important that the elements became more isolated from each other comparing to when they were inside the drop.

Qualitatively the fragmentation of the spacetime fabric could remind us of the atomization of the water drop (see Figs. 5.19 and 7.46). Figure 7.46 shows the fragmentation a drop which was vibrated in the vertical direction in its fundamental axisymmetric mode [97]. The forms of the surface waves and the small drops, shown on Fig. 7.46, are determined by surface tension and some resonant conditions.

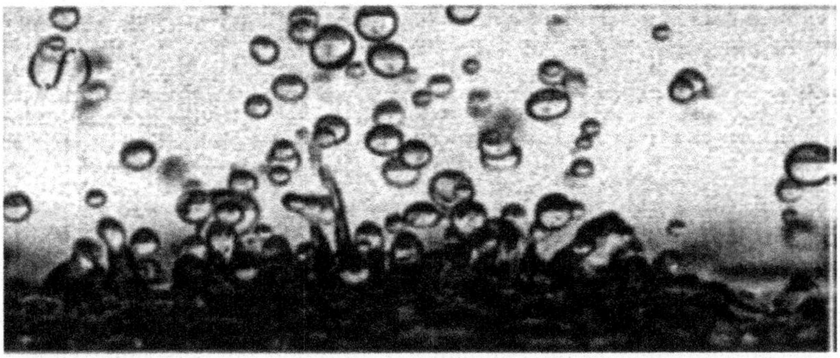

Fig. 7.46 Scheme of the rapid-ejection process and the formation of the cloud of particles

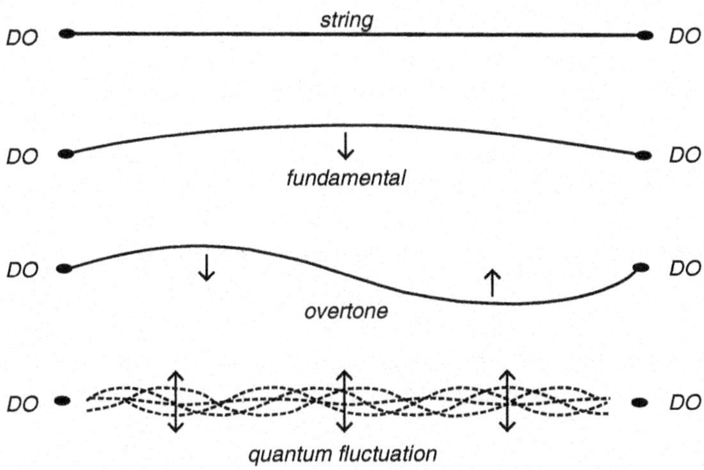

Fig. 7.47 Linear oscillations and a quantum fluctuation of a string stretched between two D0-branes [98]

Thus, the fragmentation of the spacetime fabric could resemble the atomization of the water drop (see Fig. 5.19).

As a result of the transresonant evolution the fragments (elements) gained the extremely high energy. Thus, in the course of the tunnelling the bubble (clot) of a scalar field increases strongly energy, loses the space dimensions and can strongly increase the volume. It was assumed [14] that the elements became absolutely isolated without any connections.

The process of atomization, shown in Figs. 5.19 and 7.46, corresponds to a frequency which is close to the resonance frequency of waves on the drop surface. It can serve as a model for many natural wave phenomena [13]. In particular, as a result of the transresonant evolution of the initial field a large number of small finite elements appear which possess extremely high energy. Using the terminology of the string theory these elements may be defined as strings. They begin to vibrate (Fig. 7.47).

Thus, in the course of the tunnelling the bubble (clot) of a scalar field strongly increases its energy, loses the space dimensions and can strongly increase the volume. It was assumed [14, 15, 84] that the elements became absolutely isolated without any connections (see Figs. 7.32, 7.33, 7.34, 7.36, 7.38, 7.41 and 7.43). Thus, the behaviour of the Universe's spacetime might be very surprising in the beginning.

On the whole this conclusion agrees with suggestions of Steven Carlip of the University of California, Davis [96]. In 2012 he gathered up all the theories and found that, according to many of them, the Universe had got one or two spatial dimension during the hot, dense start. In other words, geometry appears to have been radically different in the beginning. Carlip and his colleagues showed that space was split into discrete elements in the first seconds of the Universe. Each element experiences nothing outside its own existence [79, 96].

Conclusion Three fields defined by the Eq. (7.35), which was presented as (7.57) and (7.58), are normally stable. However, in certain circumstances the fields lose stability. They begin to vibrate extremely. We found the examples of parameters at which the fields begin to interact. We will call these parameters – resonant. The resulting composite field consists from the elements of those previously-independent fields. The transition of the elements from one to another is not necessarily smooth and continuous. Discontinuous transitions are possible and the jumps can be very large. We can say that in the resonant situations the field becomes fragmented.

The main thing – there are the situations when the scalar field can lose stability fully. This process can be accompanied by the disintegration of the spacetime in which the field was prior to the quantum resonant action. The disintegrated spacetime contains absolutely-isolated spatially one-dimensional string-like elements.

7.10 The Origin of the Particles of Energy and Matter as a Strongly-Nonlinear Resonant Phenomenon

We assume that a large number of small open string-like elements can appear in the scalar field. They begin to vibrate (Fig. 7.47). We can study these vibrations as a strongly-nonlinear problem which is described by NKGE (7.35). This equation has the cubic nonlinear term and the d'Alembertian operator. Near the resonant frequencies the influence of this operator is small. Using this we constructed multivalue models which describe the origin of the highly energetic particles [14, 15, 84].

The multidimensional model of the element is considered. We write the Eq. (7.35) in the form

$$\Phi_{tt} - c_*^2 \sum_{i=1}^{I} \Phi_{ii} + m^2\Phi - \lambda\Phi^3 = 0. \tag{7.60}$$

We assume at the ends of the element that

$$\Phi = 0 \quad \text{at } x_i = 0; L. \tag{7.61}$$

New coordinates r and s are introduced

$$r = ct - \sum_{i=1}^{I} k_i x_i, \; s = ct + \sum_{i=1}^{I} k_i x_i. \tag{7.62}$$

Here c is a constant. Let its value be very close to $c_* \left(\sum_{i=1}^{I} k_i^2 \right)^{0.5}$,

$$c = c_* \left(\sum_{i=1}^{I} k_i^2 \right)^{0.5} + \overline{c}, \tag{7.63}$$

where $\overline{c}$ is the perturbation of the speed c. The function Φ is represented as a sum:

$$\Phi = \Phi^{(1)} + \Phi^{(2)}, \tag{7.64}$$

where $\Phi^{(1)} >> \Phi^{(2)}$. It is assumed that

$$\Phi^{(1)} = J(r) - J(s) \text{ and } \Phi^{(2)} = j(r) - j(s). \tag{7.65}$$

Substituting (7.64) into (7.60) we write two equations

$$2\left(c^2 + c_*^2\right)\left(\Phi^{(1)} + \Phi^{(2)}\right)_{rs} = 0, \tag{7.66}$$

$$\begin{gathered}\left[2c_*\overline{c}\left(\sum_{i=1}^{I} k_i^2\right)^{0.5} + \overline{c}^2\right]\left[\left(\Phi^{(1)} + \Phi^{(2)}\right)_{rr} + \left(\Phi^{(1)} + \Phi^{(2)}\right)_{ss}\right] \\ + m^2\left(\Phi^{(1)} + \Phi^{(2)}\right) - \lambda\left(\Phi^{(1)} + \Phi^{(2)}\right)^3 = 0.\end{gathered} \tag{7.67}$$

Let

$$\begin{aligned}&\Phi^{(1)} = A\left(\sin \omega c^{-1} r - \sin \omega c^{-1} s\right) \text{and} \\ &\omega L c^{-1} = 2\pi N \left(N = \pm 1, \pm 2, \pm 3, \ldots.\right)\end{aligned} \tag{7.68}$$

It is also suggested that the function j is periodic having the period equal L. In this case the conditions (7.61) and Eq. (7.66) are satisfied. Let us now consider (7.67). We will construct the approximate solution of (7.67) which is valid where

$$\Phi^{(1)}{}_{rr} >> \Phi^{(2)}{}_{rr} \text{ and } \Phi^{(1)}{}_{ss} >> \Phi^{(2)}{}_{ss}. \tag{7.69}$$

The interaction of the opposite travelling waves is not taken into account. In this case equation (7.67) yields

$$j^3 + \overline{R}j + q\sin\omega c^{-1}r = 0, \tag{7.70}$$

where

$$q = \left[2c_*\overline{c}\left(\sum_{i=1}^{I} k_i^2\right)^{0.5} + \overline{c}^2\right]\lambda^{-1}\omega^2 c^{-2}A. \tag{7.71}$$

There are three distinct cases, when the Eq. (7.70) has real solutions:

1. Let $\overline{R} = 0$, then the Eq. (7.70) is satisfied if

$$j = (-q)^{1/3}. \tag{7.72}$$

2. Let $\overline{R} > 0$, then the function j is unique, single-valued and continuous

$$j = -2D\sinh\left[\tfrac{1}{3}\operatorname{arcsinh}\left(\tfrac{1}{2}qD^{-3}\right)\right], \tag{7.73}$$

where $D = [sign(q)](|R|/3)^{0.5}$.

3. Let $\overline{R} < 0$ and $q^2/4 + R^2/27 \le 0$. In this case there are three solutions

$$j_M = -2D\cos\left[\tfrac{1}{3}\text{arc }\cos\left(0.5qD^{-3}\right) + 2M\pi/3\right], \tag{7.74}$$

where $M = 0;1;2$. Using these continuous solutions we can construct multivalued solutions.

7.10.1 *The Origin of the Particles*

Above the model is presented for an arbitrary space dimension of the field. The interesting us result follows from it if $I = 1$ and $k_i = 1$ in (7.62). Conditions (7.69) are considered as conditions of resonance. Model calculations were made. We assumed that

$$\overline{R} = -0.0014,\ \omega = 25.5,\ c = 1 \text{ and } q = 0.0001. \tag{7.75}$$

1. First we study the nonlinear oscillations described by the Eqs. (7.70) and $\Phi^{(2)}$ (7.65). Figures 7.48 and 7.49 illustrate second and first modes of the oscillations correspondingly.

We think that Fig. 7.48 illustrates the emergence of particles of energy and matter. The particles radiate above or below the scalar field element. Namely, there are moments (for example, 0.1775 and 0.1925) when the particles separate from the energy level of the element.

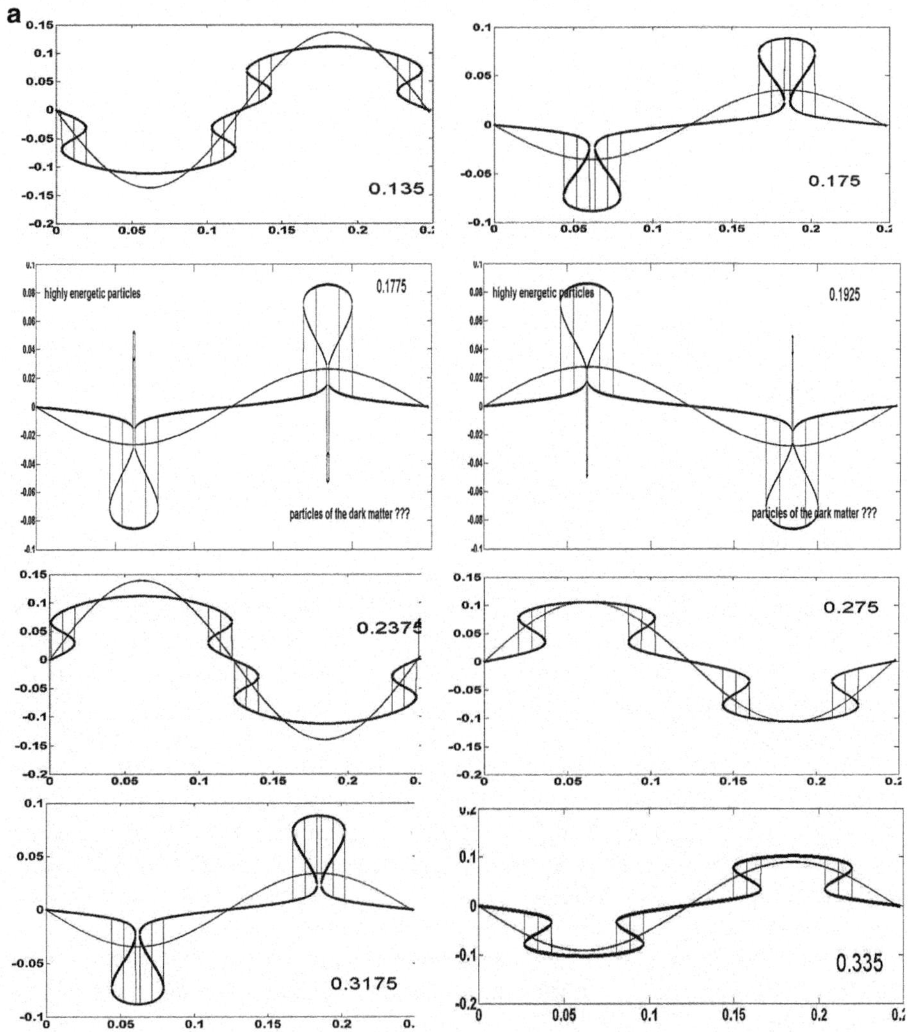

Fig. 7.48 (**a**) Resonant continuous oscillations of the scalar field elements according to the second resonant form. Periodic formation of the particles of matter. The thin smooth lines correspond to the linear oscillations [14, 15, 84]. (**b**) Resonant discontinuous oscillations of the scalar field elements according to the second resonant form. Periodic formation of the particles of matter. Curves illustrating the process of formation of the particles of matter. Curves described by the composite solution (**a**). Curves described by the discontinuous solution (**b**)

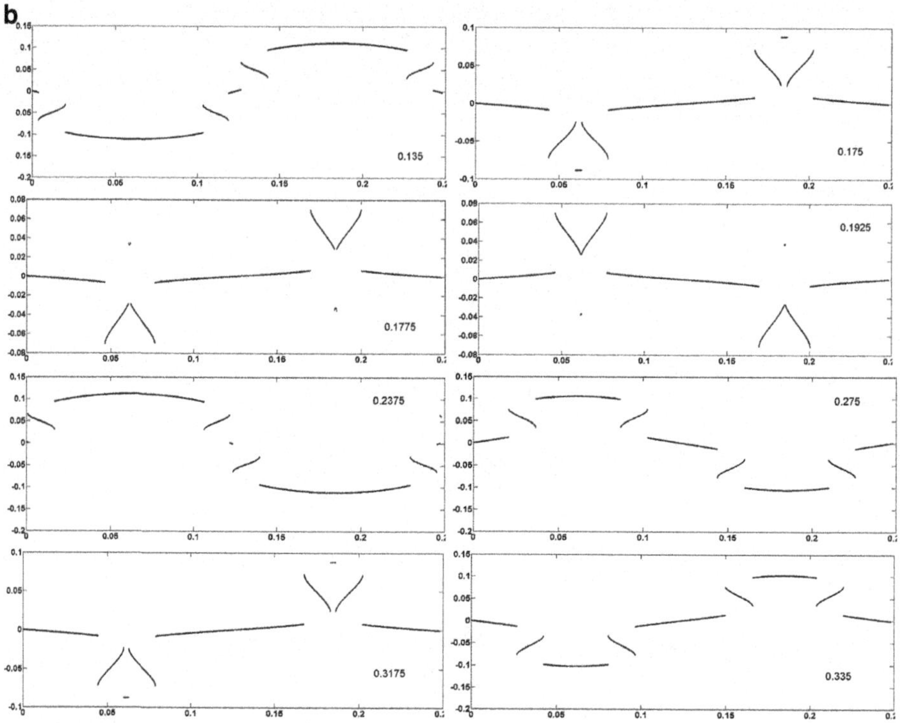

Fig. 7.48 (continued)

It is important that the appearance of the energetic particles (see Fig. 7.48) incorporates an important quantum concept – the momentary creation of pairs of particles. However, periodic radiation of a sole particle is also possible if the element oscillates according to the first resonant form (Fig. 7.49). We used (7.75) where $\omega = 26.1$.

The shapes of the nonlinear oscillations are quite different from the linear forms. In particular, the nonlinear shapes have the folds which were formed by the curved segments corresponding to the different solutions of the Eq. (7.70). Within and near these folds the scalar field can have jump discontinuities. There the gradient can be very large up to infinite values. According to (7.37) and (7.38) this gradient determines the density and the pressure of scalar fields.

Taking into account Figs. 7.48 and 7.49 we can imagine that very massive particles may be generated during the resonant oscillations of the fragments of the scalar field.

The scheme of this generation is presented in Figs. 5.19 and 7.46, where vertically-excited waves and drops on a surface of 100 millimetric volumes of water are shown. The atomization may be considered as some analogue of the scalar field fragmentation.

Qualitatively the origin of particles is shown in Fig. 7.50.

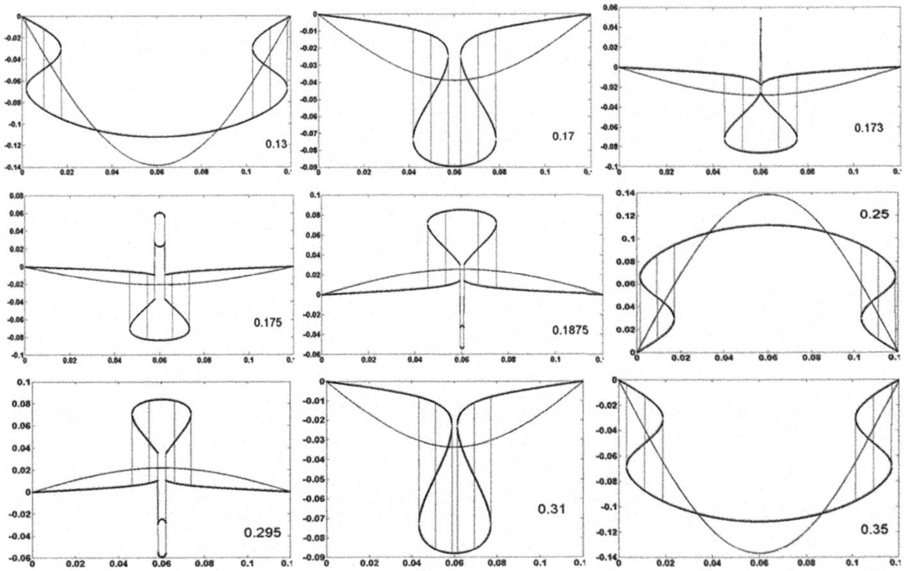

Fig. 7.49 Resonant oscillations of the scalar field elements according to the first resonant form. Periodical formation of particles of matter. The thin smooth lines correspond to the linear oscillations [14, 15]

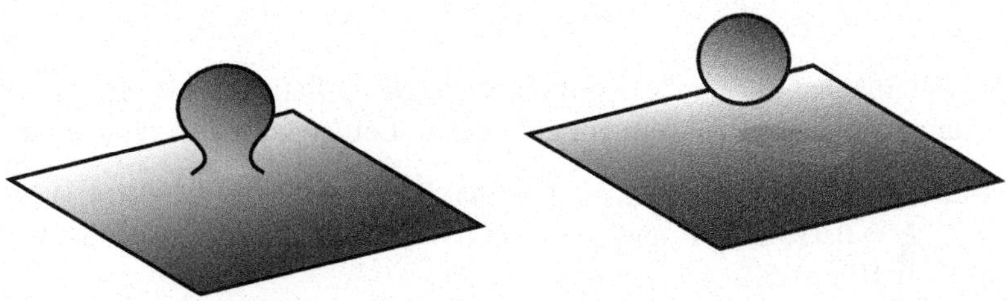

Fig. 7.50 A qualitative pattern of an appearance of the virtual particle in vacuum or a separation of new closed Universe from a big volume of space [60]

2. Then we calculate the scalar field oscillations taking into account the linear contribution. The oscillations are determined by expressions (7.64), (7.65) and (7.68):

$$\Phi = A\left[\sin\left(\omega c^{-1} r - \pi/2\right) - \sin\left(\omega c^{-1} s - \pi/2\right)\right] + j(r) - j(s). \qquad (7.76)$$

We have assumed that the shift of the linear oscillations (linear component of the solution) relative to the nonlinear oscillations (nonlinear component of the solution) at the resonance is $-\pi/2$.

Thus, the scalar field oscillations are determined by expressions (7.76) and parameters (7.75). The results of the calculation are presented in the Fig. 7.51.

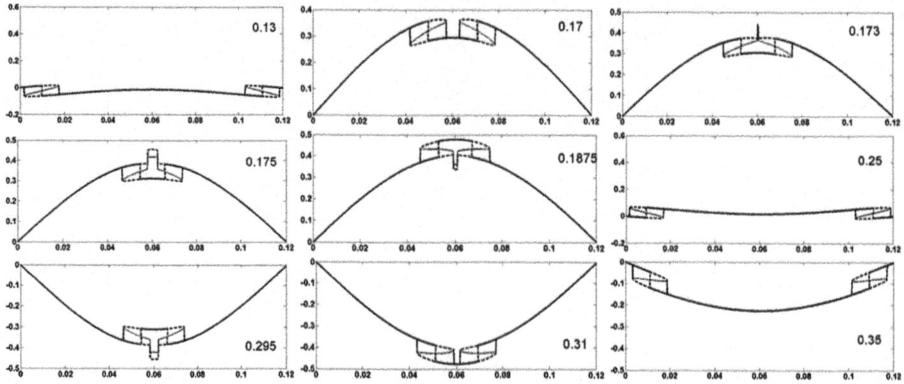

Fig. 7.51 Resonant oscillations of a fragment of the scalar field. The first resonant form [15, 84]

The elements noted above vibrate according to their resonant frequencies and can radiate particles of energy. If these particles possess some critical energy, they form matter.

7.10.2 The Origin of Dark and Normal Matters, and Dark Energy

We can imagine that during the oscillations and the radiation of the energy particles the common energy of the fragment is reduced. Let us study this phenomenon.

Dark matter particles First we assume that $\overline{R} = -m^2\lambda^{-1} = -0.0014, A = -0.2$, $\omega = 26.1$, $L = 0.12$, $c = 1$ and $q = 0.0001$. The results of the calculations are shown below (Fig. 7.52).

On the whole we have got the results which are similar to those presented in Figs. 7.48, 7.49 and 7.51. We assume that they describe the formation of very heavy particles.

Normal matter particles Increasing $\overline{R} = -m^2\lambda^{-1} = -0.00042$ we receive the results shown in Fig. 7.53.

We assume that Fig. 7.53 describes the formation of normal matter particles.

Dark energy particles One can see that Fig. 7.53 does not show the particle radiation clearly. However, we assume that the energy of the fragment continues reducing. Therefore, we assume that the coefficient $\overline{R} = -m^2\lambda^{-1} = -0.0000014$. Corresponding oscillations are shown in Fig. 7.54.

Expressions (7.37) and (7.38) show that the density and the pressure depend very strongly on the gradient of a scalar field. In points where the field changes rapidly the pressure and the density can be very large. In particular, the mass of the dark matter particles may be very near infinite, the mass of the normal matter particles may be extremely large, but the mass of the dark energy may be very small.

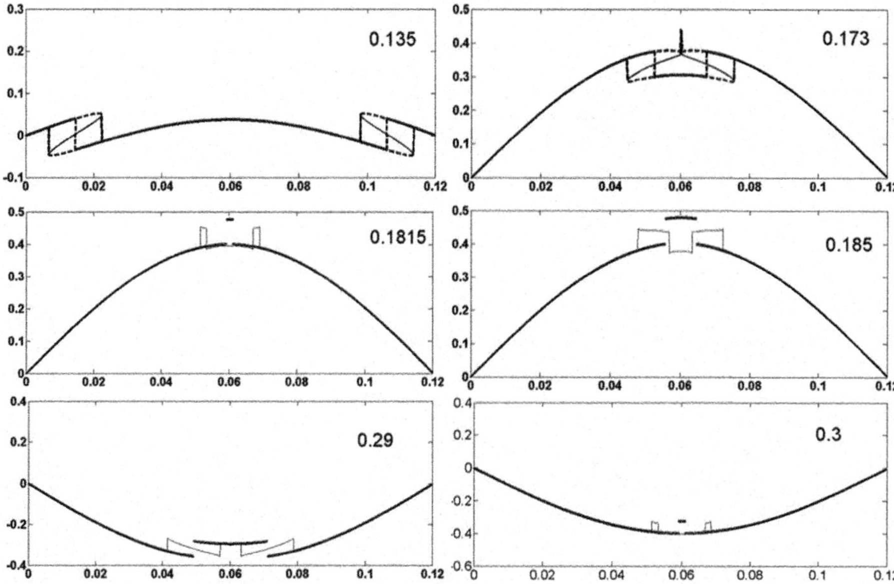

Fig. 7.52 Resonant oscillations of a fragment of the scalar field accompanied by eruptions of energetic particles. The case of $\overline{R} = -m^2\lambda^{-1} = -0.0014$ [15, 84]

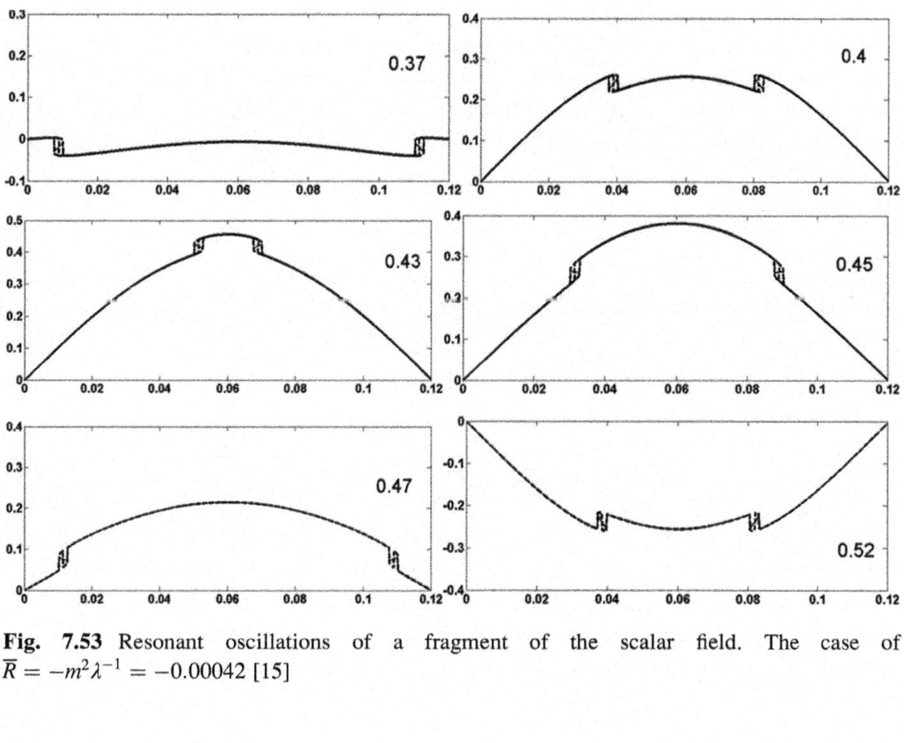

Fig. 7.53 Resonant oscillations of a fragment of the scalar field. The case of $\overline{R} = -m^2\lambda^{-1} = -0.00042$ [15]

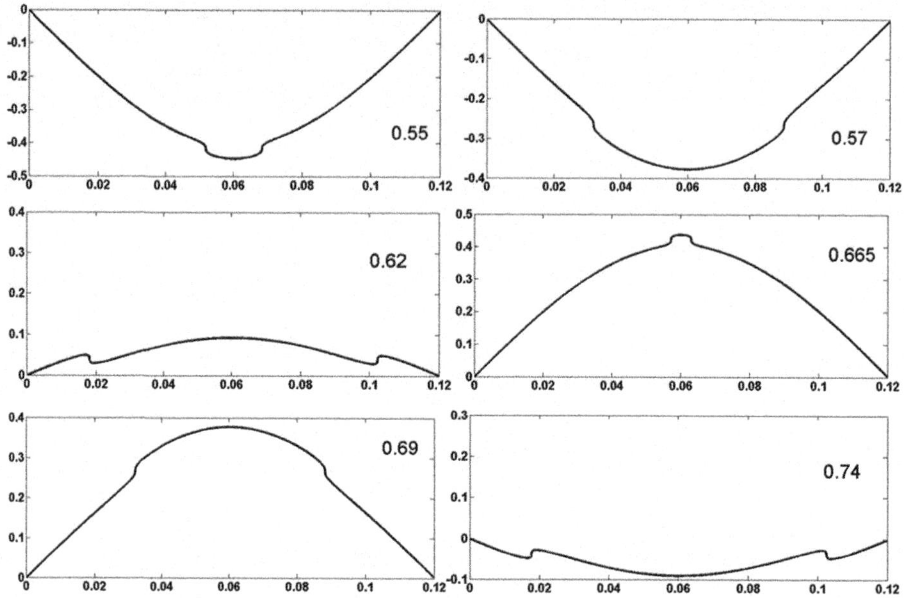

Fig. 7.54 Resonant oscillations of a fragment of the scalar field. The case of $\overline{R} = -m^2\lambda^{-1} = -0.0000014$ [15]

We assume that highly energetic particles (Fig. 7.48) could be very stable. These particles may correspond to the dark matter. We can imagine that these particles had been stable for billions of years before they begun to decay. The last process can determine the law of expansion of the modern Universe. The particles radiated near the resonances can form the normal matter. They could begin to work like a scaffold creating the cosmic structure. The dark energy is spread within this structure.

7.10.3 Boundary Conditions $\partial\Phi/\partial x_i = 0$ at $x_i = 0; L$

We now consider oscillations appearing in the fragment, if $\partial\Phi/\partial x_i = 0$ at $x_i = 0; L$.

We assume that

$$q = 0.0001,\ \omega = 25.5 \text{ and } c = 1. \tag{7.77}$$

The examples of oscillations at different values of $\overline{R}$ are shown in Fig. 7.55. The transresonant amplification of the wave takes place as we move away from 0.

We are interested in the oscillations of the fragment and the radiation of particles for the assumed boundary condition. The results of the calculations for a full period of oscillations specifically at (7.77) and $\overline{R} = -0.0021$ are shown in Fig. 7.56.

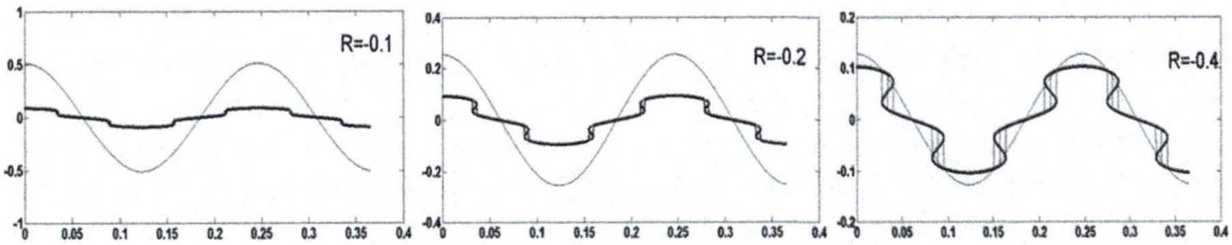

Fig. 7.55 The transresonant amplification of the wave, $R = \overline{R}/0.03$ [15]

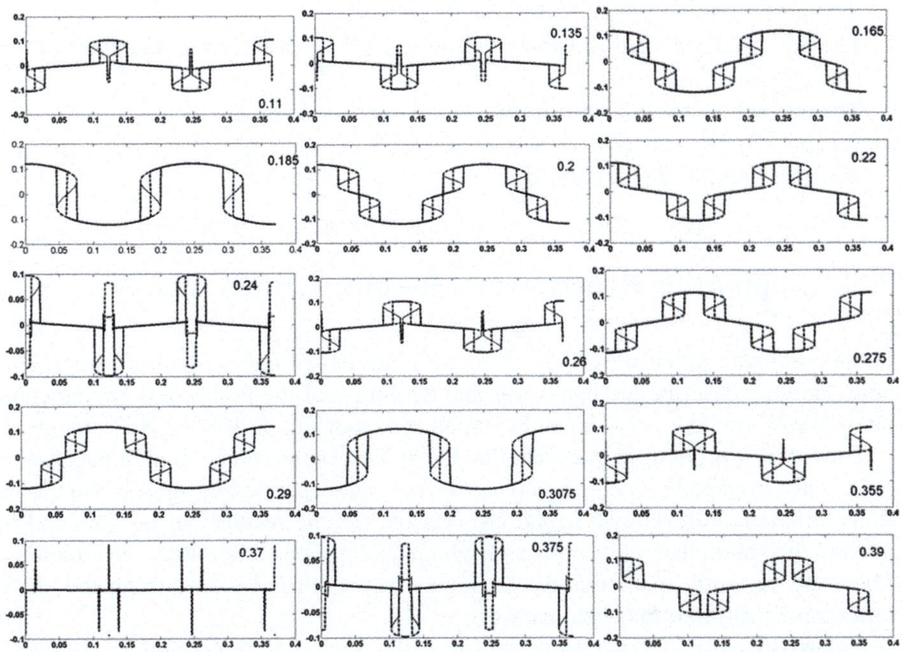

Fig. 7.56 Resonant oscillations of a fragment of the scalar field accompanied by eruptions of energetic particles. The case of $\overline{R} = -m^2\lambda^{-1} = -0.0021$ [15]

And again, just like in the earlier examples (Figs. 7.48, 7.49, 7.51 and 7.52) we are seeing the separation of energy from the main field which we interpret as emergence of various particles.

According to our theory, during the tunnelling both positive and negative particles of energy and matter appeared. However the number of the positive and negative particles might be different.

Important conclusion Thus, the calculations (Figs. 7.48, 7.49, 7.51, 7.52 and 7.56) have shown that the radiation of the particles is possible only for sufficiently small values of the transresonant parameter $\overline{R}$. They might be formed only far enough from the exact resonances. On the other hand, we get almost harmonically oscillating elements of energy if $\overline{R} \cong 0$ (Fig. 7.54). We can assume that these elements correspond to the dark energy. Thus, this energy is a remnant of the initial energy of the fragments (elements). The remnant was stored after the transresonant transition. The very energetic particles (Figs. 7.48, 7.49, 7.51, 7.52 and 7.56) correspond to the dark matter. The less energetic particles after the collapse into more elementary elements (the normal matter) could form our visible universe. According to the presented model, the transresonant process has determined the formation and contents of the matters and energies in the Universe. During this process the pure energy of the Universe was reducing, since the matter was appearing. According to this model the origin of the Universe might consist from four stages:

1. The evolution up to the resonant band;
2. The dark matter began to form when $\overline{R} = -\lambda^{-1}m^2$ crosses the boundary of the resonant band;
3. The normal matter began to form when $\overline{R}$ increases enough;
4. When $\overline{R} \approx 0$ and $\overline{R} < 0$, we have $m^2 \approx 0$. The initial energy stops the transforming into mass particles.

7.11 Supporting Experimental Results: Gravity Waves

In the previous section we presented many examples of the oscillations of a scalar field. Do they describe the behaviour and evolution of the primordial fundamental field? We do not have experimental evidence supporting or denying this statement.

However, it is possible to expect that the above results can be used in the case of other wave processes, since the wave processes are described by similar equations. Thus the method developed in the last section and the results can be extended to other highly-nonlinear wave processes which take place in other physical situations. There are certainly mathematical analogies in the formulas describing the wave processes of different physical nature.

Folded shapes of the waves presented in the Sect. 7.10 are complex. However, similar evolution was observed on the surfaces of liquids. Thus, the scalar field waves can resemble the waves in water. Let us consider this analogy at this section.

7.11.1 Resonant Particles, Drops, Jets, Surface Craters and Bubbles

Parametrically (vertically) forced surface waves, known as Faraday waves, were studied in containers, and examples of these waves are presented in Figs. 5.20, 5.21, 7.57 and 7.58 (see Sect. 5.2.3). Figure 7.57 shows the formation of jets of a liquid during the collapse of the surface depression or cavity. The calculated curves show the evolution of a resonant wave in the scalar field.

Other examples of these waves are presented in Fig. 7.58, in which the photomontages show the formation of jets of a liquid during collapse of the surface craters.

Fig. 7.57 Photographs showing the axisymmetric wave depressions (surface cavities) and jets. Results of calculations of a resonant wave in the scalar field [84]. The figures near the calculated curves determine dimensionless time

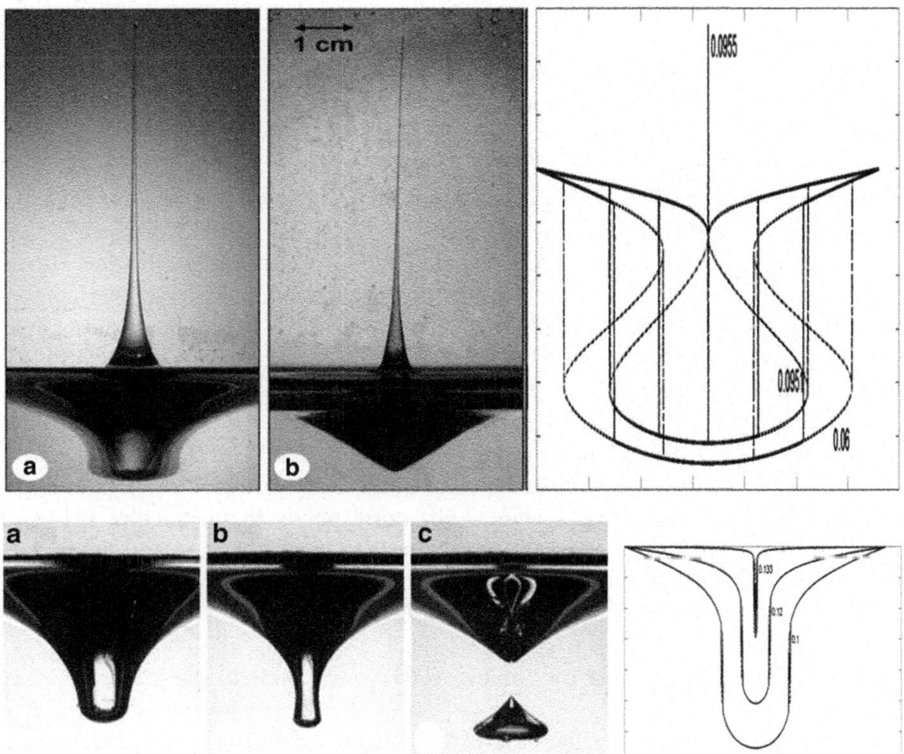

Fig. 7.58 Photomontages of the surface collapsing craters, a bubble near the liquid surface and surface jets. The calculated curves illustrate the photomontages [84]. Results of calculations of resonant waves in the scalar field. The figures near the calculated curves determine dimensionless time

The photos 7.58 demonstrate various forms of the craters arising on the liquid surface (a–c) and a bubble (c). The calculated curves are found for different dimensionless time. The curves 7.57 and 7.58 show the evolution of resonant waves in the scalar field. The general agreement of experiments and calculations is apparent.

7.11.2 Strongly-Nonlinear Faraday Waves, Interface Instability and Evolution of Waves in Vortices

Here we refer to a more conventional example of gravity surface waves to support the above findings for scalar resonant waves.

Let us compare waves of Figs. 7.52, 7.53 and 7.54 with experimental waves presented in Fig. 7.59. These waves were excited in a vertically oscillating container. Its length was 0.6 m, and the width 0.06 m. The depth of water was 0.3 m. All wave profiles (Fig. 7.59) were observed during one experiment. The forcing frequency was 1.6 Hz and the amplitude was 4.6 mm.

Very wide spectra of highly nonlinear waves are presented in Fig. 7.59. It is clear that waves of Fig. 7.59 are described qualitatively by our theory. In particular, Fig. 7.52 corresponds to Figs. 7.59a, 7.53 corresponds to Figs. 7.59b and 7.54 corresponds to Fig. 7.59c. However, there are some discrepancies between the observations and the calculations.

In our calculations we assumed that the shift of the linear oscillations (linear component of the solution) relative to the nonlinear oscillations (nonlinear component of the solution) was $-\pi/2$ (7.76). Generally speaking, this value is not known. The shift can be anything from 0 to $-\pi$. Certain discrepancies between the observations and the calculations can be explained by the variance in the value of the shift.

Another example of surface waves was presented in a dissertation of Kalinishenko [99]. His results are presented in Figs. 7.60 and 7.61 (upper rows). Results of our calculations are shown there in the bottom rows.

On the whole, the wave a (Fig. 7.60) corresponds to the picture 0.173 of Fig. 7.52. The wave b corresponds to picture 0.1875 of Fig. 7.51 and the picture 0.43 of Fig. 7.53. The wave c corresponds to the picture 0.665 and 0.69 of Fig. 7.54. The down row of Fig. 7.60 describes additionally only some particularities of waves a, b and c.

On the whole, the upper row of Fig. 7.61 corresponds to the picture 0.17 of Fig. 7.51.

And finally, we consider the case of the Richtmyer–Meshkov instability of an interface of a two-liquid system [92]. The results of instability are presented in Figs. 7.62 and 7.63 (upper rows). The results of our calculations are shown there in the bottom rows.

Results of the Richtmyer–Meshkov instability are difficult to describe analytically. However, the calculations shown in Figs. 7.62 and 7.63 allow to image the streamlines which have formed the vortices. These streamlines are shown in Fig. 7.64.

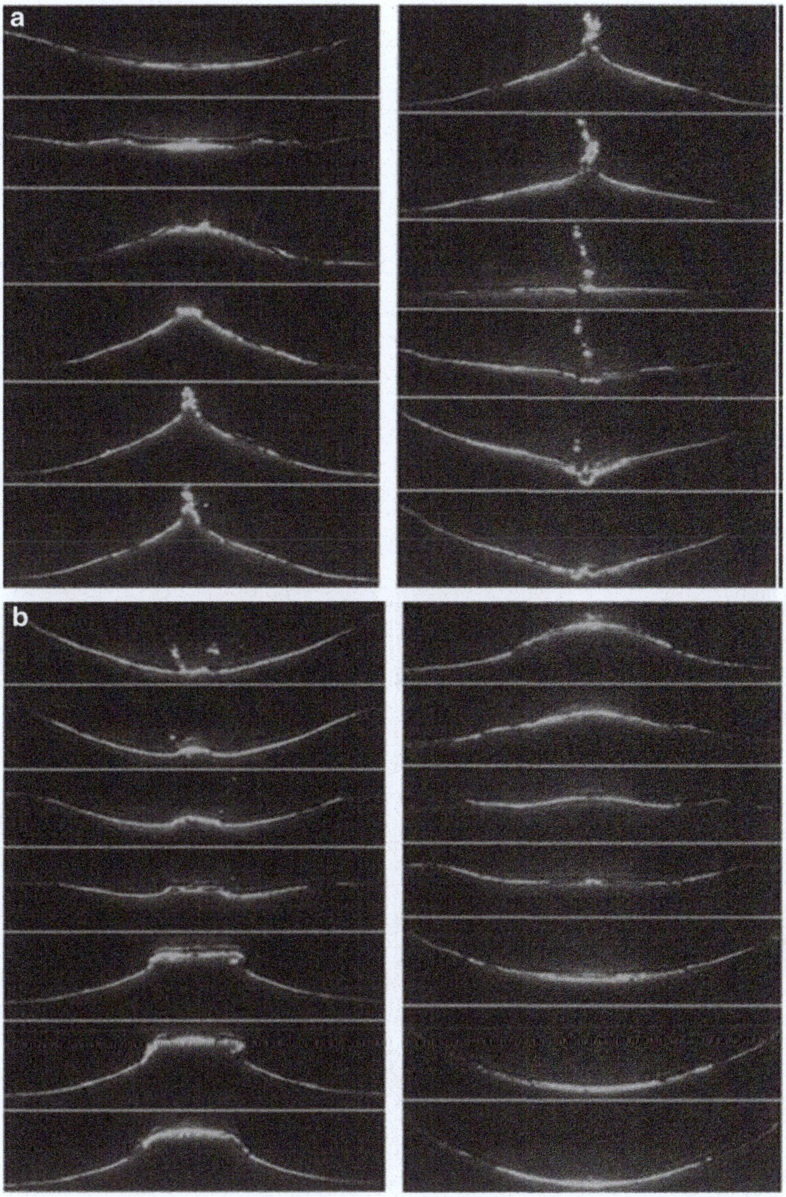

Fig. 7.59 (**a**) Gravity waves observed at the moments of time: 0, 0.04, 0.08, 0.12, 0.16, 0.2 (*left*), 0.24, 0.28, 0.32, 0.36, 0.4, 0.44 s (*right*). Similar waves may be in the scalar fields. (**b**) Gravity waves observed at the moments of time: 0.52, 0.56, 0.6, 0.64,0.68, 0.72, 0.76 (*left*), 0.8, 0.84, 0.88, 0.92, 0.96, 1, 1.04 s (*right*). Similar waves may be in the scalar fields. (**c**) Gravity waves observed at the moments of time: 1.16, 1.2, 1.24, 1.28,1.32, 1.36 (*left*), 1.46, 1.5, 1.54, 1.58, 1.62, 1.66 s (*right*). Similar waves may be in the scalar fields far enough from resonances. The transresonant nonlinear evolution of surface water waves accompanied by the strong change of the wave forms and by the eruption of the particles from wave tops (see **a**) [97]

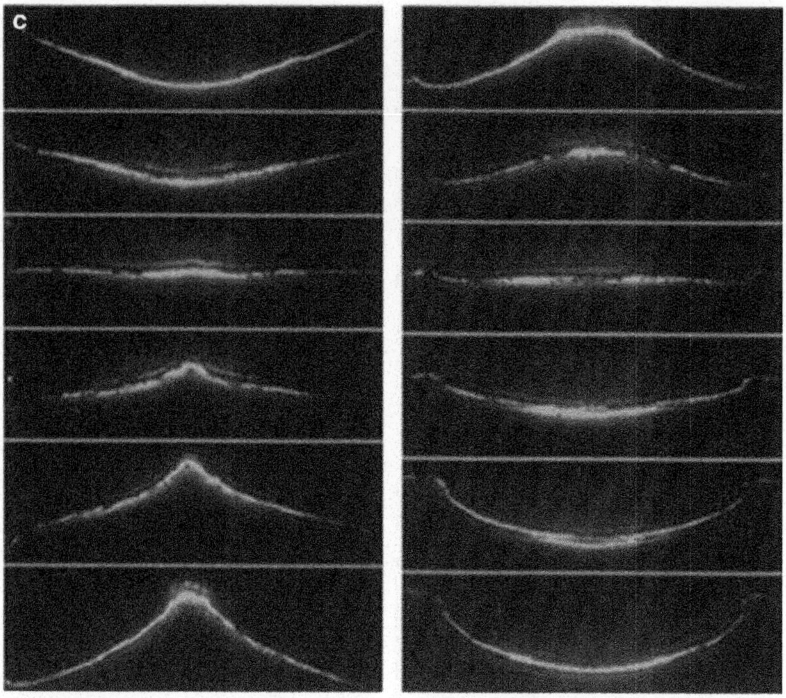

Fig. 7.59 (continued)

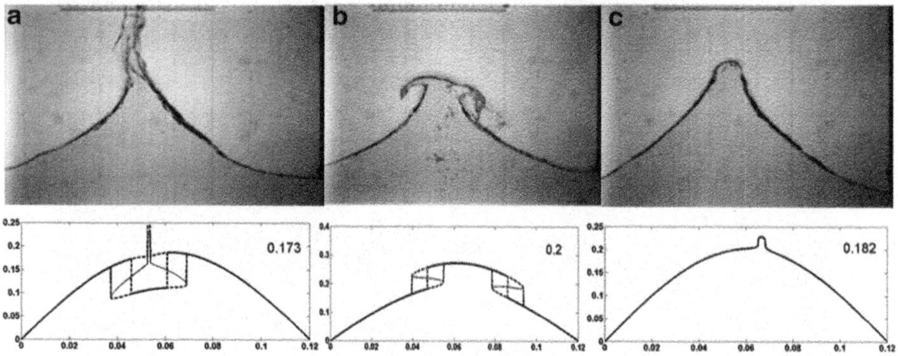

Fig. 7.60 The comparison of experimental data [99] with calculations [15, 84]

It is seen (Figs. 7.62, 7.63 and 7.64) that the calculations describe qualitatively the appearance of vortices because of the Richtmyer–Meshkov instability. We stress that R in Figs. 7.62, 7.63 and 7.64 corresponds to the transresonant parameter $\overline{R}$.

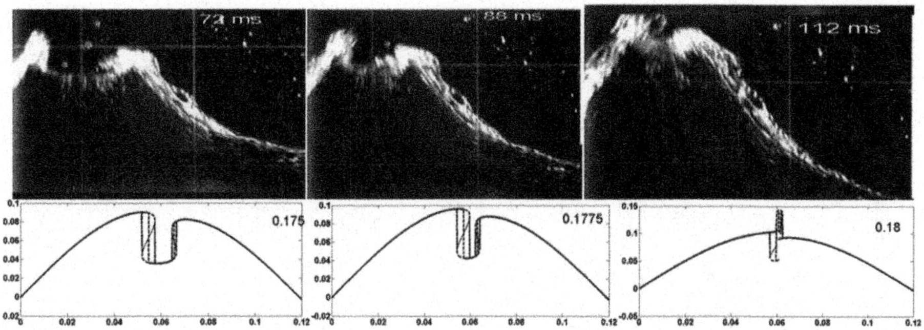

Fig. 7.61 The comparison of experimental data [99] with our calculations [15, 84]

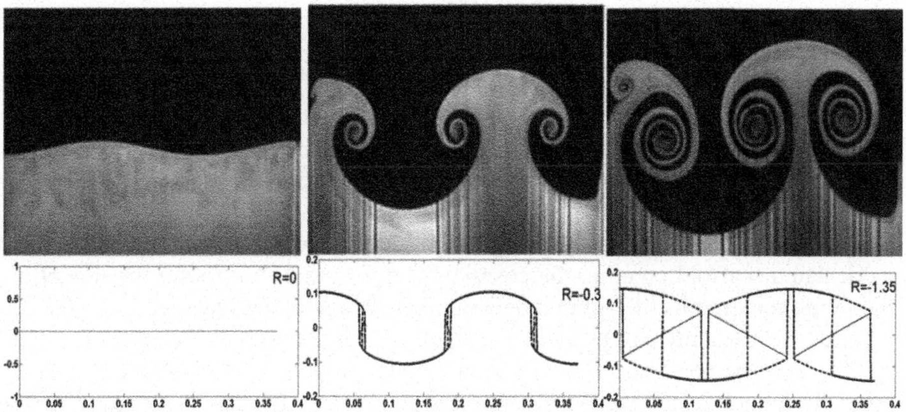

Fig. 7.62 The comparison of experimental data [92] with calculations. Here R is some transresonat parameter

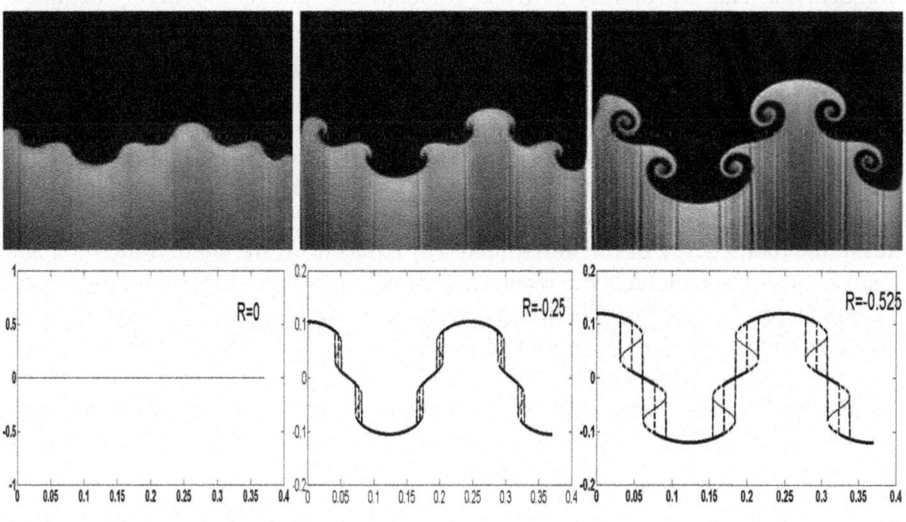

Fig. 7.63 The comparison of experimental data [92] with calculations

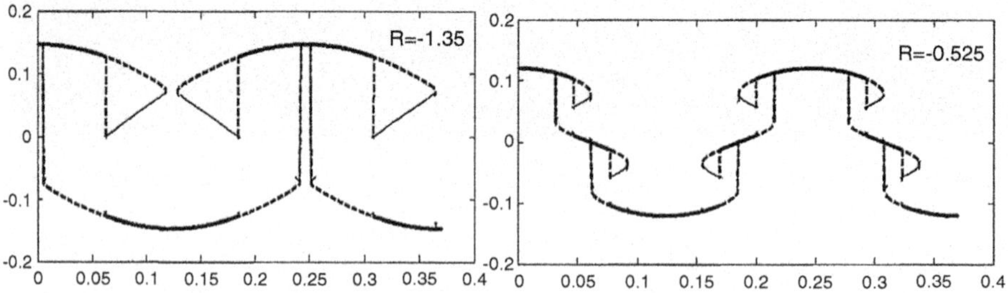

Fig. 7.64 The vortex scheme of certain experimental results [92], which follows from results of calculations (see Figs. 7.62 and 7.63 *right*)

7.12 The Origin of the Universe

The evolution scenario of the scalar field subjected to a quantum fluctuation presented above can be viewed as describing the origin of the Universe. We think that the spacetime of the bubble might undergo transformations in which its fabric was fragmented into the string-like vibrating elements. Then the elements began to emit different kinds of highly-energetic particles. Some of them could begin to work like a scaffold creating the cosmic structure. As a result after the tunnelling the fabric repaired itself within a new number of spatial dimensions.

Thus, the dynamic part of the scalar field – the bubble (clot) of energy (Fig. 7.27 right) – may be considered as the 'seed' of the Universe. The Universe appeared after the tunnelling of the bubble through the energy barrier. As a result of the quantum action the bubble obtained huge energy. This energy caused the bubble to expand. During this process the initial field was being fragmented into almost infinite quantity of highly-energetic one-dimensional elements which oscillated at resonant frequencies. The waveforms of these resonant oscillations are highly-nonlinear. These waveforms emit particles of energy and matter. The individual elements noted above are only one-dimensional 'shards' of space and time. The three-dimensional space and time emerged only after the appearance of enough matter forming this spacetime.

The origin of the Universe was not instantaneous. The Universe had enough time to become homogeneous and isotropic. After the tunnelling the one-dimensional vibrations of the elements continued to generate particles of energy and matter. As a result the total energy of the initial field was reduced. At the same time the spatial dimensions of the bubble were increasing. Finally the three-dimensional space and time emerged [14, 15, 84]. The Universe had appeared where high-energy particles, the dark matter and dark energy were floating in the four-dimensional spacetime.

The spacetime appeared as some sort of large-scale construction of more fundamental elements. The spacetime properties would 'arise' from the underlying physics of its constituents, just as water's properties emerge from the particles that comprise it [100, 101]. The construction (condensate, primordial medium) had properties that could not be found in the individual elements. In particular, waves began to form and propagate in the primordial medium.

According to the Eq. (7.57) the Universe was born due to the variation of the parameter $\overline{R}$ inside of some resonant band (transresonant process). The dark mater was born when the value of $\overline{R}$ was minimal within the resonant band. The normal energy emerged when $\overline{R}$ began to increase. The dark energy appears when $\overline{R}$ is increased up to near zero. Roughly speaking, the dark energy is no used part of the initial energy.

After the tunnelling the volume of the Universe increased so much that quantum effects became unimportant for its global evolution but the role of matter and gravitation became extremely important. After this moment the evolution may be studied using models of the early Universe [52, 53, 85–89]. The subsequent global evolution of the Universe was probably determined by the parameters of the scalar fields and the matter, in particular, by their heterogeneities which were formed during the tunnelling. The dynamic part of the scalar field which was originally approximately uniform, gained some weak heterogeneity during the tunnelling through the potential barrier. Similar heterogeneity could also arise due to quantum effects in the course of the origin of the Universe. Due to these heterogeneities waves and vortices began to form and propagate within the very early Universe.

7.13 The Evolution of the Universe After the Tunnelling

The appearance of time and the three-dimensional space did not mean that the expansion of the Universe stopped. It continued to possess a huge amount of energy. High-energy elements vibrating with resonant frequencies continued to generate particles of matter. Of course, the energy of those vibrations kept reducing. Therefore the particles appeared with less and less energy. On the other hand the high-energy particles that appeared earlier were breaking up into smaller-energy particles. As a result the Universe was being filled by the particles more and more familiar to us. The temperature and the primordial matter of the very early Universe appeared. Evolution and interaction of different particles of this matter were described by different scalar fields. The particles and new scalar fields began to interact.

Let us consider a new dynamic field χ, which interacts with a basic scalar field φ but which has no mass term. In this case according to the quantum field theory an effective potential (7.78) depends on the functions φ and χ [52]:

$$V_{eff}(\varphi) = V_0 + \frac{1}{2}m^2\varphi^2 - \lambda'\varphi^2\chi^2 - \frac{1}{4}\lambda\varphi^4 + C^*\varphi. \tag{7.78}$$

Here $C*$ is constant. The scalar field determining the temperature is considered. We can write

$$\lambda'\chi^2 = g^2T^2, \tag{7.79}$$

if the scalar field is in a thermal equilibrium. Taking (7.79) into account we rewrite (7.78) in the form

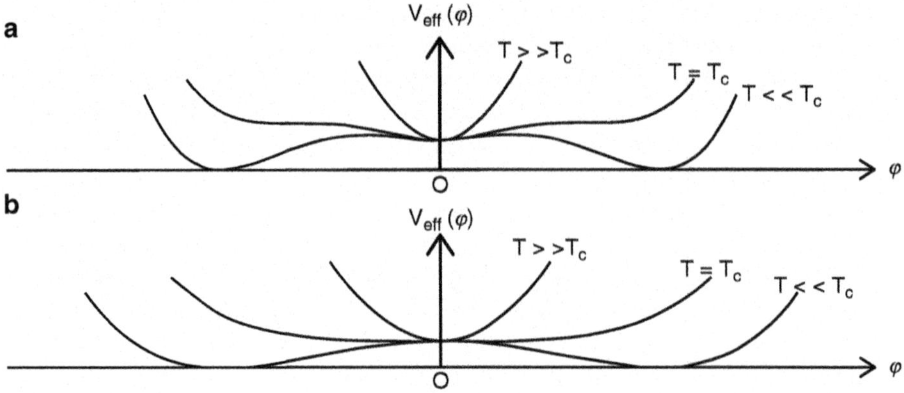

Fig. 7.65 Various representations of (7.78) [90]. The different effective potential of the field are presented. (**a**) When the temperature is high, the effective potential has a single minimum at $\varphi = 0$. As the temperature decreases, two other additional minima develop. (**b**) When the temperature is high, the effective potential has a single minimum at $\varphi = 0$. As the temperature decreases, two other minima develop

$$\varphi_{tt} - c_*^2 \nabla^2 \varphi = -\left(m^2 - 2g^2T^2\right)\varphi + \lambda\varphi^3 - C^*. \tag{7.80}$$

Thus, near

$$m^2 = 2g^2T^2, \tag{7.81}$$

the Eq. (7.80) can instantly change its solution. The condition (7.81) corresponds to the phase transitions in a physical field. In other words, if $m^2 = 2g^2T^2$ takes place, the scalar field can bifurcate into new scalar fields (Fig. 7.65).

The matter formed a continuum having density and pressure. The volume of the matter continued to increase, therefore the temperature and the density began to diminish. As a result, the system crossed different critical states when it changed very strongly. The rough scheme of noted transitions is shown in Fig. 7.66.

Similar transitions may trigger the generation of particles, waves or vortices in the field since (7.81) determines a certain resonant situation. Near this situation the Eq. (7.80) may approximately describe the evolution of smooth waves into composite waves. The evolutions of harmonic waves into mushroom-like waves and vortices were studied in [8–13]. It was found that these evolutions were determined by cubic nonlinearity and resonances.

Examples of the evolution of the unidirectional scalar waves calculated for different harmonic perturbations, different shifts in (7.76) and C^* are presented in Figs. 7.67, 7.68, 7.69 and 7.70.

On the whole the evolutions resemble Figs. 5.22 and 5.33 and pictures shown in the Sect. 7.9.

2. On the other hand the primordial matter of the very early Universe may be modelled as a inviscid liquid (gas) [101–103]. Let us assume that the liquid is in

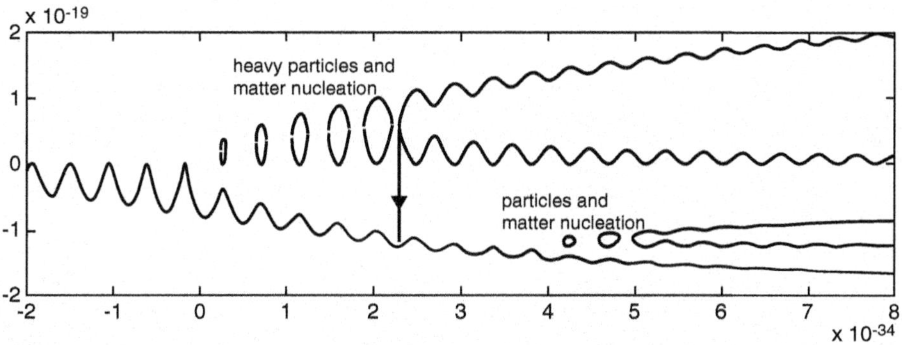

Fig. 7.66 Rough schemes of the birth of particles and the matter during the tunnelling and a cascade of transitions (bifurcations) which could form our Universe [14, 15, 84]

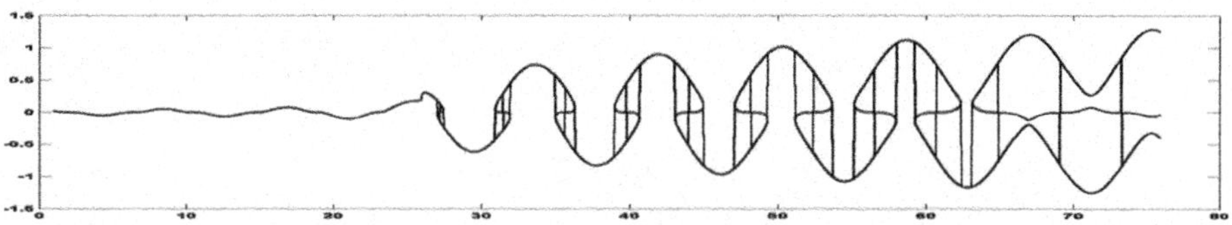

Fig. 7.67 Unidirectional waves crossing the resonant band. The evolution of small perturbations

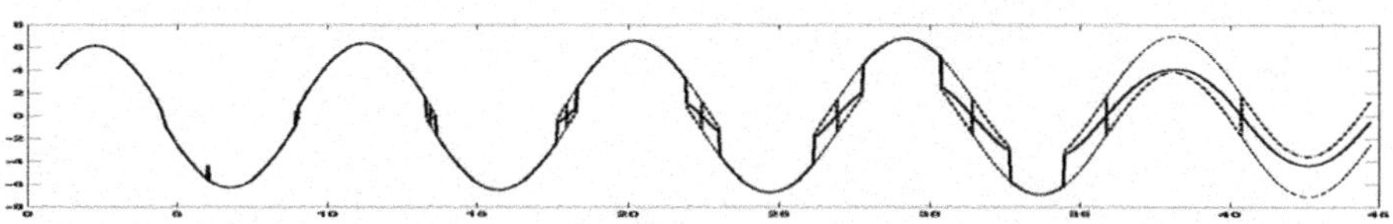

Fig. 7.68 Unidirectional waves crossing the resonant band. A process of an appearance of strongly nonlinear waves and folds of wave profile

thermal and radiation equilibrium. In this case, one-dimensional waves in this continuum may be very roughly described by the Eq. (7.2). Solutions of this equation were considered in Sects. 7.1, 7.2, 7.3, 7.4, 7.5 and 7.6. These solutions described the formation of shock waves, solitons and breaking waves. On the other hand quantum vortex lattices were observed in superfluids [104]. Similar wave structures and vortices could appear in the very early Universe too [8, 104–106].

The profiles of Figs. 7.67, 7.68, 7.69 and 7.70 contain closed loops. They may be interpreted as vortices on the wave surface. Perhaps, the loops shown in these figures could form the seeds of the galaxies and its clusters in the early Universe. Indeed, in March 2014 curly, vortex-like patterns were observed in CMB with the help of the telescope at the South Pole (Fig. 7.71).

Can we consider the patterns in Fig. 7.71 as the 'footprints' of the vortices (Figs. 7.67–7.70) which were imprinted on the sky a lot of time ago? Of course, this is

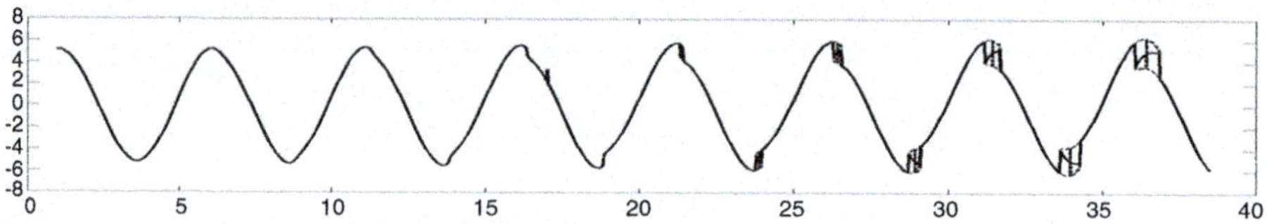

Fig. 7.69 Unidirectional waves crossing the resonant band. A process of the evolution of waves into vortices [14, 15, 84]

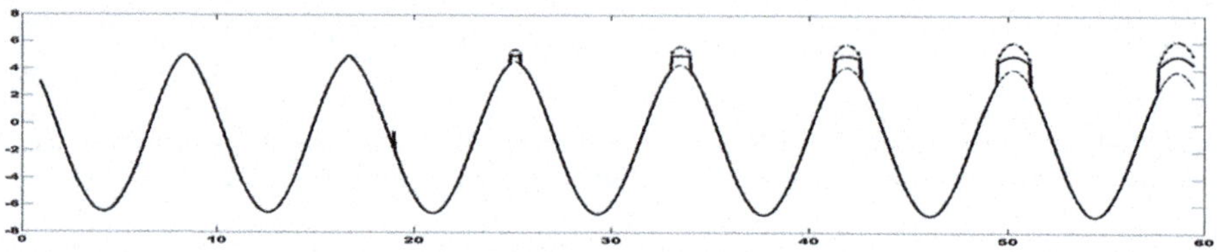

Fig. 7.70 Unidirectional waves crossing the resonant band. A process of the radiation of particle by the wave

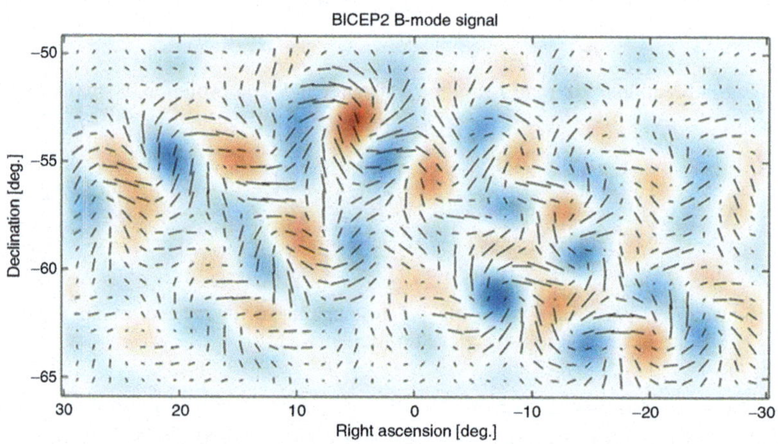

Fig. 7.71 Pinwheel-like swirls in a map of the Background Imaging of Cosmic Extragalactic Polarization [59]

only a suggestion. Modelling the birth of the Universe is a bit like playing whack-a-mole. All time the new results knock out the old theories.

Thus, according to our theory the initial harmonic wave can be distorted very strongly by nonlinearities near resonances. The wave profiles began to contain the loops which can contain a huge volume of matter and which may be interpreted as vortices. We think that during the subsequent evolution of the scalar field these volumes may be transformed into galaxies and clusters of galaxies.

7.14 Supporting Experimental Results: Interface Instability and Waves Evolving into Vortices

The generation of vortices in different physical fields as a result of the instability is a known effect. Apparently during instability both waves and vortices may appear in scalar fields. For example, in a Bose-Einstein condensate vortices were created due to the snake instability (Fig. 7.72).

On the whole, the process shown in Fig. 7.72 corresponds qualitatively to the wave evolution during the Richtmyer–Meshkov instability (see Figs. 7.42, 7.62 and 7.63).

Thus we have shown that the scalar field can form a wide spectrum of nonlinear waves which may evolve into vortices. In particular, this evolution (Fig. 7.67) is similar to the evolution presented in Fig. 7.73.

The evolution of the perturbations of interface between two environments into vortices was studied in [108]. Figure 7.73 was taken from [108]. There theoretical studies of the nonequilibrium Ising-Bloch bifurcation have utilized the FitzHugh-Nagumo reaction–diffusion model and a variant of the complex Ginzburg-Landau equation that describes amplitude modulations of forced oscillations. We can consider the perturbations of the interface as a heterogeneities in the very early Universe. They are shown by Fig. 7.73 (a, d). The initial interface (a, d) changes as a result of growth of the perturbations. Figure 7.73 (e, f) shows the positions of vortices (the spots in the figure) at the different moments of time. Thus, Fig. 7.73

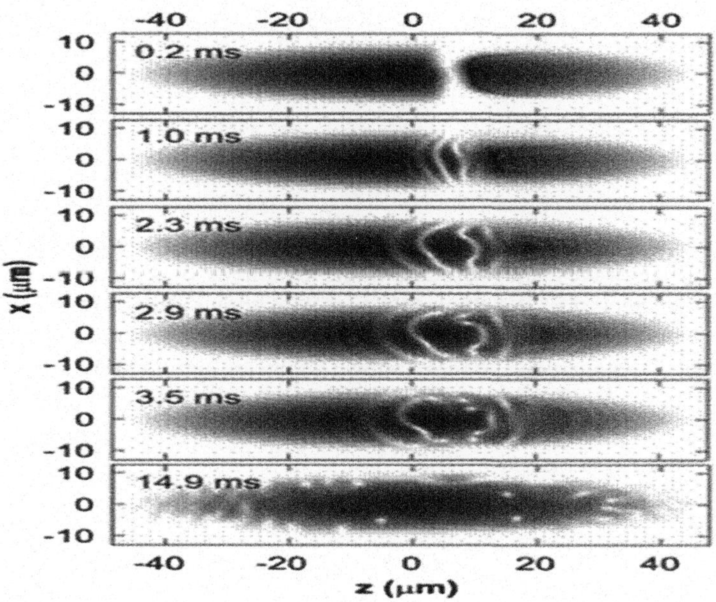

Fig. 7.72 Spontaneous formation of vortices through the snake instability [107]. The initial perturbation (0.2 ms) evolves into many vortices (14.9 ms)

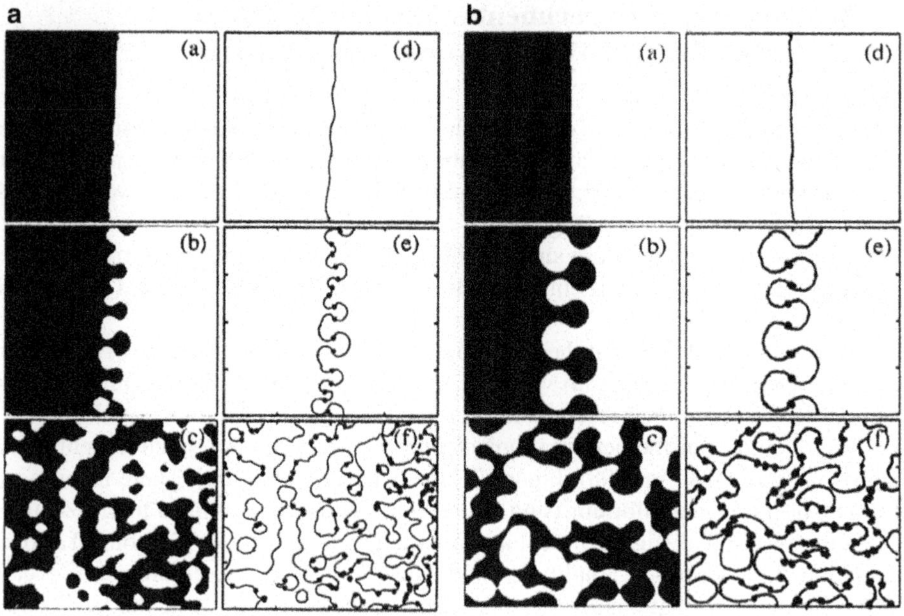

Fig. 7.73 Evolution of the initial perturbations of some interface into the elasica forms (mushroom-like waves), vortices and turbulence. The occurrence of the vortices shown by spots during the Belousov -Zhabotinsky reaction (**a**). The occurrence of the vortices and turbulence according to the numerical solution of the Ginzburg-Landau equation (**b**) [108]

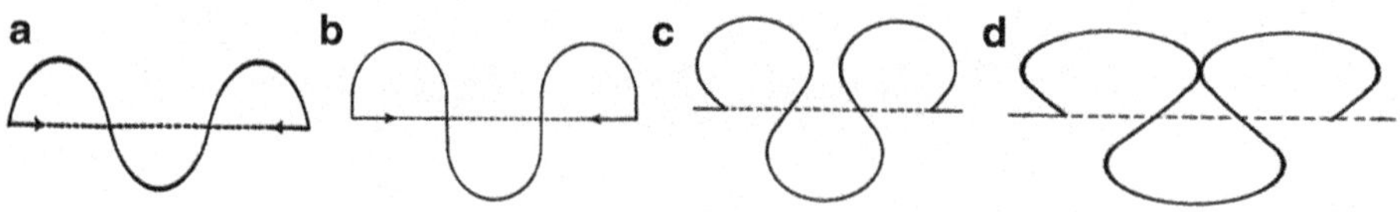

Fig. 7.74 The evolution of the elastica forms as a function of the compression parameter [109]

(d–f) demonstrates the evolution of the originally very smooth heterogeneity in mushroom-like waves and vortices. Some examples of similar evolution were shown in the Sect. 5.2. We think that Fig. 7.73 shows certain rough schemes of global evolution very early Universe and the formation of galaxies.

Let us note that many wave forms shown above are similar to the forms shown in Figs. 7.73 and 7.74. They can form during longitudinal compression of thin rods.

We stressed earlier that the elastica forms can be formed in many highly-nonlinear wave systems.

7.15 Resume of the Sects. 7.7, 7.8, 7.9, 7.10, 7.11, 7.12, 7.13 and 7.14

We considered a version of the nonlinear Klein-Gordon equation and the approximate solutions, which describe the evolution of the Universe as an evolution of the clots (the bubbles) of energy (Fig. 7.75). Almost all energy of these bubbles is concentrated on their surface. Inside the bubble the energy is almost uniform. However, there are traces of non-uniformity. Such traces could appear in the course of the tunnelling of the energy bubble through the energy barrier (Fig. 7.44) which resulted from a quantum fluctuation.

During the tunnelling the bubble is being filled by matter. Some spacetime elements begin to interact forming new four-dimensional space-time. As a result our Universe appears. Its size is much greater than the Planck size.

On the whole, the central part of Fig. 7.75 resembles a Mexican hat-shaped energy landscape (see Fig. 7.76). The ball sitting on the top of the sombrero looks symmetrical. But it is unstable. The slightest nudge and the ball falls to a lower energy state in the 'valley' of the hat. Then the ball begins to oscillate between the valley walls – creating the particles. At the same time, our theory is quite different from the scheme of Fig. 7.76. In particular, the oscillations of the clot energy in the crater of the hill are stable (Figs. 7.27 and 7.28). The stability is lost only after the action of quantum perturbations. The tunnelling and the fall can be faster in one direction (Figs. 7.44 and 7.75). Thus, they can depend on the direction of the motion. Nonsymmetrical fields could be generated during the fall, while they could not appear according to Fig. 7.76. It could explain why the Universe has the 'axis of evil' (Fig. 7.45). The particles appear during the tunnelling and the falling. According to the model, it is possible, that the Universe had not reached the bottom of the valley. It continues to fall.

The tunnelling process leads to the transformation of the multi-dimensional spacetime into our familiar three-dimension space and time. The moment this process finishes is the moment of the origin of our Universe. The Universe is then

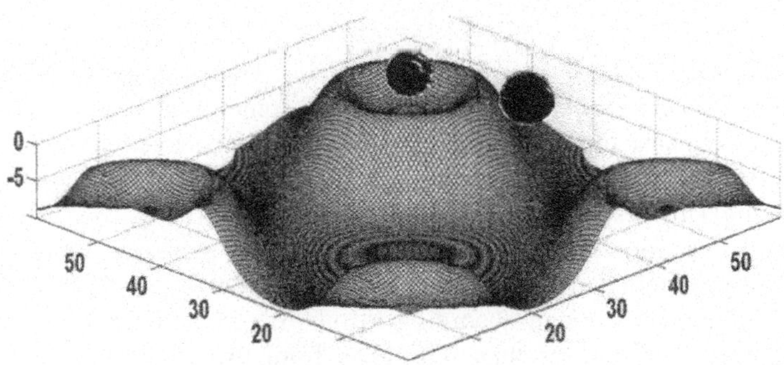

Fig. 7.75 The energy clot tunnels through the potential wall. As a result the clot of energy rapidly changes. At the same time the new three-dimensional space and the directed arrow of time (the unidirectional time) are being formed

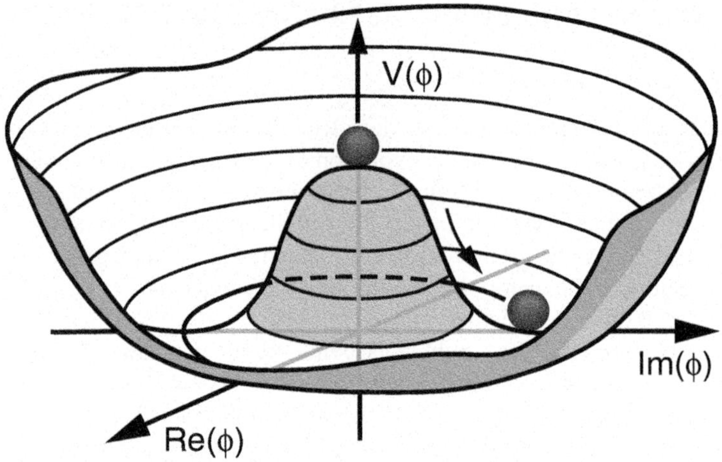

Fig. 7.76 The rolling ball represents the sliding value of the energy density of the Higgs fields. Internet. http://sites.uci.edu/energyobserver/2013/02/24/why-create-misleading-pictures-of-how-the-higgs-provides-mass/

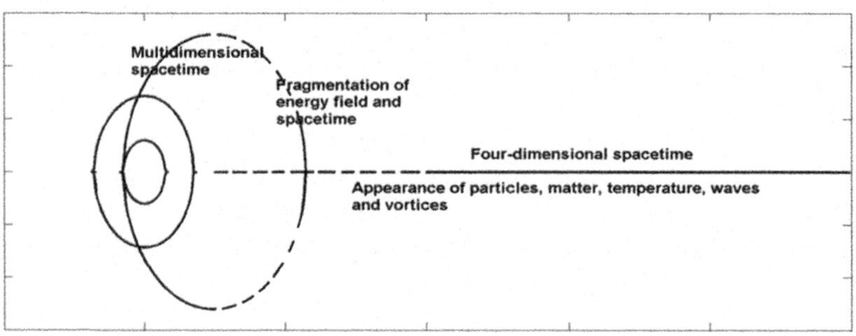

Fig. 7.77 Schemes demonstrating the evolution of the initial bubble of energy and the appearance of particles, matter, waves and vortices. The appearance of the matter forms the four-dimension spacetime

filled by different kinds of energies and particles. High-energy particles float in four-dimensional spacetime. Generally speaking, some of these particles can have multidimensional structure inherited from the multidimensional spacetime.

The production of the particles continues after the origin of the Universe. Inside the expanding Universe, the heavy highly-energetic particles decay into lighter particles and radiation. The Universe size continued to increase. According to the well-known terminology the described fast expansion of the Universe may be called the inflation. Although according to our representations the speed of the expansion was much slower than it follows from the known inflation models. According to our model the Universe could form rather large, almost evenly-filled by the particles of energy and matter. Therefore the special stage of the superfast expansion is not required for our model (Fig. 7.77).

The speed at which these particles are produced reduces all the time. However, it is possible that particles of matter form in the depths of cosmos even now. Certainly, the volume of their production is not comparable to the volumes in the first moments of the existence of the Universe. This production may support and can even accelerate the continuing expansion of the Universe.

Thus, we explain qualitatively the emergence of the Universe and describe its initial evolution. It is assumed that the tunnelling was a process at which the multidimensional spacetime collapsed and the newly-formed spacetime elements began to vibrate with resonant frequencies creating energy and matter particles. Then the waves and the vortices were formed (see Figs. 7.67, 7.68, 7.69 and 7.70).

Our Universe appeared with its four-dimensional spacetime. The Universe size was much greater than the Planck size and it was substantially evenly filled with highly-energetic particles of matter.

Resume of the research

1. The pre-Universe exists within a multidimensional spacetime. This pra-Universe is described by a scalar field. The field is roiled by the tiny quantum fluctuations;
2. At any moment the pra-Universe gives birth to billions of 'seeds' of rapidly evolving Universes, one of which accidentally evolved into our Universe;
3. The Universe sprang into existence due to quantum fluctuations;
4. During the earliest stage of the evolution the multidimensional scalar field was fragmented into vibrating elements having very high energy. The fragments are modelled as one-dimensional strings;
5. Highly-nonlinear oscillations of those elements emitted very heavy particles of mass and energy which formed the four-dimensional spacetime;
6. Our Universe appeared with huge energy, mass and finite size;
7. The spacetime began to stretch very rapidly as more and more particles appeared and the heavy particles began breaking up into lighter particles and the energy continued to transform in mass. It was the Universe's rapid growth spurt;
8. Waves and vortices began to form in matter.

According to the developed model the birth of the Universe was the pure nonlinear process. In particular, the dark mass emerged as the very stable clots of the initial energy which were generated due to the nonlinearity. The normal mass was formed by very small parts of the instable clots of the initial energy which were generated due to the nonlinearity. The dark energy is some remnant of the initial energy remaining after radiation of high-energetic mass particles. Figuratively speaking, the dark energy is coals and ashes from a bonfire which was recently burnt down. The normal and dark matters are sparks and flames of the long-vanished bonfire.

7.16 Discussion and Comments

In this Chapter we developed certain ideas which had been collected in [8]:

'The amplitude and the form of the resonant waves is defined by nonlinear, dispersive, dissipative and spatial effects. In the trans-resonant frequency band the balance between these effects varies along with the frequency. Therefore within the band unexpected phenomena can be generated. For example, the trans-resonant transformation of harmonic smooth waves into shock waves were observed in tubes. In mechanical systems, only the first resonances can typically be observed, because of the damping and the narrowing of the resonant band with the increase of the resonance order. However, in electronic, optic, crystal and quantum systems the damping may be very small and higher order resonances can occur. Electrons and atoms can also form resonators. Interesting elliptic standing wave patterns were recently observed formed by a single atom and a 2D free electron gas. It was found that the Bose-Einstein condensate can have unprecedentedly large nonlinearity. As a result, a resonantly tuned light pulse travels at a velocity of only 17 m/s and is strictly compressed in the condensate. This effect might explain various anomalous observations, in particular, in space. Indeed, the varying-speed-of-light theory of the cosmos was developed during recent years. The acoustical, resonant and nonlinear effects may be important for cosmology. It has been discovered that the background radiation has large-size peaks (a fundamental mode) and smaller peaks ('overtones'). The early Universe rang like a spherical resonator after the big bang. Thus, recent observations show that the acoustic model of the early Universe is correct. The rapid expansion of the Universe produced nonlinear pressure and density waves, since the matter of the early Universe was highly nonlinear and very dense.

One of the goals of this paper is to study highly nonlinear wave phenomena in various dissipative-dispersive resonant systems. These systems can be surface layers, microresonators, the early Universe, etc. We develop the theory of trans-resonant wave phenomena in these systems. The perturbed wave equations are studied. It is known that the physical processes of generation and transformation of waves can differ dramatically, nevertheless equations and analytical solutions which describe these processes are often similar. For example, shock-, soliton- and cnoidal-like solutions are well known in nonlinear dynamics. In particular, it has been recently found that similar solutions describe waves in spherical resonators and different anomalous wave phenomena, and wave patterns observed in water and granular layers. It has been shown that the perturbed Maxwell wave equation and the perturbed Klein-Gordon field equation (φ^4 field) have similar solutions.'

We considered the origin and the evolution of the Universe using ideas noted above. We assumed that the origin of the Universe may be described by the well-known scalar nonlinear equation. The approximate solutions of this equation describe the strict sequence of stages of the evolution of the very early Universe. It is seemed that every stage has some analogue in the modern theories of cosmology. We only tried to construct the continuous picture of the origin. We understand that some our results may be considered as crazy. Therefore we tried to illustrate them, when it was possible, by experimental data and results of different authors.

The great tragedy of science The standard model of the universe is based on an oversimplification of Einstein's equations of general relativity. There are two big assumptions. First, that the universe is homogeneous, looking roughly the same in all locations. Second, it is isotropic, looking roughly the same in all directions from any standpoint. These assumptions led to the Big Bang model and the cosmic inflation theory. The Einstein's equations become very complex and the Big Bang and the inflation become slippery concepts if these assumptions are not valid.

It is found today that galaxies exist in clusters and filaments of matter distributed around the boundaries of huge, bubble-shaped voids. These voids have roughly one-tenth of the clusters' matter density, but account for more than 60 % of the universe's volume. 'Everyone knows that the universe is inhomogeneous,' says Thomas Buchert of the University of Lyon in France. 'To idealize such a complex structure with a homogeneous solution is a bold idealization'. The problem is, we don't have a bird's-eye view of the universe on such scales. David Wiltshire, cosmologist at the University of Canterbury in New Zealand, thinks that what is going on right in front of our eyes might be distorting our view [110].

The idea of the cosmic primeval fireball was born in the mind of George Gamow [111–115]. This idea is fantastically successful. On that basis Gamow and his co-authors presented brilliant predictions about chemical element abundances when they have come out of the cosmic fireball and about the cosmic microwave background (CMB). This idea was very successful since Gamow's main interest was in the cosmic history after the Big Bang (approximately starting with 1 s after the Big Bang). For example, the hot Big Bang theory predicted that no galaxy in the Universe contains less than 23 % of helium. This amount was confirmed by observations. But these and other successes had not shed light on the mystery which is the origin of the Universe. Indeed, according to the hot Big Bang theory we have the singularity in the beginning where the laws of physics are not valid. Therefore the Gamow model does not work very close to the beginning.

The inflation theory is trying to correct this situation.

Observations tell us that CMB is remarkably isotropic. However, there is a tiny bit of anisotropy in CMB (see Fig. 7.45). Planck's map (Fig. 7.45) greatly improves cosmologists' understanding of the universe, but it does not solve lingering mysteries over unusual patterns in the CMB. These include a 'preferred' direction in the way the temperature of light varies, dubbed the cosmic 'axis of evil', as well as an inexplicably cold spot that could be evidence for universes beyond our own. Perhaps, it is not very surprising. The last data give the theorists something from which to build mathematical models about contents of the early Universe. Some physicists wonder whether the axis of evil requires a rethink of our ideas about the

Universe. They include Joao Magueijo at Imperial College in London, who coined the term 'the axis of evil'. According to Fig. 7.45, there may be something seriously wrong with Big Bang and inflation models.

'The great tragedy of science,' as Thomas Huxley observed, is 'the slaying of a beautiful hypothesis by an ugly fact'. He was talking about the origins of life, but it may be true for the starting point of the Big Bang model and the cosmic inflation theory.

Some notes about the inflation models: pros and cons This explains high interest in the BICEP2 data.

Using a radio telescope at the South Pole, the US-led team has detected the first evidence of primordial gravity waves (Fig. 7.71). It may be considered as a long awaited confirmation that the origin of the Universe was accompanied by a very fast expansion. Cosmologists knew that inflation would have a distinctive signature: the brief but violent period of expansion would have generated gravitational waves, which compress space in one direction while stretching it along another. Although the primordial waves would still be propagating across the Universe, they would now be too feeble to detect directly. But they would have left a distinctive mark in the CMB: they would have polarized the radiation in a curly, vortex-like pattern known as the B mode.

The results were declared during March 2014. In particular, the journal Science [56] published an exciting paper 'First wrinkles in spacetime confirm cosmic inflation'. 'This is a totally new, independent piece of cosmological evidence that the inflationary picture fits together,' says theoretical physicist Alan Guth of the Massachusetts Institute of Technology (MIT) in Cambridge, who proposed the idea of inflation in 1980. He adds that the study is 'definitely' worthy of a Nobel prize. 'If the BICEP2 data are confirmed, apostles of inflation may celebrate the dramatic conformation of the general idea of inflation' says Aleksei Starobinsky of the Landau Institute for Theoretical Physics in Moscow, Russia, who had published his own theory of inflation in December 1979 [57]. Several physicists are suggesting that the data of the BICEP2 will change the face of cosmology and particles physics.

Then Journal Nature collected reactions of leading researchers about the BICEP2 data [116]. We quote a few of them: 'It's just amazing because many great intellectual discoveries are never confirmed at the time when the authors are still alive. I'm not dead yet and they are already seeing this gravitational-wave signal. So I was just flying there on the airplane thinking is this going to break? If my life ends right now I'm not going to know.' Andrei Linde, theoretical physicist at Stanford University in California, and one of the founders of inflation theory.

'This is Nobel prize material, no question. It's not everyday that you wake up and learn something fundamentally new about the Universe, a telegram from the very earliest moments of the Universe. It also is the first detection of the effects of a gravitational wave on matter: that's a great landmark achievement just in time for the one-hundredth birthday of Einstein's general [theory of] relativity next year.'

Marc Kamionkowski, cosmologist at Johns Hopkins University in Baltimore, Maryland.

'There's no doubt that the gravitational waves, if confirmed, would kill the cyclic and ekpyrotic models that my collaborators and I proposed. We have always been very clear about that. But that is a big if.' Paul Steinhardt, theoretical physicist Princeton University in New Jersey.

However, although further observations may yet confirm the findings, independent researchers now say they no longer think that the original data constituted significant evidence. In the intervening months, the Planck satellite has reported new measurements that indicate the Milky Way may contain more dust than assumed by the BICEP2 team. Several groups have concluded that it is possible that dust could reproduced all (or most of) the claimed BICEP2 polarization signal.

Paul Steinhardt stressed [69]: 'The BICEP2 instrument detects radiation at only one frequency, so cannot distinguish the cosmic contribution from other sources.... Now a careful reanalysis by scientists at Princeton University and the Institute for Advanced Study, also in Princeton, has concluded that the BICEP2 B-mode pattern could be the result mostly or entirely of foreground effects without any contribution from gravitational waves. .... The inflationary paradigm is fundamentally untestable, and hence scientifically meaningless.'

In the book [62] Paul Steinhardt and Neil Turok critique the Big Bang model using following arguments: 'We were motivated by the fact that, over the last few decades, more and more elements have had to be added to the Big Bang Theory to make it consistent with what we observe. To explain why the universe is so uniform and isotropic, we had to a new feature called inflation. To explain the formation of structure, we had to add dark matter. The recent discovery that the expansion of the universe has begun to speed up has required the addition of something called dark energy. Each of these elements have been added one by one to make today's Big Bang Theory a kind of patchwork of disconnected ideas. Furthermore, in the background, there has been the disturbing notion that the big bang is the beginning of everything – something that has never been successfully explained by fundamental physics.'

'These things aren't connected in a coherent theory. What's disturbing is when you have a theory and you make a new observation, you have to add new components' –Steinhardt think. 'And they're not connected...There's no reason to add them, and no particular reason to add them in that particular amount, except the observations. The question is how much you're explaining and how much you're engineering a model. And we don't know yet.' [117].

Alfonso the Wise and the Big Bang The second century witnessed the culmination of Greek cosmology. Its author was Claudius Ptolemaeus (Ptolemy). In his book Ptolemy launched his unique and crowning achievement: the first complete and rigorous account of the planets' erratic motions. Epicycles, deferents, eccentrics and equants were introduced to describe the motions. The Arabs, in their efforts to improve the accuracy of Ptolemaic theory, had added more epicycles. As result the theory became scandalously complex. For example, Alfonso the Wise, the thirteenth-century king of Castile (1221–1284), was rather sceptical about the theory 'If the Lord Almighty had consulted me before embarking upon creation, I should have recommended something simpler'.

King Alfonso could have make a similar remark about the Big Bang model. This model explains many of the detailed features of the Universe seen today. But the model assumes that the Universe had to start with certain properties that have never been successfully explained by the fundamental physics. So that to explain new observational data, researchers are introducing new and new suggestions which are not fully supported by observations, experiments and the results of fundamental physics. As a result the initially simple Big Bang theory became quite complex.

Singularity of the Big Bang and the metaverse Of course, it is impossible to overthrow the main results of the Big Bang theory. However, the theory says nothing about the explosion itself. On closer examination it was found that this explosion was very sensitive to very small perturbations.

Alex Vilenkin [60] compared the stability of the expansion with the stability of a pin balancing on its point. Nudge it slightly in any direction and it will fall. A tiny perturbation from the required power level results in a cosmological disaster, such as the fireball collapsing under its own weight or the Universe being nearly empty. At the same time, all properties of every object in the Universe are determined, in the final analysis, by a small number of constants of nature. The Standard Model of particle physics, which describe strong, weak, and electromagnetic interactions of all known particles, contains twenty-five 'adjustable' constants. Research in different areas of physics has shown that many essential features of the Universe are sensitive to the precise values of some of the constants. Any change to any of these constants could change our Universe. And most likely neither we, nor any other living creatures, would be around to admire this new universe.

Thus, our presence in the Universe depends on a precarious balance between different tendencies. In particular, the balance would be destroyed if the constants of nature were to deviate from their present values. What are we to make of this fine-tuning of the constants? We can suggest that there are infinite number of universes and that one of them corresponds to our Universe [60].

A few notes about our theory We presented above the coherent model for the emergence and initial evolution of the Universe. In particular, we found that a scalar field can give birth to the billions of 'seeds' of rapidly evolving Universes. Most of them are Planck-sized 'flicker' universes, which blink in and out of existence. But then, there are some universes that evolve to a large size. One of them accidentally formed our Universe with the fundamental parameters which we have.

We reject the starting point of the Universe as the explosion or the singularity. The model without the initial singularity is introduced. The pra-history of the Universe is introduced instead of the singularity.

1. We have developed an approximate model of the origin of the Universe. The model is common and has potential for development. This is not an engineering-type model like the Big Bang theory or the Big Bang. It is based on the equations

of the quantum scalar field. We found the number of solutions. Using these solutions one can calculate the pressure and density of the field, and then the temperature. Of course, these calculations may be quite complex. But through these calculations we can align our model to the theory of the Big Bang. From the point in time when the density and temperature, calculated according to our model, increase up to the density and temperature, calculated according to the Big Bang theory, further calculation must be carried out using the model of the Big Bang.

2. We attempted to develop the strongly-nonlinear model for the origin of the Universe in which all the elements fit together in a tight and natural way. According to the theory the vacuum with a small positive energy (this condition can be associated with dark energy or cosmological constant) tunnels into a state of negative energy. The origin of the particles, mass and temperature is directly related to strongly nonlinear properties of the quantum scalar fields. In result, our theory opens the way for the development and further generalizations to explain the existence of dark matter and dark energy.

    2.1. During the tunnelling and subsequent time a portion of the energy, which can be qualitatively evaluated as the value of the second order relative to the initial energy of the entire field, is converted into particles and mass (see, for example, Figs. 7.48, 7.49, 7.51 and 7.53). However, the bulk of the energy determined by pure harmonic oscillations (see Fig. 7.54) remained unused. We can assume that this energy was locked into the formed four-dimensional spacetime. This energy may correspond to the dark energy.

    2.2. Some highly energetic particles (see, for example, Fig. 7.48) generated in the beginning could be very stable. They could work like a scaffold creating the cosmic structure. We can imagine that these particles had been stable for billions of years before they begun to decay. This process can determine the law of expansion the modern Universe. On the other hand, these particles may correspond to the dark matter.

3. We stress that our universe must be asymmetric since the landscape around the potential wall was asymmetric during the tunnelling (Figs. 7.27, 7.28 and 7.75). It is also interesting that the origin of the particles is described by the solutions of the wave equation. Therefore it is not surprising that they have wave properties in accordance with the theory of Louis de Broglie.
4. We do not have any precise data about the physical processes near the beginning. These processes could be very different from those familiar to us. Near the beginning the gravitation could be absent, the Einstein theory may not be applicable and the sound speed might be much larger than the light speed. We assumed that the matter of the very early Universe could have unprecedentedly large nonlinearity, much greater than nonlinearity of the Bose-Einstein condensate [8]. This nonlinearity was often ignored. We stress that the interaction of particles is the interaction of the fields. Results of the interactions are determined by the nonlinearity of the fields.

Nonlinearity and resonances In the book we tried to unify the nonlinear wave dynamics using idea of resonance. Highly-nonlinear equations were used. Simple examples of these equations were presented in Chaps. 6 and 7. The equations contain the d'Alembert operator. In the general case, they include in addition dissipative and dispersive terms. Traditional solutions of these equations centre on perturbation methods. However, in the natural resonators the resonant growth in linear waves occurs. As a result, the interaction of the nonlinear, dissipative, dispersive and spatial effects in these resonant systems becomes very complex. However, very close to resonances the influence of the nonlinearities becomes dominant. This phenomenon allows us to simplify strongly the analysis of the resonant waves. We explained the appearance of the extreme waves by certain resonances.

The resonances provide a common platform to reduce disciplinary boundaries and to construct the unified theory of the extreme waves.

We presented the scheme of the theory describing strongly nonlinear waves both in continuum and scalar fields. The theory describes the wide spectrum of the wave phenomena which have a resonant nature. The theory describes many elastica forms, which attracted the attention of many of the brightest minds in the history of mathematics and physics, including Galileo Galilei, James Bernoulli, Leonhard Euler, Gustav Kirchhoff, Max Born and others (see the reference [43] in Chap. 5). We have shown the similarities of nonlinear phenomena which take place from oscillations of gas in tubes through to the origin of the Universe.

References

1. Galiev ShU (1999) Topographic effect in a Faraday experiment. J Phys A Math Gen 32:6963–7000
2. Darwin C (1839) Journal of researches into the geology and natural history of the various countries visited by H.M.S. Beagle, under the command of Captain FitzRoy, R.N. from 1832 to 1836, 1st edn. Henry Colburn, London
3. Galiev ShU (2003) The theory of nonlinear trans-resonant wave phenomena and an examination of Charles Darwin's earthquake reports. Geophys J Int 154:300–354
4. Galiev ShU (1988) Нелинейные волны в ограниченных средах (Nonlinear waves in bounded continua). Naukova Dumka, Kiev (in Russian)
5. Galiev ShU (1997) Resonant oscillations governed by the Boussinesq equation with damping. In: Proceedings of 5th international congress on sound and vibration, Adelaide, pp 1785–1796
6. Galiev ShU (1998) Topographic amplification of vertical-induced resonant waves in basins. Adv Hydrosci Eng 3:179–182
7. Galiev ShU, Galiev TSh (1998) Resonant travelling surface waves. Phys Lett A 246:299–305
8. Galiev ShU, Galiyev TSh (2001) Nonlinear transresonant waves, vortices and patterns: from microresonators to the early universe. Chaos 11:686–704
9. Galiev ShU (2003) Transresonant evolution of wave singularities and vortices. Phys Lett A 311:192–199
10. Galiev ShU (2005) Strongly–nonlinear two–speed wave equations for coastal waves and their application. Phys D 208:147–171

11. Galiev ShU (2008) Strongly-nonlinear wave phenomena in restricted liquids and their mathematical description. In: Perlidze T (ed) New nonlinear phenomena research. Nova, New York
12. Galiev ShU (2009) Modelling of Charles Darwin's earthquake reports as catastrophic wave phenomena. https://researchspace.auckland.ac.nz/handle/2292/4474
13. Galiev ShU (2011) Геофизические Сообщения Чарльза Дарвина как Модели Теории Катастрофических Волн (Charles Darwin's geophysical reports as models of the theory of catastrophic waves). Centre of Modern Education, Moscow (in Russian)
14. Galiev ShU, Galiyev TSh (2013) Scalar field as a model describing the birth of the Universe. InterNet. Galiev – Galiyev, Dec 18 2013
15. Galiev ShU, Galiyev TSh (2014) Nonlinear scalar field as a model describing the birth of the Universe. Известия Уфимского Научного Центра Российской Академии Наук (Herald of Ufa Scientific Center, Russian Academy of Sciences (RAS)) 2:7–33
16. Lamb H (1932) Hydrodynamics, 6th edn. Dover Publications, New York
17. Galiev ShU (1972) Вынужденные продольные колебания нелинейно-упругого тела (Forced longitudinal oscillations of a nonlinear elastic body). Izv Acad Nauk USSR, Mech Solid Body 4:80–87
18. Nigmatulin RI (1991) Dynamics of multiphase media. Parts I and II. Taylor & Francis, London
19. Galiev ShU, Panova OP (1995) Periodic shock waves in spherical resonators (survey). Strength Mat 10:729–746
20. Carcione JM, Poletto F (2000) Sound velocity of drilling mud saturated with reservoir gas. Geophysics 65:647–651
21. Iga Y, Nohml M, Goto A, Ikohagi T (2004) Numerical analysis of cavitation instabilities arising in the three-blade cascade. J Fluids Eng 126:419–429
22. Chester W (1964) Resonant oscillations in closed tubes. J Fluid Mech 18:44–64
23. Verhagen JHG, van Wijngaarden L (1965) Nonlinear oscillations of fluid in a container. J Fluid Mech 22:737–751
24. Натанзон МС (1965) Вынужденные разрывные колебания жидкости в трубах. (Forced discontinuous oscillations of water in tubes). Izv Acad Nauk USSR, Mechanics 2:33–42
25. Chester W, Bones JA (1968) Resonant oscillations of water waves. II. Experiment. Proc R Soc A 306:23–39
26. Natanzon MS (1977) Продольные Автоколебания Жидкостной Ракеты (Longitudinal self-excited oscillations of a liquid-fuel rocket). Mashinostroenie, Moscow
27. Cox E, Mortell MP (1986) The evolution of resonant water-wave oscillations. J Fluid Mech 162:99–116
28. Ockendon H, Ockendon JR, Johnson AD (1986) Resonant sloshing in shallow water. J Fluid Mech 167:465–479
29. Ilgamov MA, Zaripov RG, Galiullin RG, Repin VB (1996) Nonlinear oscillations of a gas in a tube. Appl Mech Rev 49:138–154
30. Chabchoub A, Hoffmann NP, Akhmediev N (2011) Rogue wave observation in a water wave tank. Phys Rev Lett 106:204502
31. Ibrahim RA (2005) Liquid sloshing dynamics. Cambridge University Press, Cambridge
32. Faltinsen OM, Timokha AN (2009) Sloshing. Cambridge University Press, Cambridge
33. Kobine JJ (2008) Nonlinear resonant characteristics of shallow fluid layers. Philos Trans R Soc A 366:1331–1346
34. Nayfen AH, Mook DT (1995) Nonlinear oscillations. Wiley, New York
35. Galiev ShU, Ilgamov MA, Sadikov AV (1970) Periodic shock waves in a gas. Izv Acad Nauk USSR, Mech Fluid Gas 2:57–66
36. Prejean SG, Haney MM (2014) Shaking up volcanoes. Science 345:39
37. Alidibirov M, Dingwell DB (1996) Magma fragmentation by rapid decompression. Nature 380:146–148

38. Ichihara M, Ohkunitani H, Ida Y, Kameda M (2004) Dynamics of bubble oscillation and wave propagation in viscoelastic liquids. J Volcanol Geotherm Res 129:37–60
39. Gonnermann HM, Manga M (2007) The fluid mechanics inside a volcano. Annu Rev Fluid Mech 39:321–356
40. Galiev ShU (1994) Influence of cavitation on transient deformation plates and shells by the liquid. In: Zhemin Z, Qingming T (eds) Proceedings of IUTAM symposium on impact dynamics. Peking University Press, Beijing
41. Galiev ShU (1997) Influence of cavitation upon anomalous behavior of a plate/liquid/ underwater explosion system. Int J Impact Eng 19(4):345–359
42. Kimmoun O, Branger H (2007) A particle image velocimetry investigation on laboratory surf-zone breaking waves over a sloping beach. J Fluid Mech 588:353–397
43. Taylor GI (1954) An experimental study of standing waves. Proc R Soc A 218:44–59
44. Kutschera U (2010) Darwin's geological time dilemma. Nat Geosci 3:71–72
45. Repcheck J (2003) The man who found time. Perseus Publishing, New York
46. Brake ML (2009) Revolution in science: how Galileo and Darwin changed our world. Palgrave Macmillan, New York
47. Verschuuren GM (2012) Darwin's philosophical legacy. Lexington Books, New York
48. Kragh H (1999) Cosmology and controversy. Princeton University Press, Princeton
49. Glegg B (2009) Before the big bang: the prehistory of the universe. St. Martin's Press, New York
50. Brockman J (2014) The universe: leading scientists explore the origin, mysteries, and future of the cosmos. HarperCollins, New York
51. Linde A (2005) Particle physics and inflationary cosmology. arXiv:hep-th/0503203vl
52. Lyth DY, Liddle AR (2009) The primordial density perturbation. Cambridge University Press, Cambridge
53. Serjeant S (2010) Observational cosmology. Cambridge University Press, Cambridge
54. Guth AH (1981) Inflationary university: a possible solution to the horizon and flatness problems. Phys Rev D 23(2):347–356
55. Linde A (1982) A new inflationary universe scenario: a possible solution of the horizontal, flatness, homogeneity, isotropy and primordial monopole problems. Phys Lett 108B (6):389–393
56. Cho A, Bhattacharjee Y (2014) First wrinkles in spacetime confirm cosmic inflation. Science 343:1296–1297
57. Cowen R (2014) Telescope captures view of gravitational waves. Nature 507:281–283
58. Ade PAR, Aikin RW, Barkats D et al (2014) Detection of B -mode polarization at degree angular scales by BICEP2. Phys Rev Lett 112:241101
59. Greene B (2014) Listening to the big bang. Smithsonian Magazine http://www.smithsonianmag.com/science-nature/history-big-bang-theory-18095116
60. Vilenkin A (2006) Many worlds in one. Hill and Wang, New York
61. Krauss LM (2012) A universe from nothing. Atria Paperback, New York
62. Steinhardt PJ, Turok N (2007) Endless universe. Doubleday, New York
63. Penrose R (2010) Cycles of time: an extraordinary new view of the universe. Bodley Head, London
64. Bojowald M (2007) What happened before the big bang? Nat Phys 3:523–525
65. Brandenberger R (2012) The matter bounce alternative to inflationary cosmology. arXiv:1206.4196v1 [astro-ph.CO] 19 Jun 2012
66. Bojowald M (2008) Follow the bouncing universe. Sci Am 299:44–51
67. Bernard C (ed) (2007) Universe or multiverse? Cambridge University Press, Cambridge
68. Barrau A (2007) Physics in the multiverse: an introductory review. CERN Cour 47(10):13–17
69. Steinhardt P (2014) Big band blunder bursts the multiverse bubble. Nature 510:9
70. Grossman L (2013) Cosmic baby snaps hint at bubbly birth. New Scientist 2935:12
71. Ashtekar A, Pawlowski T, Singh P (2006) Quantum nature of the big bang. Phys Rev Lett 96:141301

72. Ashtekar A, Corichi A, Singh P (2008) Robustness of key features of loop quantum cosmology. Phys Rev D 77(2):024046
73. Barrau A, Linsefors L (2014) Our universe from the cosmological constant. ArXiv:1406.3706 [gr-qc]
74. Wetterich C (2013) A universe without expansion. ArXiv:1303.6878 [astro-ph.CO]
75. Afshordi N, Mann RB, Pourhasan R (2014) The black hole at the beginning of time. Sci Am 311(2):26–33
76. Morton J (2014). Learn more about the birth of the Universe. NZ Herald. May 27
77. Musser G (2014) Gravitational waves reveal the universe before the big bang: an interview with physicist Gabriele Veneziano. Sci Am 310(4). http://blogs.scientificamerican.com/critical-opalescence/2014/04/03/gravitational-waves-reveal-the-universe-before-the-big-bang-an-interview-with-physicist-gabriele-veneziano/
78. Quantum Universe (2003) U.S. Department of Energy and the National Science Foundation. http://www.interactions.org/pdf/Quantum_Universe_GR.pdf
79. Reich ES (2012) Theorists bridge space-time rips. Nature 491:19
80. Maggie M (2013) Pop-up universe. New Scientist 2937:38–41
81. Merali Z (2013) Theoretical physics: the origins of space and time. Nature 500:516–519
82. Cowen R (2013) How to see quantum gravity in big bang traces. Nature. doi:10.1038/nature.2013.13834
83. Ellis G, Silk J (2014) Scientific method: defend the integrity of physics. Nature 516:321–323
84. Galiev ShU, Galiyev TSh (2014) Coherent model of the births of the universe, the dark matter and the dark energy. https://researchspace.auckland.ac.nz/handle/2292/23783
85. Harrison ER (2000) Cosmology: the science of the universe. Cambridge University Press, Cambridge
86. Weinberg S (2008) Cosmology. Oxford University Press, Oxford/New York
87. Gorbunov DS, Rubakov VA (2011) Introduction to the theory of the early universe. Hot big bang theory. World Scientific, Singapore/Hackensack
88. Gorbunov DS, Rubakov VA (2011) Introduction to the theory of the early universe. Cosmological perturbations and inflationary theory. Hot big bang theory. World Scientific, Singapore/Hackensack
89. Bronnikov KA, Rubin SG (2013) Black holes, cosmology and extra dimensions. World Scientific, Singapore/London
90. Mo H, van den Bosch F, White S (2011) Galaxy formation and evolution. Cambridge University Press, Cambridge/New York
91. Stock S, Bretin V, Chevy F, Dalibard J (2004) Shape oscillation of a rotating Bose-Einstein condensate. Europhys Lett 65:594–600
92. Niederhaus CE, Jacobs JW (2003) Experimental study of the Richtmyer-Meshkov instability of incompressible fluids. J Fluid Mech 485:243–277
93. Mureika J, Stojkovic D (2011) Detecting vanishing dimensions via primordial gravitational wave astronomy. Phys Rev Lett 106:101101
94. Stojkovic D (2013) Vanishing dimensions: theory and phenomenology. arXiv:1304.6444 (hep-th)
95. Stojkovic D (2013) Vanishing dimensions: a review. Mod Phys Lett A 28(37):1330034
96. Brooks M (2014) Silence is olden. New Scientist 2973:34–37
97. Jiang L, Perlin M, Schultz WW (1998) Period tripling and energy dissipation of breaking standing waves. J Fluid Mech 369:273–299
98. Gubser SS (2010) The little book of string theory. Princeton University Press, Princeton
99. Kalinichenko VA (2009) Nonlinear effects in surface and internal waves Faraday. Dissertation, Russian Academy of Sciences, Ishlinsky Institute for Problems in Mechanics
100. Liberati S, Maccione L (2014) Astrophysical constraints on Planck scale dissipative phenomena. Phys Rev Lett 112:151301

101. Moskowitz C (2014) If spacetime were a superfluid, would it unify physics – or is the theory all wet? http://www.scientificamerican.com/article/superfluid-spacetime-relativity-quantum-physics/
102. Martinez G (2013) Advances in quark gluon plasma. arXiv:1304.1452v1 [nucl-ex]
103. Galiev ShU (2000) Unfamiliar waves excited due to parametric and resonant effects. Proc Am Inst Phys 511:361–367
104. Gomez LG, Ferguson KR, Cryan JP et al (2014) Shapes and vorticities of superfluid helium nanodroplets. Science 345:906–909
105. Shandarin SF, Zeldovich YaB (1989) The large–scale structure of the universe: turbulence, intermittency, structures in a self–gravitating medium. Rev Mod Phys 61(2):185–220
106. Miniati F et al (2000) Properties of cosmic shock waves in large-scale structure formation. Astrophys J 542:608–621
107. Dutton Z, Budde M, Slowe C, Hau LV (2001) Observation of quantum shock waves created with ultra-compressed slow light pulses in a Bose-Einstein condensate. Science 293:663–668
108. Marts B, Hagberg A, Meron E, Lin AL (2004) Bloch-front turbulence in a periodically forced Belousov-Zhabotinsky reaction. Phys Rev Lett 93:108305
109. Love AEH (1944) The mathematical theory of elasticity. Cambridge University Press, Cambridge
110. Clark S (2014) Ghost busters. New Scientist 2975:32–35
111. Gamow G (1946) Expanding universe and the origin of elements. Phys Rev 70:572–573
112. Gamow G (1948) The evolution of the universe. Nature 162:680–682
113. Alpher RA, Bethe H, Gamow G (1948) The origin of chemical elements. Phys Rev 73:803–804
114. Gamow G (1948) The origin of elements and the separation of galaxies. Phys Rev 74:505–506
115. Gamow G (1952) The creation of the universe. Viking, New York
116. Cowen R (2014) Experts hail the gravitational-wave revolution. Nature. doi:10.1038/nature.2014.14902
117. Than K (2007) Greatest mysteries: how did the universe begin? http://www.livescience.com/1774-greatest-mysteries-universe.html

Chapter 8
Last Comments on Charles Darwin's Geophysical Observations

> *... A few naturalists, endowed with much flexibility of mind, ...may be influenced by this volume; but I look with confidence to the future, to young and rising naturalists...*
> *(**Charles Darwin**. On the Origin of Species by Means of Natural Selection, London, John Murray, 1859, p. 311.)*

We tried to show in the book that Charles Darwin's geophysical observations continue to arouse great interest up to this moment.

8.1 Darwin's Discoveries and the Instability of Nature

Darwin was the first naturalist who witnessed a huge natural disaster and published a detailed scientific description of this awful phenomenon. He described the seismic and ocean waves and the explosions of volcanoes. He wrote ... *the earth, considered from our earliest childhood as the type of solidity, has oscillated like a thin crust beneath our feet; and in seeing the laboured works of man in a moment overthrown, we feel the insignificance of his boasted power* ... [1, pp. 601–602]. Darwin's words can be interpreted very widely and applicable for all areas of human activity. At the same time grandiose and terrible natural catastrophes may excite a research enthusiasm, and inspire further research.

Darwin's description strikes our imagination, just as much as does the description by Pliny the Younger of the shivering of the earth and ocean, during the eruption of Vesuvius in 79. There is even a stronger analogy between Darwin's accounts and the events described in the epic about Gilgamesh, which has come down to us from Ancient Sumer and was probably written nearly 5,000 years ago.

Below we will present some extracts from this great literature.

The hearts of the Great Gods moved them to inflict the Flood.

Just as dawn began to glow
there arose from the horizon a black cloud.
Adad rumbled inside of it,
before him went Shullat and Hanish,
heralds going over mountain and land.
Erragal pulled out the mooring poles,

Sh.U. Galiev, *Darwin, Geodynamics and Extreme Waves*,
DOI 10.1007/978-3-319-16994-1_8

forth went Ninurta and made the dikes overflow.
The Anunnaki lifted up the torches,
setting the land ablaze with their flare.
Stunned shock over Adad's deeds overtook the heavens,
and turned to blackness all that had been light.
The... land shattered like a... pot.
All day long the South Wind blew ...,
blowing fast, submerging the mountain in water,
overwhelming the people like an attack.
No one could see his fellow,
they could not recognize each other in the torrent.
The gods were frightened by the Flood,
and retreated, ascending to the heaven of Anu.
The gods—those of the Anunnaki—were weeping with her,
the gods humbly sat weeping, sobbing with grief(?),
their lips burning, parched with thirst.
Six days and seven nights
came the wind and flood, the storm flattening the land.
When the seventh day arrived, the storm was pounding,
the flood was a war—struggling with itself like a woman writhing (in labor).
The sea calmed, fell still, the whirlwind (and) flood stopped up.
(http://www.ancienttexts.org/library/mesopotamian/gilgamesh/tab11.htm
Translated by Maureen Gallery Kovacs)

It is possible to interpret in various ways the information in the epic, but it appears probable that a description is given there of the joint action of a major earthquake (The... land shattered like a... pot), of volcanoes (The Anunnaki lifted up the torches,/setting the land ablaze with their flare) and a huge tsunami (Adad rumbled inside of it,/before him went Shullat and Hanish,/heralds going over mountain and land./ Erragal pulled out the mooring poles,/forth went Ninurta and made the dikes overflow). As a result, Southern Mesopotamia was shattered and flooded.

The land had been flooded during many years or, perhaps, many centuries, before the Sumerians created a state on this land. The flood is connected also with the history of a ship described in the epic. Animals and Ziusudra's family were saved from the flood in this ship. The history of Ziusudra and his family was reproduced in the Old Testament in slightly changed form, although the time distance between these outstanding literary works is about 1,500–2,500 years

It is necessary to note that in that period the evolution of society was very slow, so that many generations during many hundreds of years could preserve the memory of great events. Only during recent decades an information has started to be created, collected and accumulated very quickly. As a result, in the mass consciousness there is no stable memory, and an important event may be erased in a literal sense within a few years.

Homer's poems were initially transmitted orally from generation to generation, but the epic of Gilgamesh was written on clay tablets and translated into several languages over many centuries. Many dozens of generations were brought up on those tablets and on earlier legends. In them, the long-term memory of society was preserved. In particular, it is known from Sumerian tablets that in Sumer there was a tradition of the ritual murder of a specially-appointed ruler, in expiation of sins of all society. The sacrificial ruler reigned only for 5 days. Later, in Christian religion, Jesus acts also as the redeemer. It is interesting that he carried the title – Jesus of Nazareth, Ruler of the Jews. The analogy with the Sumerian tradition is clear.

The basic events in the epic were described by Darwin using other literary means. Certainly, the scales of these catastrophes were different – but the basic components of the descriptions are identical. They are earthquake, eruption of volcanoes and tsunami. According to Darwin, all were the result of a local loss of stability in the earth crust. Darwin wrote that . . . *for, in order to break up and throw over portions of very thick crust, as in Diagram 3, there must have been great horizontal extension, and this, if sudden, would have caused as many continuous outbursts of volcanic matter* ... and ... *if the force had acted suddenly, these portions of the earth's crust would have been absolutely blown off*, ... [2, p. 79].

Catastrophic events are called that, because they are unexpected and are accompanied by transient wave processes. Such events seem to be occurring more frequently. According to the Center for Research on Environmental Decisions (Columbia University), during the last 40 years there has been a fast growth in numbers of natural catastrophic phenomena [3]. We can agree with this conclusion. Probably, we have entered into a new phase of the Earth's evolution, when its climate and surface begin to change intensively.

Certainly, before reaching far-reaching conclusions, it is necessary to consider and study attentively those cases of rapid change of the face of the Earth, which have been recorded in its geology and in the history of mankind.

But it is possible that human activity is a reason for future possible natural cataclysms. Let us explain what we mean. We can assume that a system consisting of the Earth, ocean and atmosphere is initially in a condition of a very fragile balance. Then our mines, gas and oil bore-holes, artificial thermal fields and streams, technological and industrial processes, huge water basins – despite the rather small total effect of them, – can strongly disturb the system. Small disturbances can break the unstable balance of Nature. The near future will show, whether that is to.

In this context I recall a fantastic story "When the Earth Screamed" by Arthur Conan Doyle (1929), which I read as a child. In that story, some people drilled an extremely deep bore-hole. When they had drilled through the Earth crust, the Earth started to shake. It was found that the Earth is a live being – and even the rather weak disturbances caused by people could arouse the Earth, with catastrophic consequences.

Are we approaching such a situation? Recent Chinese data are supporting this opinion [4]. A catastrophic earthquake in Ludian County (3 August, 2014) started to rise just as two giant reservoirs on upper Yangtze were being filled with water. In particular, seismicity may be connected with an injection of different fluids at ground depth. Analysis of numerous shallow earthquake sequences induced by fluid injection at depth reveals that the maximum magnitudes are determined by the total volume of fluid injected. Activities involving fluid injection include (1) hydraulic fracturing of shale formations or coal seams to extract gas and oil, (2) disposal of wastewater from these gas and oil activities by injection into deep aquifers, and (3) the development of enhanced geothermal systems by injecting water into hot, low-permeability rock. The maximum magnitudes sometimes exceed 5. However, the maximum magnitudes might be up to 7. Induced seismicity is currently the largest obstacle for using above technologies near urban centers (DOI: 10.1002/2013JB010597).

8.2 Catastrophic Earthquakes and Tsunamis of Recent Years

Many of the world's cities are built on sedimentary basins (for example, Los Angeles, Tokyo, New York, Shanghai, Calcutta, Jakarta, Delhi) or on the tops of ridges. From the standpoint of earthquake risk, however, basins (valleys) and hilltops are often the least-desirable places to build because of the topographic effect. It is now recognized that hills, sedimentary basins, lakes, and continental shelf are natural resonators, where seismic-induced waves may be amplified.

A scheme of a city located on a sediment layer is given in Fig. 1.7 (left). Seismic vertical waves can be trapped by a sedimentary layer and amplified due to the resonance. The ruin of cities can be also connected with another acoustical effect of the amplification of seismic waves or the amplification of their influence on buildings. First, the amplitude of waves amplifies at their transition into the sediment layer. The amplification is proportional to the ratio of densities of the materials. In the case of the weak top layer, the amplification can be sufficient for the destruction of buildings that Darwin observed in Chile. The seismic wave can be reflected from the free ground surface. The city may be in ruins, if a frequency of the reverberations is equal to the resonant frequency of any building or a group of buildings.

However, the worst situation is when the forcing frequency of the rocky breed can correspond to the resonant frequency of the top layer. It can lead to a strong growth of the amplitude of the trapped wave. As a result, the city can be destroyed. The scheme of Fig. 1.7 shows the very dangerous case when an earthquake source is located at a small depth under a city. Vertical rarefaction waves may be generated in this case. Probably, the waves which have destroyed the city of Port-au-Prince were the rarefaction waves.

Let us remember that the 2010 Haiti earthquake (with magnitude 7.1) was the most destructive in the region during all recorded history. The state capital, Port-au-Prince, was almost destroyed.

The Haiti 12 January 2010 Earthquake The earthquake source had been located approximately 13 km below the surface and 22 km to the southwest from the capital. The earthquake was generated by the interaction of the Caribbean and North American plates. Thus, the source of the Haiti 2010 earthquake was in upper-lying layers of the ground. Therefore, the strongly-nonlinear waves of the vertical acceleration could be excited within the upper-lying sediment layers and the hills. One reason for the catastrophic results of the Haiti 12 January 2010 earthquake, which subjected about 3 million people to 'severe shaking', may be the relatively loose ground. The configuration and mechanical properties of the upper ground layers amplified the seismic waves. Another important reason is the seismic centre, which was close to the capital. Therefore, the seismic wave could not weaken rapidly when it approached the ground surface. Then the seismic wave transmitted into the loose sediment layer, where it was trapped and amplified due to resonance. The most dramatic damages were also associated with the topographic effects, namely, with the amplification of seismic waves within hills and small ridges. These circumstances and the weakly-structured buildings led to the death of about 240,000 people.

The 27 February 2010 Chilean Earthquake On 27 February 2010, around 03.34 in local time, an earthquake with magnitude 8.8 occurred in Chile (the coast of the Maule region). It was 120 km northeast of the city of Concepcion and 410 km south from Santiago, and the epicentre was approximately 35 km below the ground surface. The earthquake occurred on the same section of the fault that had caused the earthquake which Darwin observed in 1835. More information about the Chilean earthquake 2010 can be found at www.geerassociation.org/GEER.../Maule.../Section_4_GeologyVer7.pdf

A tsunami warning was issued for the coasts of Chile and Peru, east coast of Australia, and some Pacific Islands. In particular, a tsunami hit the coast of Chile with height up to 1.6 m. Within 200 km radius of the earthquake epicentre there are many volcanoes, but they were calm during the earthquake.

Comparing Darwin's data and the results of the 27 February 2010 Chilean earthquake, we can formulate two questions. The earthquake in Chile was one of the biggest in recent time – so why was the tsunami smaller than might be expected? On the other hand, we can consider the last Chilean earthquake as a test of the connection of earthquakes and volcanoes. Why did the earthquake not waken the volcanoes?

Here we are again emphasizing that the active Earth is a very complex system [5]. There are a lot of factors involved in the earthquake-volcano connection and a tsunami formation: location, depth, preparedness, energy, propagation, and interaction of waves, and so on. The earthquake-induced events may be compared only if we know these factors. However, we do not know them.

Earthquake in Christchurch (22 February 2011, New Zealand) An example of the catastrophic influence of vertical seismic shock is the earthquake in Christchurch (Fig. 1.1). As in case of Port-au-Prince the earthquake magnitude was not very large – about 6.3. However, its effects were catastrophic. This resulted from the fact that the earthquake source settled down at a depth of only 5 km, almost under the city centre. Therefore the earthquake-induced vertical movement prevailed over the horizontal movement, namely, the vertical acceleration down reached 2.2 g. If a similar event took place in 1835 it would be enough to explain all of the seismic phenomena described by Darwin.

Japanese Earthquake on March, 11th, 2011 In some measure the events described by Darwin were repeated during an earthquake which devastated some part of Japan territory on March, 11th. The earthquake centre settled down in the sea on a distance of 130 km from the coast at depth about 24 km. Hence, the distance approximately corresponded to Darwin's reports. The magnitude was about 9, which

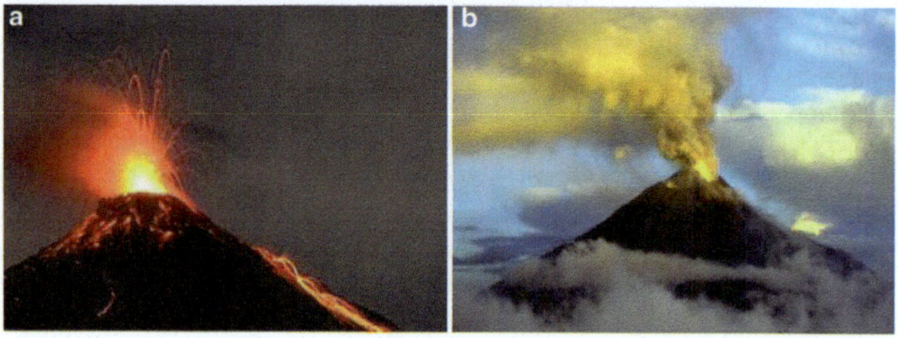

Fig. 8.1 A volcano on the mountain Karangetang in Indonesia (**a**) and the volcano of Shiveluch (**b**). They began intensive eruption very soon after the Japanese earthquake on March, 11th 2011. Internet. http://focus.ua/country/174139/ and http://newsland.com/news/detail/id/964275/

also corresponds approximately to modern estimations of the magnitude of the Chilean 1835 earthquake. Therefore, it is not surprising that amplitudes of tsunamis in both cases were similar. In Japan they reached more 10 m, and in Chile, according to Darwin, the tsunami reached 20–24 ft (6–7 m).

Even more importantly for us, that earthquake did awake volcanoes. In Indonesia, the volcano Karangetang awoke almost after that earthquake. A column of emissions formed, and on the slopes some lava started to flow down (Fig. 8.1a).

Three volcanoes started to erupt on Kamchatka. The volcano Shiveluch threw out a huge column of ashes to the height of 8 km (Fig. 8.1b). Later, the volcano Karymsky ejected ash columns to the height of 4 km, and then the volcano Kizimen awoke. On March 15th in the southwest of Japan, the volcano Sinmoe started to throw out stones and ashes to the height of 4 km. Before January 2011 this volcano had not been active for 52 years. We notice that Sinmoe is located 1,500 km from the earthquake area.

Let us stress that the specified connection of that earthquake with the beginning of volcanic eruptions corresponds to Darwin's reports, and also to data from satellites [6, 7]. According to the news agency "Xinhua", 13 volcanoes in Japan have started to become more active. An earthquake of magnitude 6.4 occurred near the volcano Fujiyama on March 15th. The epicentre was very close to the magma chamber of that famous volcano. Thus, after the strong earthquake, the earth crust of the east part of Japan became unstable [8, 9].

Earthquake in Ludian County (Yunnan), China (3 August, 2014) The Ludian earthquake occurred on 3 August 2014. The epicentre was at a depth of 10.0 km. This shallow earthquake (magnitude 6.1) was weaker than the earthquake in Christchurch. Its depth was larger than in Christchurch. However, we think that the vertical acceleration was there very large like in Christchurch. The earthquake killed nearly 600 people. The earthquake also triggered multiple landslides that have blocked rivers.

8.3 Closing Remarks

It is necessary to note that depths of the Earth and dynamic processes occurring there have been studied much less than the depths of cosmic space. Therefore, often we can only make assumptions about reasons for different underground phenomena. In particular, although the theory of linear seismic waves is classical and it describes well many observations, sometimes it is not sufficient. The linear theory cannot describe complex underground phenomena. For example, recently the assumption has been formulated [10–13] that some cracks on the Earth surface may be generated by discontinuous nonlinear waves propagating along an underground fault.

Underground Mach cone At a weak point of an old fault, a critical tension can form a new crack, which can begin to propagate along the fault. As a result, nonlinear seismic waves are excited. If the crack speed is great enough, these waves form a Mach cone – a cone surface on which wave parameters change discontinuously (Fig. 8.2).

Is it possible that the unexpected character of some destructions of a ground surface is determined by a Mach cone surface propagating near the Earth surface? That is a very complicated question, to which we do not have any answer.

We can also ask about the possibility of generation of a large Mach cone surface in a solid body, due to wave deformations. The character of the failure on this surface is a question too. There are only a few experimental researches of the formation of the Mach cones connected with earthquakes [10–13]. Apparently, those are only the beginning of research on this topic. We emphasize that Mach cones have been well studied only for gas-like media.

However, the possibility of the formation of similar cones was clearly shown more than 20 years ago [14–16]. The process of perforation of metal targets was

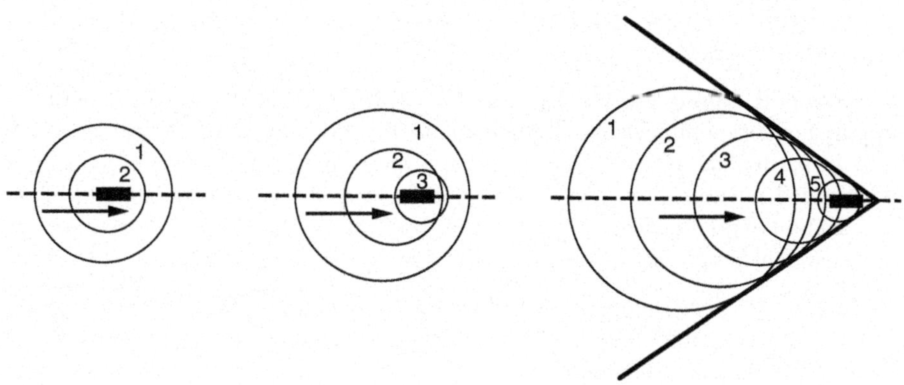

Fig. 8.2 The Mach cone arising when the speed of a body (or crack) exceeds some critical speed [12]

studied. The penetration and perforation of plates of two aluminium alloys by a cylindrical steel projectile were examined experimentally and simulated numerically. The influence of the impact velocity upon the penetration depth and the deformation and perforation modes were investigated. Particular attention was given to the cone formation in the targets, due to temperature rising and material melting at the sites of intensive plastic shear deformations. An attempt was made to simulate numerically the dynamics of the initiation and development of adiabatic shear bands in plates.

Some results of this research are presented in Fig. 8.3 for different moments of intrusion of a solid body (the projectile with a flat penetrating end) into softer material.

The deformation patterns in plate sections which form during the penetration are shown in Fig. 8.3. At velocities close to the ballistic limit (corresponding to complete perforation) the materials' response is different. The Al-Zn-Mg alloy is characterized by the initiation of internal spallation cracks (Fig. 8.3g). Conversely, in the AMg6 alloy such cracks are absent for all interaction velocities. There is also a notable difference between the failure modes in the cases of complete perforation (Fig. 8.3d, h). The formation of the Mach cone is shown clearly in photos b, f and g. In particular, an example of the Mach cone and cracks ahead of it are shown clearly in Fig. 8.3g, an example of a Mach cone is shown in Fig. 8.3f, and the generation of the Mach cone is shown in Fig. 8.3b.

At the same time, Fig. 8.4 shows that the plug forced out after the AMg6 plate complete perforation is separated into two parts, one of which is cone-shaped crater. The crater surface is smooth and shiny. Such specific failure modes can naturally be associated with the material melting in the sites of intensive plastic shear deformations. The second is formed by material taken from the crater (at the right).

The test calculation was performed for AMg6 alloy. The grid distortion in the target at two different instants are presented in Fig 8.5. The grid in the projectile is not shown in view of its minor distortion.

The developed information on these experiments can be found in [14–16].

We think that Figs. 8.4 and 8.5 illustrate qualitatively the development of supershear ruptures during shallow and deep earthquakes [17–19].

Darwin and Nonlinear Geodynamics There is increasing recognition of the importance of nonlinear wave phenomena in many areas of geophysics [20–25]. In particular, the nonlinear local amplification of seismic waves is a more widespread phenomenon than seismologists assumed earlier. An example of the localization of nonlinear seismic effects is given in Fig. 8.6. Nonlinear waves can be formed, if the amplitude of forced oscillations is sufficiently large.

Almost 200 years ago Darwin described many catastrophic local phenomena. We can now interpret those as results of strongly-nonlinear wave processes forced by the Concepcion earthquake.

During the voyage Darwin was exposed to the challenging influence of many new scientific facts. His talent for observations and his genius for logical explanations of them were developed. Darwin explained the earthquake-induced

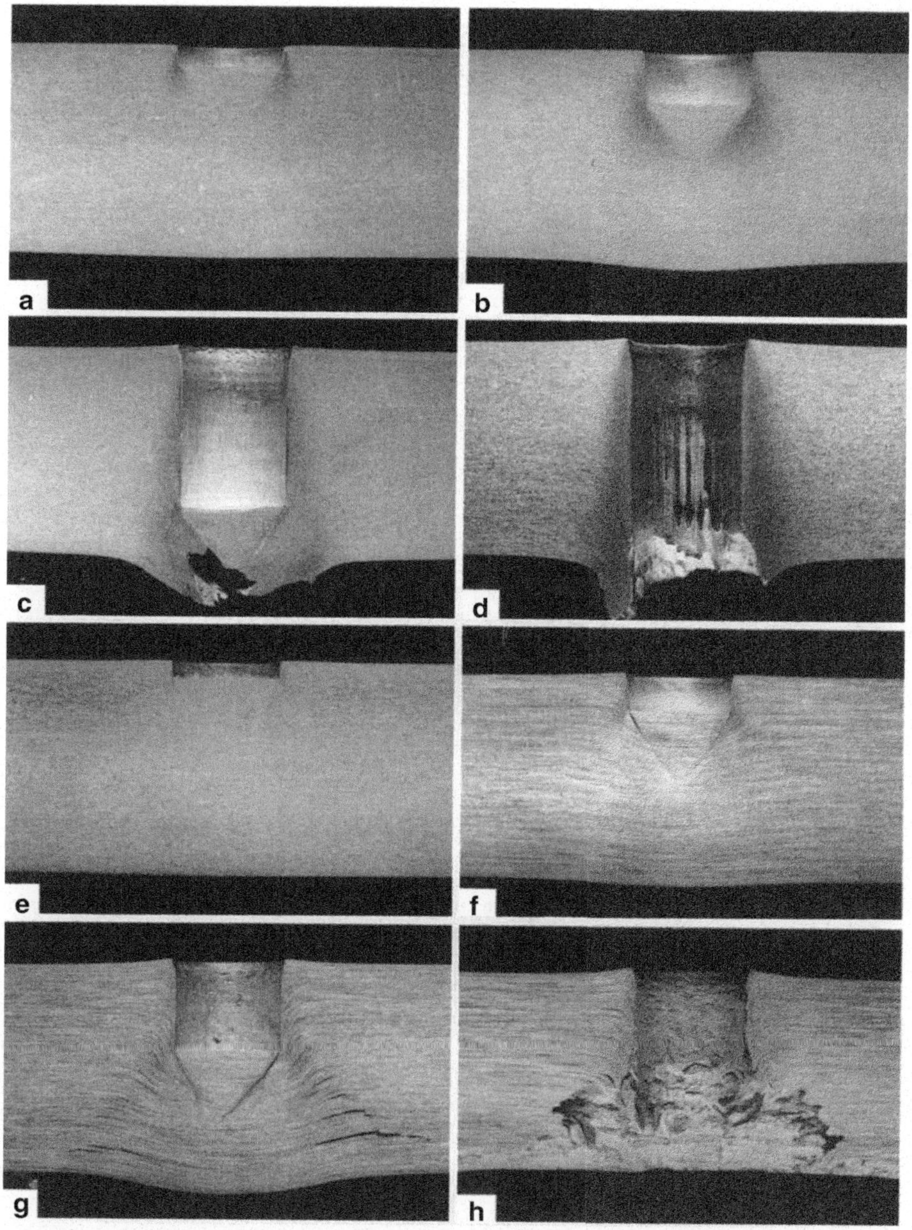

Fig. 8.3 Cross-section of specimens of AMg6 (**a–d**) and Al-Zn-Mg (**e–h**) alloys after tests (the surfaces were polished and pickled with hydrofluoric acid): (**a**) $V_0 = 138$ m/s; (**b**) $V_0 = 224$ m/s; (**c**) $V_0 = -390$ m/s; (**d**) V_0 $^{= 437}$ m/s; (**e**) $V_0 = 157$ m/s; (**f**) $V_0 = 240$ m/s; (**g**) $V_0 = 340$ m/s; (**h**) $V_0 = 440$ m/s [14]

Fig. 8.4 The typical plug and cone after complete perforation of 60 mm thick aluminium alloy target. Melting cone surfaces are formed by intensive plastic shear deformations. The projectile velocity $V_0 = 660$ m/s. Plug and cone heights are 15 and 8 mm, respectively [14]

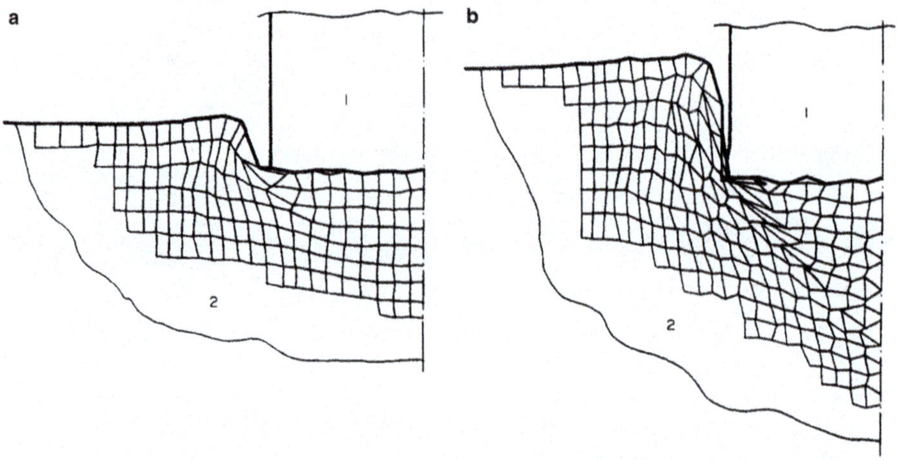

Fig. 8.5 Lagrangian grid distortions near the contact boundary between projectile (*1*) and 30-mm thick AMg6 alloy plate (*2*) for $V_0 = 224$ m/s at instants $10.8{*}10^{-6}$ (**a**) and $32.2{*}10^{-6}$ s (**b**)

Fig. 8.6 The eastern Bay of Plenty (in New Zealand) was struck by a severe earthquake in March 1987. These photographs of Edgecumbe railway station, with track bent in opposite directions, show the magnitude of the forces that were unleashed. (*Left*) – Ground level view, (*right*) – birds-eye view [26]

catastrophic effects, which he observed in Chile 1835, by the fast elevation of the land and by the local circumstances.

The author has focused his attention on these ideas. The local effects of the sharp vertical elevation of the land were analysed. These effects occur where the geology of surface layers or/and the land surface change quickly. Thus, Darwin observed local topographic effects of the earthquake-induced elevation of the land. Due to the topographic effect, strong amplification of seismic waves can occur. This amplification can be accompanied by different local nonlinear effects; namely, by earthquake/volcano interaction, fracture of the land, elevation of islands and tsunamis. Darwin explained the catastrophic results of that earthquake by strong vertical shocks at the surface of the Earth. We think that those ideas of Darwin are supported by the material of this book.

Apparently, Darwin's main idea is valid when the earthquake source is close enough to the Earth surface and the vertical acceleration is greater than g. The existence of very strong vertical seismic accelerations has been established within the past few years, whereas this possibility had not been considered.

Catastrophic events on the Earth's surface attract more and more attention [27, 28]. It is impossible to survey comprehensively the huge literature which is devoted to these events and various aspects of the development of geophysics connected with extreme waves. But even that which we have examined allows us to conclude that Darwin's ideas about earthquakes, eruptions of volcanoes and dynamics of the Earth are not well known, and his priority is not recognized [29]. For example, Darwin's work is not mentioned in a very interesting and detailed review of the history of geology [30].

This book is intended to change that situation.

References

1. Darwin C (1890) Journal of researches into the natural history and geology of the countries visited during the voyage of H.M.S. Beagle around the world, under the command of Captain FitzRoy, R.N. T. Nelson and Sons, London
2. Darwin C (1840) On the connexion of certain volcanic phenomena in South America; and on the formation of mountain chains and volcanoes, as the effect of the same power by which continents are elevated. (Read March 7, 1838) In: Barrett PH (ed) (1977) The collected papers of Charles Darwin. The University of Chicago Press, Chicago
3. Schiermeier Q (2012) Disaster toll tallied. Nature 481:124–125
4. Qiu J (2014) Chinese data hint at trigger for fatal quake. Nature 513:154–155
5. Watt SFL, Pyle DM, Mather TA (2009) The influence of great earthquakes on volcanic eruption rate along the Chilean subduction zone. Earth Plan Sci Lett 277:399–407
6. Geoscience: volcanoes respond to earthquakes (2010) Nature 467:887
7. Donne DD (2010) Earthquake-induced thermal anomalies at active volcanoes. Geology 38 (9):771–774
8. Prejean SG, Haney MM (2014) Shaking up volcanoes. Science 345:39
9. Brenguier F, Campillo M, Nakeda T et al (2014) Mapping pressurized volcanic fluids from induced crustal seismic velocity drops. Science 345:80–82

10. Rosakis AJ, Samudrala O, Coker D (1999) Cracks faster than the shear-wave speed. Science 284:1337–1340
11. Xia K, Rosakis AJ, Kanamori H (2004) Laboratory earthquakes: the sub-Rayleigh-to-supershear rupture transition. Science 30:1859–1861
12. Fisher R (2009) Seismic boom. New Sci 2719:32–35
13. Rousseau CE, Rosakis AJ (2009) Dynamic path selection along branched faults: experiments involving sub-Rayleigh and supershear ruptures. J Geophys Res 114:808303
14. Astanin VV, Galiev SU, Ivashchenko KB (1991) Cone formation in targets beneath a penetration projectile. Int J Impact Eng 11(4):515–525
15. Astanin VV, Galiev SU, Ivashchenko KB (1988) Pecularities of the deformation and failure of aluminum targets in interaction with a steel projectile along the normal. Strength Mat 12:52–58
16. Astanin VV, Galiev SU, Ivashchenko KB (1987) Numerical-experimental study of the elastoplastic interaction of a striker with an obstacle. Strength Mat 11:97–100
17. Wang D, Mori J (2012) A Moderate earthquake with supershear rupture. Bull Seismol Soc Am 102:301–308
18. Yue H, Lay T, Freymueller JT et al (2013) Supershear rupture of the 5 January 2013 Craig, Alaska (M_w 7.5) earthquake. J Geophys Res 118:5903–5919
19. Zhan A, Helmberger DV, Kanamory H et al (2014) Supershear rupture in a Mw 6.7 aftershock of the 2013 Sea of Okhotsk earthquake. Science 345:204–207
20. Anderson JG (2007) Physical processes that control strong ground motion. In: Schubert G (ed) Treatise on geophysics. Elsevier, Amsterdam/Boston
21. Singh SK, Mena E, Castro R (1988) Some aspects of source characteristics of the 19 September 1985 Michoacan earthquake and ground motion amplification in and near Mexico City from strong motion data. Bull Seism Soc Am 78:451–477
22. Aki K (1993) Local site effects on weak and strong ground motion. Tectonophysics 218:93–111
23. Oglesby DD, Archuleta RJ (1997) A faulting model for the 1992 Petrolia earthquake: can extreme ground acceleration be a source effect? J Geophys Res 102(B6):11877–11897
24. Field EH, Johnson PA, Beresnev IA, Zeng Y (1997) Nonlinear ground-motion amplification by sediments during the 1994 Northridge earthquake. Nature 390:599–604
25. Aoi S, Kunugi T, Fujiwara H (2008) Trampoline effect in extreme ground motion. Science 322:727–730
26. Eiby GA (1989) Earthquakes. Heinemann Reed, Auckland
27. Keller EA, Devecchio DE (2014) Natural hazards: earth's processes as hazards, disasters, and catastrophes. Pearson Education, Auckland
28. Castaños H, Lomnitz C (2012) Earthquake disasters in Latin America. Springer, Dordrecht/New York
29. Greene MT (1982) Geology in the nineteenth century. Cornell University Press, Ithaca
30. Bodissey B (2009) A history of geology and planetary science. Internet

Index

Sh.U. Galiev, *Darwin, Geodynamics and Extreme Waves*,
DOI 10.1007/978-3-319-16994-1

The manufacturer's authorised representative in the EU is Springer Nature Customer Service Centre GmbH, Europaplatz 3, 69115 Heidelberg, Germany. If you have any concerns regarding our products, please contact ProductSafety@springernature.com

Printed and bound by CPI Group (UK) Ltd, Croydon, CR0 4YY
22/09/2026
02221065-0006